Lecture Notes in Computer Science

Lecture Notes in Artificial Intelligence   **15835**

Founding Editor

Jörg Siekmann

Series Editors

Randy Goebel, *University of Alberta, Edmonton, Canada*
Wolfgang Wahlster, *DFKI, Berlin, Germany*
Zhi-Hua Zhou, *Nanjing University, Nanjing, China*

The series Lecture Notes in Artificial Intelligence (LNAI) was established in 1988 as a topical subseries of LNCS devoted to artificial intelligence.

The series publishes state-of-the-art research results at a high level. As with the LNCS mother series, the mission of the series is to serve the international R & D community by providing an invaluable service, mainly focused on the publication of conference and workshop proceedings and postproceedings.

Shuhan Yuan · Fragkiskos Malliaros · Xin Zheng
Editors

Trends and Applications in Knowledge Discovery and Data Mining

PAKDD 2025 Workshops, ADUR
FairPC, GLFM, PM4B and RAFDA
Sydney, NSW, Australia, June 10–13, 2025, Proceedings

 Springer

Editors
Shuhan Yuan ⓘ
Utah State University
Logan, UT, USA

Fragkiskos Malliaros ⓘ
Université Paris-Saclay
Gif-sur-Yvette, France

Xin Zheng ⓘ
Griffith University
Gold Coast, QLD, Australia

ISSN 0302-9743   ISSN 1611-3349 (electronic)
Lecture Notes in Artificial Intelligence
ISBN 978-981-96-8196-9   ISBN 978-981-96-8197-6 (eBook)
https://doi.org/10.1007/978-981-96-8197-6

LNCS Sublibrary: SL7 – Artificial Intelligence

This Springer imprint is published by the registered company Springer Nature Singapore Pte Ltd.
The registered company address is: 152 Beach Road, #21-01/04 Gateway East, Singapore 189721, Singapore

If disposing of this product, please recycle the paper.

Foreword

It is our great pleasure to present the workshops held in conjunction with the 29th Pacific-Asia Conference on Knowledge Discovery and Data Mining (PAKDD 2025). PAKDD is one of the longest-established and leading international conferences in the areas of data mining and knowledge discovery. It provides an international forum for researchers and industry practitioners to share their latest developments, new ideas, original research results, and practical development experiences from all KDD-related areas including data mining, statistical and symbolic machine learning, databases, knowledge acquisition and automatic scientific discovery, data visualization, and knowledge-based systems.

This book constitutes the proceedings of the following workshops that were held in conjunction with the 29th Pacific-Asia Conference on Knowledge Discovery and Data Mining, PAKDD 2025, which took place in Sydney, Australia, during June 10–13, 2025:

– Workshop on Advanced Data-Driven Techniques for Urban Resilience (ADUR 2025)
– Workshop of Foundational AI for Pervasive Computing (FairPC 2025)
– Workshop on Graph Learning with Foundation Models (GLFM 2025)
– Workshop on Pattern Mining and Machine Learning for Bioinformatics (PM4B 2025)
– Workshop on Research and Applications of Foundation Models for Data Mining and Affective Computing (RAFDA 2025)

For ADUR 2025, the Workshop on Advanced Data-Driven Techniques for Urban Resilience, 10 submissions were received, and 6 papers have been accepted. For FairPC 2025, the Workshop of Foundational AI for Pervasive Computing, 2 survey papers and one full paper were accepted from a total of 5 submissions. For GLFM 2025, the Workshop on Graph Learning with Foundation Models, 7 qualified submissions were received, and 4 papers were accepted. For PM4B 2025, the Workshop on Pattern Mining and Machine Learning for Bioinformatics, 6 full papers were accepted from a total of 14 submissions. For RAFDA 2025, the Workshop on Research and Applications of Foundation Models for Data Mining and Affective Computing, 18 submissions were received, consisting of 5 invited papers and 13 regular submissions. Eventually, 12 papers (5 invited papers and **7** regular papers) were accepted.

We express our sincere gratitude to all workshop organizers, workshop program committee members, and authors whose diligent efforts and commitment contributed significantly to the high quality and success of the workshop program.

We hope participants found the workshop sessions engaging and insightful, providing opportunities for new collaborations and further research directions in data mining and related fields.

June 2025

Shuhan Yuan
Fragkiskos Malliaros
Xin Zheng

Contents

Workshop on Research and Applications of Foundation Models for Data Mining and Affective Computing (RAFDA 2025)

Workshop on Advanced Data-Driven Techniques for Urban Resilience (ADUR 2025)

ADUR 2025 Preface

Advanced Data-Driven Techniques for Urban Resilience (ADUR), a workshop of the 29th Pacific-Asia Conference on Knowledge Discovery and Data Mining (PAKDD), addressed the escalating challenges faced by modern urban environments. Urban areas are increasingly facing significant shock events and chronic stress. Shock events include natural disasters such as earthquakes and severe storms, which may damage infrastructure and displace populations, as well as cyberattacks that disrupt critical services and public health crises that strain healthcare systems. At the same time, chronic stresses such as income inequality, housing unaffordability, and inadequate public transportation gradually erode a city's capacity to maintain stability. As a result, there is an increasing need to understand and respond to how these challenges impact our economy, communities, and environment.

In this context, urban resilience—the ability of individuals, communities, businesses, governments, and urban systems to survive, adapt, and thrive—becomes crucial. Ensuring resilience is vital for maintaining functionality and supporting the well-being of residents during these disruptive events, enabling cities to withstand both acute shocks and ongoing stresses. To effectively enhance urban resilience, advanced data-driven techniques play a pivotal role by empowering cities to proactively enhance infrastructure, promote social equity, and optimize access to essential services. By leveraging machine learning, predictive analytics, and data mining, cities can gain actionable insights to anticipate disruptions, allocate resources efficiently, and implement timely interventions. These data-driven strategies not only strengthen cities' capacity to adapt to evolving challenges but also ensure sustainable development and improved well-being for all residents. The objective of this workshop was to provide a collaborative platform for discussing and showcasing advanced data-driven techniques that enhance urban resilience. It focused on methodologies that support urban areas in responding effectively to acute shocks and chronic stresses, ensuring sustainability, adaptability, and long-term urban well-being.

We extend our gratitude to the PAKDD Program Committee, organizers, and workshop chairs for their invaluable support.

The workshop received papers that cover one or several of the following topics:

1. Infrastructure and Maintenance

   - Data-driven infrastructure health and maintenance
   - Machine learning for electricity and power system optimization
   - Traffic analysis using data mining and predictive modelling
   - Communication network resilience using machine learning

2. Urban Planning and Population Analysis

   - Data mining techniques for dwelling and population analysis
   - Machine learning for urban planning and capacity management

- Urban computing for data-driven urban planning
- Urban computing for transportation systems using predictive analytics

3. Environmental Monitoring and Sustainability

- Weather event prediction through advanced data analytics
- Urban computing for environmental monitoring with machine learning
- Data-driven approaches for urban energy consumption management
- Climate change impact assessments through data analytics

4. Big Data and Smart Sensing

- Urban sensing technologies integrated with data analytics
- Urban data management technologies leveraging big data techniques
- Managing urban big data in the cloud with data mining
- Anomaly detection and event discovery in urban areas through AI

5. Social Applications and Public Safety

- Machine learning in urban social applications
- Economic analysis in urban areas using data-driven techniques
- Public safety and security enhanced by machine learning data-driven analysis of housing affordability and social equity

6. Advanced Machine Learning and AI Techniques

- Foundation models and large language models for urban data applications
- Federated learning for decentralised urban data analysis
- Predictive modelling and AI-driven simulations
- Deep learning architectures for urban resilience

ADUR 2025 received 10 submissions, among which 6 papers were accepted for inclusion in the proceedings as well as oral presentation at the workshop. All papers were peer-reviewed double-blind by 3 reviewers from the Program Committee and selected on the basis of these reviews. The review process focused on the technical quality, originality, significance, and clarity of the paper, as well as its relevance to data mining. The accepted papers represent a compelling mix of advanced data-driven techniques tailored to address the complex challenges of urban resilience. They showcase innovative applications of methods such as supervised learning, meta-learning, and multi-task learning to enhance cities' capacity to anticipate, respond to, and recover from both acute shocks and ongoing stresses. Their acceptance reflects a rigorous review process, marked by thorough discussions and consensus among the reviewers.

We extend our sincere gratitude to the organizers, reviewers, and authors for their unwavering dedication and hard work, which significantly contributed to the success of this workshop. We also express our appreciation to the Organizing Committee, Program Committee members, especially the workshop chairs of the PAKDD 2025 Organizing Committee, and the technical staff for their pivotal roles in ensuring the success of ADUR 2025. Special thanks are extended to Springer for their invaluable assistance in publishing the proceedings. Lastly, we acknowledge the invaluable contributions of all participants and speakers at ADUR 2025, whose collective support made the workshop a dynamic, engaging, and triumphant event.

June 2025   Bin Liang
                                                                                Zhidong Li
                                                                                 Rui Wang

ADUR 2025 Organization

Program Committee Chairs

Bin Liang	University of Technology Sydney, Australia
Zhidong Li	University of Technology Sydney, Australia
Rui Wang	Shanghai University, China

Program Committee

Bahman Javadi	Western Sydney University, Australia
Xu Zhang	China University of Mining & Technology, China
Feng Zhou	Renmin University of China, China
Sara Shirowzhan	University of New South Wales, Australia
Xunhui Fan	Macquarie University, Australia
Xiaohan Su	University of Technology Sydney, Australia
Rongchen Wu	University of Technology Sydney, Australia

Deep Surrogate Model to Predict the Effectiveness of Substrates for Absorbing CO_2 from Air

Nina Ghanbari Ghooshchi⬛, Weihong Wang⬛, and Ashfaqur Rahman$^{(\boxtimes)}$⬛

CSIRO's Data61, Eveleigh, Australia
ashfaqur.rahman@data61.csiro.au

Abstract. Due to air pollution, CO_2 is commonly present in the air. Substrates can be designed to absorb CO_2 to make air free from pollutant. Fluid dynamics model based computer simulations are commonly used to design substrates and understand the amount of absorption the substrate can do. In this paper, we explored the use of deep learning model as an efficient surrogate for Computational Fluid Dynamics (CFD) simulations. CFDs are often constrained by high computational costs. Our investigation centers on a specific scenario involving the flow of air containing CO_2 through a tube with a metal scaffold (i.e., substrate) designed for selective CO_2 adsorption.

We developed a deep learning model capable of accurately predicting transport (i.e., absorption) and mixing properties from the geometric characteristics of the substrate, thereby circumventing the need for computationally intensive CFD simulations. Our model employs a 3 Dimensional (3D) Convolutional Neural Network (CNN) to process the discretized 3D shapes of the tube and scaffold. We validated this approach using data generated across multiple runs of CFD as part of an optimization process. Our findings demonstrate that our deep learning models can predict transport and mixing properties with high accuracy, significantly reducing computation time compared to traditional CFD methods.

Keywords: Computational Fluid Dynamics (CFD) emulation · Computational Fluid Dynamics (CFD) surrogate · 3 dimensional convlutional neural network (3DCNN) · digital twin

1 Introduction

Pollution is a key environmental concern that refers to the introduction of harmful substance to the environment. Different industries across the world produce waste containing harmful materials that end up in the water ways. Similarly burning of fossil fuel in the industries produces CO_2 contaminating the air. Effective filters/substrates are commonly designed that can absorb pollutants from fluid (waste water/air). The interaction between such substrates and fluid are commonly modeled using Computational Fluid Dynamics (CFD).

S. Yuan et al. (Eds.): PAKDD 2025 Workshops, LNAI 15835, pp. 7–18, 2025.
https://doi.org/10.1007/978-981-96-8197-6_1

CFD involves usage of numerical methods and algorithms to simulate fluid flow behavior. Real-world applications often require testing fluid flow under diverse physical conditions. CFD provides an economical way to recreate these physical scenarios mathematically and analyze fluid behavior in various test situations. However, CFD models are computationally demanding and can take a long time to generate results, even on advanced computing systems. Machine Learning (ML) based surrogate models present a quicker alternative to traditional CFD models if they can deliver accurate predictions. This paper examines a data-driven ML model to assess its potential in accurately mimicking CFD simulations.

This paper focuses on a CFD model designed to simulate the flow of air containing CO_2 through a tube equipped with a metal scaffold (the method can be generalized to any fluid containing pollutant). This scaffold is engineered to selectively adsorb CO_2 from the air. The primary goal of the research is to design a metal scaffold that maximizes both absorption and fluid mixing. The process is iterative and employs an evolutionary optimization method using a Genetic Algorithm [1]. This algorithm generates various shape parameters over multiple generations [2]. These parameters are input into a custom geometry generator to create a solid body with a specific shape. Subsequently, a grid generator uses this geometry to produce a computational grid.

In this approach, the CFD model begins by taking the computational grid as its initial input. It proceeds to solve the Lattice Boltzmann equations [3], calculating two critical metrics: the transportation rate (indicating CO_2 - carbon-dioxide absorption by the metal substrate) and fluid mixing. Achieving uniform deposition on the substrate requires effective fluid mixing. These metrics serve as the fitness criteria for the Genetic Algorithm (GA). Initially, the GA initializes a population with random shapes, and the CFD evaluates their fitness scores. Subsequent generations of shapes undergo modifications based on these fitness values provided by the CFD. This iterative process continues until the GA converges on an optimal solution. The optimization process, over the generations, produces a large number of alternative shapes of the substrates and their corresponding fitness scores (absorption and mixing). This provides us with a dataset that can be used for training and testing supervised machine learning models.

The entire optimization process (based on CFD) is notably time-intensive. The key bottleneck is the CFD part (that produces the fitness scores i.e., absorption and mixing) takes a large time (approx. 2 h on high performance computing platforms). If trained well and accurately, supervised ML models can make such predictions within few seconds - significantly reducing the time needed by the optimization process giving the opportunity to explore more alternative shapes. This motivated us to investigate -"Can supervised ML models accurately predict substrate fitness?".

Several research efforts ([4–6]) are observed in the literature that aim to develop ML models as a surrogate for CFD. The one close to our problem is presented in [7] where a vanilla ML (decision tree, neural network etc.) is used as a surrogate for CFD to predict absorption and mixing based on Shape Parameter

Vector (SPV). SPV is an encoded parametric representation of the substrate's shape that is used by an algorithm in a recursive manner to produce the gridded representation of the substrate. SPV based prediction of fitness values produced accurate results in [7]. A key problem with SPV based approach is that the length of the vector varies across different substrate families. As a result ML models trained on one shape family cannot be generalized to another substrate family.

To address the generalization problem, we investigated ML models leveraging deep learning, where the input consists of the 3D grid housing the scaffolds and through which the fluid flows. Since the grid size remains consistent across all scaffold families examined in the research, this approach allows us to test models across different scaffold families addressing the generalization problem. Given the substantial grid dimensions ($52 \times 52 \times 400$), sophisticated methods are necessary to extract features from the grid. Our exploration led us to employ a 3DCNN that integrates both feature extraction and prediction components.

The key research questions addressed in this paper are:

1. Can we develop a deep learning model (3D CNN) to infer fitness values (absorption and mixing) from grid representation of substrate
2. How accurate is the 3D CNN model compared to vanila vanilla ML (decision tree, neural network etc.) model presented in [7]
3. How well does the 3DCNN based deep learning model generalize across different families

Note that the improvement of computational complexity of ML models (vanilla ML/3DCNN) w.r.t. CFD model is very obvious (hours vs seconds). Hence we did not provide any results on computational complexity of the two approaches in this paper.

1.1 CFD Modelling

In this paper, the CFD model focuses on calculating the steady-state velocity field for fluid flow within a lengthy tube. This involves solving the Lattice Boltzmann (LB) equations, a microscopic model known to yield results equivalent to the Navier-Stokes equations. The LB method offers advantages in handling complex geometries efficiently and is highly parallelizable, enhancing both computational speed and accuracy.

Once the velocity field is established, fictitious tracers are introduced into the tube. Their paths are determined by solving a simple equation ($dr/dt = v(r)$), where r represents the position of a tracer particle at time t, and v denotes the corresponding velocity at that position. To ensure statistical significance, thousands of tracers are simulated, providing data to quantify key parameters such as metal scaffold transport and fluid mixing.

2 Proposed Deep ML Approach

To enable the training of versatile ML models capable of handling various substrate families, our approach involves directly training these models on the grid that contains the filter or substrate. Since the grid dimensions remain consistent across all substrate families, this uniformity facilitates seamless training and testing processes across different families (i.e., it addresses the generalization problem faced in [7]).

In a grid of dimensions $X \times Y \times D$, each grid point is indexed as (x, y, d). Points in the grid marked as 0 represent empty spaces through which fluid can flow, while points marked as 1 denote substrate or filter locations where fluid flow is obstructed. Fig 1 presents a hypothetical scenario of a substrate being placed in a grid and it's cross sections

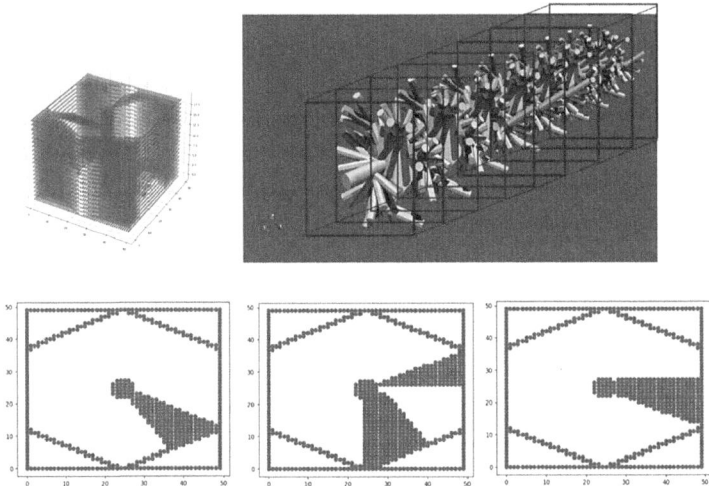

Fig. 1. A grid view of the substrate that is being placed inside a hexagonal tube. top-left: the discretiation of a 3D structure as a tensor; top-right: conceptualisation of the 3D substrate as a series of planes through which fluid must flow sequentially; and bottom: cross sections of the data tensor that represent binary grids of solid and void pixels.

Convolutional 3D (Conv3D) networks expand the principles of 2D (2 dimensional) convolutional neural networks [9] into three dimensions, making them particularly effective for managing volumetric data and, in some cases, spatiotemporal data. Originally developed for applications like video recognition [16] and medical imaging [10,11], Conv3D networks have shown exceptional ability in capturing intricate spatial and temporal patterns. [16,17]. The fundamental concept involves applying 3D kernels to 3D input data, enabling the network to simultaneously learn features across width, height, and depth. This direct processing of 3D data provides substantial advantages in situations where spatial and temporal dimensions are interconnected, such as in fluid dynamics [12,13], climate modelling [14,15], and biomedical image analysis. [18,19].By utilising 3D convolutions, these networks can effectively capture intricate dependencies and achieve high predictive accuracy in regression tasks that involve volumetric data.

In this paper, we employed Conv3D neural networks to tackle a regression task aimed at predicting complex relationships within volumetric voxel grid data. Conv3D architectures are well-suited for processing 3D spatial data, leveraging their ability to capture complex spatial features within volumetric voxel grids. Each Conv3D layer convolves a 3D kernel across the input volume, extracting hierarchical representations of features. These networks excel in capturing intricate patterns within dynamic sequences, making them particularly effective for our task of predicting continuous-valued outputs from volumetric inputs. The Conv3D model architecture used in our research includes multiple convolutional layers followed by activation functions, culminating in fully connected layers for regression output. Our approach demonstrates the efficacy of Conv3D networks in handling volumetric spatial data and showcases their potential in advancing predictive modelling tasks across various domains.

2.1   Structure

The deep learning network architecture comprises a feature extraction module and a prediction module. The feature extraction module consists of three 3D convolutional blocks, each incorporating a Conv3D layer followed by a ReLU (Rectified Linear Unit) activation and a MaxPool3D (3D max pooling) layer. On the other hand, the prediction module is implemented as a Multi-layer Perceptron (MLP) consisting of three fully connected layers.

We trailed with different hyper parameters for the 3D CNN and present here the ones that produced best results. All 3D convolutional layers feature kernels of size (3,3,3) and a stride of (1,1,1). The number of filters is indicated in each respective box. 3D max pooling layers utilize kernels of size (2,2,2). The output of the final pooling layer is flattened and serves as input to the fully connected layers. The number of nodes for each fully connected layer is denoted within the box. This network receives input in a 3D shape of size (400,52,52) and yields two fitness values as output. Fig 2 presents the structure of the proposed CNN architecture.

3   Experiments

To evaluate the effectiveness of our proposed method, we implemented this model in pytorch and we set the parameters for training as follows. Batch size = 32, number of epochs = 10, learning rate = 0.001, loss function = L1, optimizer = Adam. These hyper parameters presented in this paper are the ones that produced best results from trial-and-error.

We obtained data for training and testing the deep ML models on three different shapes of substrate famillies: Tree, Ribbon, and Hexrain [2]. We trained and tested four instances of the structure: The first model named modelTree is trained on Tree family, second model named modelRibbon is trained on Ribbon

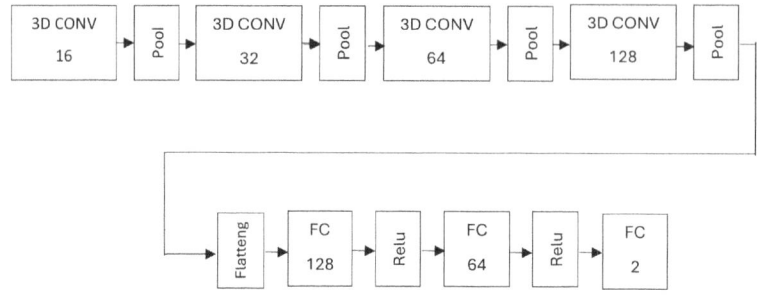

Fig. 2. the structure of the neural network used for training.

data, third model named modelHexrain is trained on Hexrain family and the fourth model named modelAll is trained on the combination of all of these families. Given our limited dataset, we employed 5-fold cross-validation for reporting. This involved randomly partitioning the dataset into five folds, training on four of these folds and testing on the remaining fold, repeating this process for every fold combination. Following data splitting, we augmented the training set ensuring augmentation was performed separately to prevent potential data leakage. Subsequently, we assessed the trained model on the test data. The results of this experiment for each of the models on a representative fold (Fold 1) is depicted in the following figures.

Figure 3 represent the test results of a representative fold based on modelTree experiments that is trained on data from Tree family and tested on the same family. In these figures Target 1 represents absorption and Target 2 represents mixing. The average error across the five folds demonstrate good predictions, with a mean error of 5.57 for absorption and 4.81 for mixing.

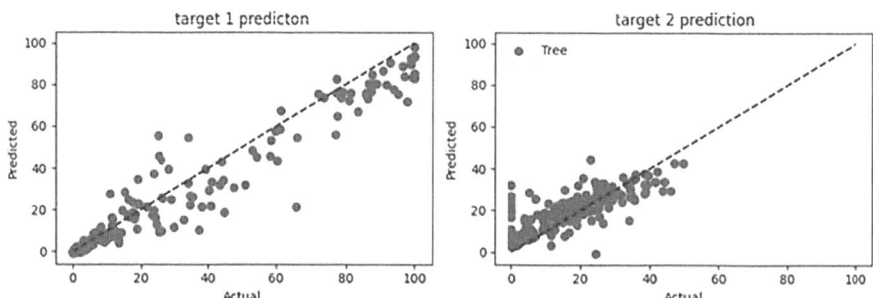

Fig. 3. Prediction Result on test set of a representative fold on Tree family substrates. Here Target 1 represents absorption and Target 2 represents mixing.

Figure 4 represent the test results of a representative fold based on model-Ribbon experiments that is trained on data from Ribbon family and tested on the same family. The average error across the five folds demonstrate good predictions, with a mean error of 6.56 for target1 and 8.73 for target2. Since this family has the fewest data points, the resulting error is larger compared to the prediction errors of other families, which is expected.

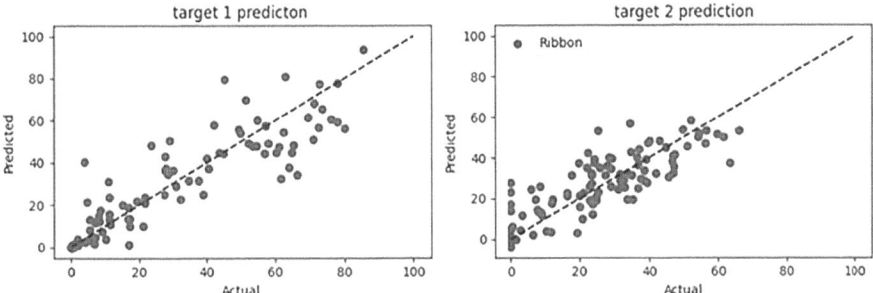

Fig. 4. Prediction Result on test set of a representative fold on Ribbon family substrates. Here Target 1 represents absorption and Target 2 represents mixing.

Figure 5 represent the test results of a representative fold based on model-Hexrain experiments that is trained on data from Hesrain family and tested on the same family. The average predictions are good, with a mean error of 7.55 for target1 and 5.72 for target2 across the five folds.

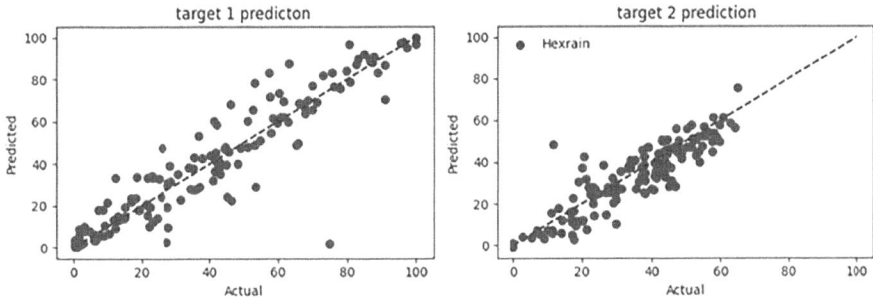

Fig. 5. Prediction Result on test set of a representative fold on Hexrain family substrates. Here Target 1 represents absorption and Target 2 represents mixing.

We also trained the model on data from all three families and tested on each family separately. Figure 6 present the prediction result on a representative fold based on experiments modelAll that was trained and tested on data from all three families. The mean errors for each family are as follows: Tree (5.41, 4.61), Ribbon (6.42, 8.65), and Hexrain (7.30, 5.18) across the five folds.

The performance of modelAll (trained on combined data from three families) is better than the models trained individually on each family. This suggests that combining data from different families to train deep ML model improves the prediction accuracy. Note that this combination is not possible in [7] as the parametric shape representation of the three families have different dimensions.

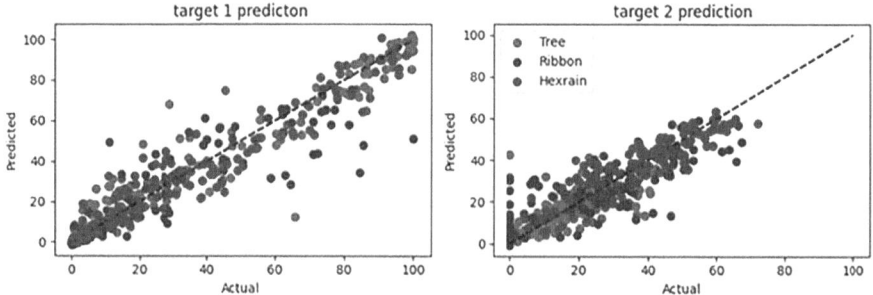

Fig. 6. Prediction Result on test set of a representative fold on all three families of substrates. Here Target 1 represents absorption and Target 2 represents mixing.

3.1 Cross Family Results

To test the transferability of the model between different families, we also tested the models trained on single families on the data from other families. The results of this experiment are depicted in the following figures.

Table 1 presents a comprehensive summary of our findings across different experimental scenarios. This includes details on the training and testing setups, specifying the families used for training and testing. Additionally, we provide the mean absolute error for the predictions of two fitness values. In calculating errors for models trained and tested within the same family, and for the 'modelAll' tested across three families, we utilized cross-validation. The reported error represents the mean error across all folds. However, in the cross-family scenario, where the model is trained on one family and tested on others, we did not employ cross-validation. Instead, we trained and tested the model using all available data from the respective families.

Based on the table analysis, it is evident that for the model to accurately predict fitness values within a specific family of designs, training on data from that same family is crucial. The model's performance diminishes significantly when attempting predictions on designs it hasn't been trained on. However note that, when the trained model has some knowledge of the new family (as in modelAll that is trained with samples from all families), the situation improves as evident from the last three rows of Table 1. We conclude that generalization can be achieved using the proposed model in two ways:

1. The proposed approach offered an uniform feature space (i.e., input to the model - the whole 3D grid) across all substrate families resulting in a position where same deep ML architechture can be trained and tested across all families,

2. The deep ML model (3D CNN) performs well on a new family, if it has some knowledge of the new family (as evident from last three rows of Table 1). We suggest retraining (or transfer learning) the deep ML model with some samples from the new family that will assist the model generalize well on a new family.

We have also provided a comparison between the proposed ML model (3D CNN) and SPV (Shape Parameter Vector) based vanilla ML models as presented in [7]. The models in [7] were trained and tested on Hexrain family only and we provide a comparison on that only. The error results are presented in 2. We can observed that the proposed 3D CNN outperforms Multiple Linear Regression, Support Vector Regression, and Neural Network Regression. Decision Tree Regression performs better than any other model. Note that the proposed approach (3D CNN) can be generalized across different substrate families but the vanialla ML based approach in [7] can not.

Table 1. Comparison of performance of different ML models to predict target variables in the test set. Here Target 1 represents absorption and Target 2 represents mixing.

Model	Trained	Tested	Target1 error	Target2 error
modelTree	Tree	Tree	5.57	4.81
modelTree	Tree	Ribbon	13.44	17.15
modelTree	Tree	Hexrain	32.17	19.26
modelRibbon	Ribbon	Tree	11.40	12.45
modelRibbon	Ribbon	Ribbon	6.56	8.73
modelRibbon	Ribbon	Hexrain	28.19	15.58
modelHexrain	Hexrain	Tree	21.06	12.29
modelHexrain	Hexrain	Ribbon	12.21	13.64
modelHexrain	Hexrain	Hexrain	7.55	5.72
modelAll	All	Tree	5.41	4.61
modelAll	All	Ribbon	6.42	8.65
modelAll	All	Hexrain	7.30	5.18

4   Conclusions and Future Directions

In this paper, we investigated the application of deep learning models to directly predict transport and mixing quantities that traditionally require running a time-consuming CFD model to simulate air flow when a substrate is placed in a tube to absorb CO_2. The method can be generalized to design substrates for absorbing metals from waste water as well. Our machine learning models predict transport and mixing quantities directly from substrate shape properties. Compared to

Table 2. Comparison of performance of proposed 3DCNN model and vanialla ML models in [7] on Hexrain substrate family. Here Target 1 represents absorption and Target 2 represents mixing.

Model Name	Target 1 Error	Target 2 Error
Multiple Linear Regression	18.61	13.23
Support Vector Regression	9.55	5.38
Neural Network Regression	9.58	6.56
Decision Tree Regression	1.41	0.71
Proposed 3D CNN	7.55	5.72

other ML methods that base their predictions on a fixed number of shape properties ([7]), the deep learning approach is more general. The input to this method is the discretized 3D shape of the tube and the substrate, which is consistent across all substrate families. Empirical results demonstrated the effectiveness of the method in predicting transport and mixing quantities with reasonable error.

Based on the research question we addressed, we found that

1. 3D CNN based deep learning model can be used as a surrogate for CFD and can predict absorption and mixing with error (5.41, 4.61) on Tree family, (6.42, 8.65) on Ribbon family, and (7.30, 5.18) on Hexrain family.
2. The proposed 3D CNN based method performs better than Multiple Linear Regression, Support Vector Regression, and Neural Network Regression but not Decision Tree when compared to models in [7] on Hexrain family. Note that the proposed method can generalize across substrate families that the methods in [7] are unable to do.
3. The deep ML model (3D CNN) performs well on a new family, if it has some knowledge of the new family. We suggest retraining (or transfer learning) the deep ML model with some samples from the new family that will assist the model generalize well on a new family.

Considering the structure of the substrate, we only accounted for the solid and empty spaces in the tube. For future work, we plan to include velocity fields, which can be calculated relatively quickly using CFD, as additional features. We hope that incorporating velocity fields will further improve the predictive accuracy of our method.

We also aim to explore the feasibility of applying semi-supervised learning techniques. Additionally, we intend to investigate the application of generative AI models as another potential direction for enhancing our approach.

Acknowledgments. We would like to acknowledge the FDFM (Future Digital Manufacturing) fund from CSIRO for their support in this work.

References

1. Katoch, S., Chauhan, S.S., Kumar, V.: A review on genetic algorithm: past, present, and future. Multimedia Tools Appl. **80**, 8091–8126 (2021)
2. Pereira, G.G., Howard, D., Lahur, P., Breedon, M., Kilby, P., Hornung, C.H.: Freeform generative design of complex functional structures. Sci. Rep. **14**(1), 11918 (2024)
3. Benzi, R., Succi, S., Vergassola, M.: The lattice Boltzmann equation: theory and applications. Phys. Rep. **222**(3), 145–197 (1992)
4. Wang, H., et al.: Recent advances on machine learning for computational fluid dynamics: a survey. arXiv abs/2408.12171 (2024)
5. Azfarizal, M., Shah Hizam Md Yasir, A., Fariz Mohamed Nasir, M.: A machine learning-based comparative analysis of surrogate models for design optimisation in computational fluid dynamics. Heliyon **9**(8) (2023)
6. Vinuesa, R., Brunton, S.L.: Enhancing computational fluid dynamics with machine learning. Nat. Comput. Sci. **2**, 358–366 (2022)
7. Rahman, A., Pereira, G., Kilby, P., Lahur, P.: Surrogate model for CFD based on machine learning In: 25th International Congress on Modelling and Simulation, pp. 711–717. MODSIM, Darwin, NT, Australia (2023)
8. Tran, D., Bourdev, L., Fergus, R., Torresani, L., Paluri, M.: Learning spatiotemporal features with 3D convolutional networks. In: Proceedings of the IEEE International Conference on Computer Vision, pp. 4489–4497. IEEE, Santiago, Chile (2015)
9. Krizhevsky, A., Sutskever, I., Hinton, G.E.: ImageNet classification with deep convolutional neural networks. Commun. ACM **60**(6), 84–90 (2017)
10. Ker, J., Wang, L., Rao, J., Lim, T.: Deep learning applications in medical image analysis. IEEE Access **6**, 9375–9389 (2017)
11. Singha, A., Thakur, R.S., Patel, T.: Deep learning applications in medical image analysis. Biomed. Data Min. Inf. Retrieval: Methodol. Tech. Appl., 293–350 (2021)
12. Guo, X., Li, W., Iorio, F.: Convolutional neural networks for steady flow approximation. In: Proceedings of the 22nd ACM SIGKDD International Conference on Knowledge Discovery and Data Mining, pp. 481–490. ACM, San Francisco California USA (2016)
13. Morimoto, M., Fukami, K., Zhang, K., Nair, A.G., Fukagata, K.: Convolutional neural networks for fluid flow analysis: toward effective metamodeling and low dimensionalization. Theoret. Comput. Fluid Dyn. **35**(5), 633–658 (2021). https://doi.org/10.1007/s00162-021-00580-0
14. Chattopadhyay, A., Hassanzadeh, P., Pasha, S.: Predicting clustered weather patterns: a test case for applications of convolutional neural networks to spatiotemporal climate data. Sci. Rep. **10**(1), 1317 (2020)
15. Haidar, A., Verma, B.: Monthly rainfall forecasting using one-dimensional deep convolutional neural network. IEEE Access **6**, 69053–69063 (2018)
16. Tran, D., Bourdev, L., Fergus, R., Torresani, L., Paluri, M.: Learning spatiotemporal features with 3D convolutional networks. In: Proceedings of the IEEE International Conference on Computer Vision, pp. 4489–4497. IEEE, Santiago, Chile (2015)
17. Hara, K., Kataoka, H., Satoh, Y.: Can spatiotemporal 3D CNNs retrace the history of 2D CNNs and ImageNet?. In: Proceedings of the IEEE Conference on Computer Vision and Pattern Recognition, pp. 6546–6555. IEEE, Salt Lake City, UT, USA (2018)

18. Abraham, G.K., Jayanthi, V.S., Bhaskaran, P.: Convolutional neural network for biomedical applications. Comput. Intell. Appl. Healthc., 145–156 (2020)
19. Trajanovski, S., et al.: Towards radiologist-level cancer risk assessment in CT lung screening using deep learning. Comput. Med. Imaging Graph. **90**, 101883 (2021)

Inferring Sensor Metadata Based on Machine Learning for Portable Building Applications

Ashfaqur Rahman[1], Mashud Rana[1(✉)], Mahathir Almashor[2], and John McCulloch[2]

[1] Data61, CSIRO, Eveleigh, Australia
{ashfaqur.rahman,mdmashud.rana}@data61.csiro.au
[2] Energy, CSIRO, Eveleigh, Australia
{mahathir.almashor,john.mcculloch}@csiro.au

Abstract. Modern buildings integrate a multitude of interconnected sensors and control devices for efficient energy management. Different conventions are adopted by vendors, technicians, and operators for naming and creating metadata for these sensors and devices. Analytical tools for understanding system status, fault diagnosis, and optimizing energy consumption in such complex environments necessitate automated categorization or classification of sensors and their metadata according to a standard ontology like Brick or Haystack. Previous research has predominantly focused on automatically inferring only the classes of the sensors from their text name strings. However, there exists an evident void in research concerning the automated *inference of sensor metadata* (i.e., properties) and in this paper we aim to address this *novel* and challenging problem. Our *contribution* lies in developing data-driven approaches that combine NLP based context aware representation of sensor names with machine learning to accurately infer sensor properties. The effectiveness of the developed approaches is assessed using a proprietary dataset comprising 129 buildings. The results of this study hold implications for improving the portability and adaptability of energy analytics applications within smart building systems.

Keywords: Metadata inference · Sensor properties · Sensor classification · Energy efficiency · Energy analytics

1 Introduction

Commercial buildings account for a quarter of energy consumption in Australia [7] and contribute to global struggle to address climate change [10]. To deal with this problem, numerous sensors are nowadays placed in these modern buildings to monitor and efficiently manage energy usage. Each space, equipment, or sensor within buildings is defined as a *point* and associated metadata are called *properties*.

Lowering energy usage and detecting the faulty equipment in buildings require engineers to understand the representative semantic model of the points, so that various data-driven analysis, visualisation, and predictive action tools can be brought to bear on the task. Simply put, we can't train a forecasting algorithm to predict cooling patterns if we don't even know which building points and accompanying data streams to feed into it. The fact that this process must be done *retroactively* on the millions of existing buildings only exacerbates the issue. Buildings are a complex maze of original and retrofit equipment, installed over time by a procession of vendors each with their own naming conventions and commercial sensitivities. Brick [3], Haystack [1] provides a generic name space for all possible points and their properties in commercial buildings. But challenge remains in mapping proprietary point names and their properties into the generic name space offered by the standard ontology like Brick or Haystack.

Regardless of the diverse naming conventions adopted to name the heterogeneous points across buildings, the name strings of the points generally contain useful information that can help the inference of their types, properties, location, and relationships [21]. For instance, from the dataset we consider in this study, a point name string (from a anonymised site and building) *dsapi-normal-splendid-section-site_K._K_B000.B2DbE3BlkDLvl6.ActivePowerB* indicates that the point is located at the level *6* of the building *B000* at site *K*. It has the following electrical {property, value} pairs as per Brick ontology: {*electricalPhaseCount, 1*}, {*electricalPhases, B*}, {*powerFlow: net*}, and {*powerComplexity: real*}. Another illustrative point name string *dsapi-thick-gusty-rest-site_N.site_N_Bld_B001.LIBBIBatteryStorageSystem.V_AC*
refers that the point belong to the building *B001* at site *N*. Moreover, the point is a Battery Storage System and has the following electrical properties: {*electricalPhaseCount, 2*}, and {*electricalPhases, AC*}.

The novel research question we seek to address in this paper is if it is possible to infer the properties (i.e., metadata) of points from their name strings utilising Machine Learning (ML) based approaches, and make the approaches generalized enough to be effective across the multitude of buildings each with their own naming conventions. Doing so would expedite the inclusion of millions of buildings onto analytical platforms such as the Data Clearing House (DCH)[1], which can help the stakeholders to monitor, manage, and ultimately reduce the power profiles of their buildings.

Our proposed approach eschews large, complex and compute-intensive models in favour of direct use of pre-trained and tested Natural Language Processing (NLP) models such as BERT [8], MPNET [22], RoBERTa [18] etc. The building point names are fed into these pre-trained models without costly additional retraining on textual data germane to each building context. We then utilise output embedding from these language models as inputs of the classical ML models in two novel and complementary pathways: i) a two-staged hierarchical approach which first identifies the properties/metadata associated with a point based on a set of binary classifiers (one for each property) and then predicts their respective values using a set of multi-class classifiers; ii) a single step approach which adopts a set of multi-class classifiers to predict the values of the properties directly where a value 'NA' for a property indicates the property does not

---

[1] https://research.csiro.au/dch/.

belong to a point. We show that our amalgamation of text embedding with classical ML models such as Support Vector Machines [6] can yield high accuracy (about 90%) when tasked with predicting the correct property values purely on the point names. More importantly, this is done with relatively straight-forward ML methods, with none of the computational excesses of more modern AI-based methods, including Generative AI tools like Large Language Models.

2   Related Work

Previous studies on points classification can be categorised into two broad groups based on the type of the data sources they exploited [21]: text and time series. The first group of approaches (e.g., [2,16,19]) primarily consider deciphering the point names by counting the frequency of different words of varying length (called k-mers [9]) in text name strings of the points. Although such a *bag-of-words* method can help to extract the encoded information from point names to some extent, it does not consider the contextual meaning or semantic relationship of the words in the points names. Almashor et al. [2] applied an Auto-encoder (a neural-network architecture) to infer the Brick classes of the points using frequency of word counts in the name strings as inputs. Balaji et al. [4] applied Random Forest (RF) model with feedback from domain experts to classify the points using similar type of inputs. Likewise, [16,17] developed text data-driven approaches based on ML models including neural networks. The effectiveness of these approaches relies heavily on the inherent similarities between point names in both the source and target buildings. As a consequence, portability of approaches focusing on exploiting text point names across multitudes of buildings are limited due to the diverse naming conventions. To address this issue, integrating domain expertise [4] or leveraging knowledge specific to the target buildings [13] becomes necessary. However, this may not always be feasible.

Unlike the diverse naming conventions adopted by different buildings, the Time Series (TS) data from similar types of points is anticipated to exhibit some consistency across different buildings. This data encapsulates discernible patterns that can serve as signatures for effectively classifying various point types, irrespective of their deployment sites or buildings [21]. Hence, the second group of approaches (e.g., [12,14]) emphasises on utilising the statistical features extracted from the TS data recorded over time by the points as inputs to the prediction models. Gao et al. [12] employed various statistical features (such as mean, mode, quantiles, and deciles) from TS data as inputs to train multiple ML models (including RF, SVM, and k-Nearest Neighbor (kNN)) to classify the points according to Haystack ontology. Hong et al. [14] investigated the clustering of TS data using a novel similarity metric (cross-predictability) and explored its application in grouping different types of points. Moreover, the utilisation of TS data in conjunction with text name strings to transfer the knowledge from source to target buildings is investigated in [15]. Although the TS based approaches enhance the portability of accurate classification of the data-driven approaches, they might not be the appropriate choice for the on-boarding of new buildings and existing buildings with occasional limited availability of time series data.

In summary, the predominant methods primarily tackle the points classification problem relying on information extracted from their names without considering the sequencing of the words in the name strings. In contrast, we aim to utilise NLP models to generate latent embedding from the point name by considering the semantic relationships of the words in the entire string. In addition, prior work relating to classification of building points primarily focuses on mapping the points according to standard ontology like Brick or Haystack as opposed to our focus on classifying the *properties* of the points. Waterworth et al. [23] also used pre-trained text embedding libraries, and Almashor et al. [2] went with a similar lightweight autoencoder architecture model. However, the focus there was on assigning Brick labels instead of predicting the property values.

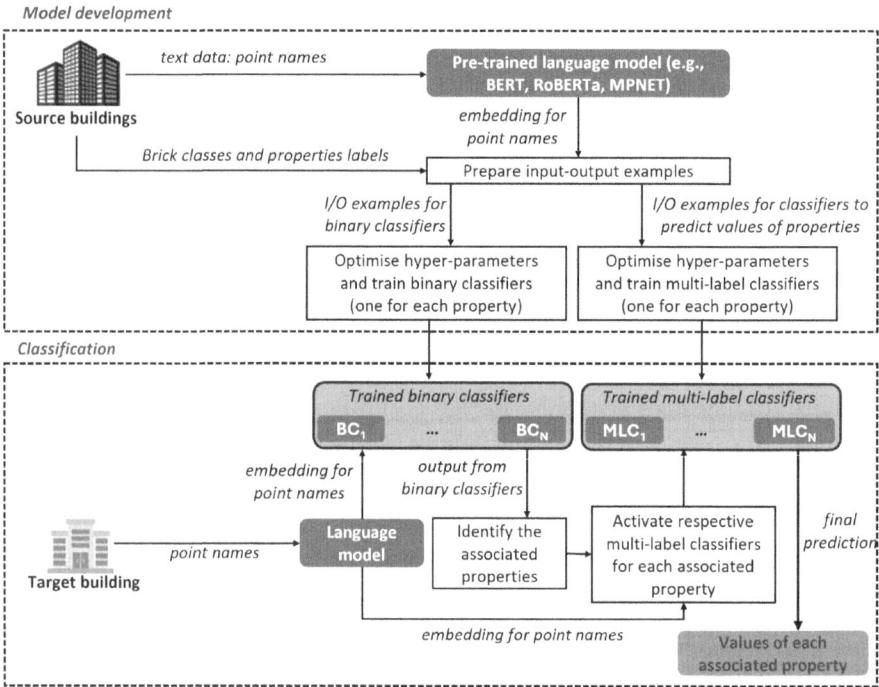

Fig. 1. Overview of the proposed hierarchical approach for classifying the properties of points in the buildings.

3   Methodology

3.1   Problem Statement

Given the point names $NS_B^P = \{m_1, m_2, ..., m_{N_s}\}$ of a set of N_s points $S_B^{N_s} = \{s_1, s_2, ..., s_{N_s}\}$ in a building B. Each point $s_{i \in \{1 \text{ to } N_s\}}$ has certain properties from the set $P = \{p_1, p_2, ..., p_{N_p}\}$ as defined in Brick ontology [3] and each property $p_j \in P$ has a set of M_j values $\{v_1^{p_j}, v_2^{p_j}, ..., v_{M_j}^{p_j}\}$. The goal is to infer the values of the properties associated with each point in S_B^N.

We propose two different approaches for classifying the properties of points in buildings: i) a hierarchical approach which adopts a set of binary and multi-class classifiers, and ii) a single step approach that only utilises multi-class classifiers. The details of these two approaches are provided below. Figure 1 presents a schematic diagram outlining both hierarchical and multi-class classification approaches.

3.2   Hierarchical Approach

This approach comprises of two main steps: *model development* and *classification*.

Model Development. The hierarchical approach addresses the properties classification tasks by deploying two different types of classifiers in series: i) a set of binary classifiers – one for each property to check if a point has a certain property associated with it, and ii) a set of multi-class classifiers where each of them aims to predict the values of a certain property.

Both set of classifiers utilise the same inputs – the embedding generated from the point names based on a language model. The points in different buildings (even within the same building) are usually named by following different conventions and by different parties (e.g., vendors, contractors) that cause the length of name strings to vary. The variable length point names are encoded into fixed length numeric vectors (known as embedding) by applying a language model (e.g., BERT, MPNET, RoBERTa). Through pre-training on extensive text data, these language models acquire the ability to produce comprehensive, context-sensitive word embedding that allow it to comprehend the complexities of language intricacies and execute text classification tasks like ours. We feed the point names from source buildings into a language model which generates an embedding vector of a fixed length for each point name string.

The embedding vectors for all points from the source buildings are then used as inputs to train both sets of classifiers (binary and multi-class) to predict the properties in a hierarchical manner. Although both sets of classifiers use the same embedding as inputs, they have different outputs. Since the purpose of utilising the binary classifiers is to identify the types of properties associated with a point, each binary classifier has output of $True(1/+ve)$ if a point has certain property or $False(0/-ve)$ otherwise. On the other hand, the outputs of a multi-class classifier vary on the specific property the classifier considers. For examples, the classifier focusing on predicting the values of the *Power Complexity* property can have a output from $\{real, apparent, reactive, import, export\}$ whereas the classifier that is design for predicting values of $ElectricalPhaseCount$ property can have output from $\{0, 1, 2, 3\}$.

Both set of classifiers are generic and can adopt any ML algorithms. In this study, for each classifier type (either binary or multi-class) we investigated several supervised ML algorithms with different pattern learning capabilities that include Decision Tree (DT) [20], Random Forest (RF) [5], AdaBoost [11], and Support Vector Machine (SVM) [6]. These ML algorithms showed promising performance for buildings' points classification in previous studies (e.g., [4,15,21]). The main reason to include different ML algorithms is to bring diversity into learning the relationship between inputs (i.e., embedding provided by language model) and output (the labels for point properties from Brick ontology). The application of multiple ML algorithms will also help to study the performance of the proposed approach comprehensively.

Moreover, we optimize different hyper-parameters for each ML algorithm through grid-search. Specifically, it involves exploring a predefined grid of hyper-parameter values. For each combination the hyper-parameter values within the grid, the model undergoes training and evaluation using k-fold cross-validation (we used $k = 10$) of the training data. This technique splits the dataset into k subsets, iteratively trains the model on $k - 1$ subsets, and evaluates its performance on remaining subset which differs in each iteration. This process helps to ensure generalizability of the model and the performance is not overly influenced by any specific data splits. After assessing all combinations, grid search selects the best-performing set of hyper-parameters based on a chosen metric, like accuracy or F-score. This set is then utilized to train the final model on the entire training dataset from source buildings for deployment.

Classification. The transfer of knowledge from source to target building is done through the trained classifiers sets. Specifically, to classify the properties of the points from a target building, we first generate the embeddings from their name strings based on a language model. These embeddings are then provided to the set of binary classifiers to determine whether a point possesses certain properties. If the output of a certain binary classifier is $True(1/+ve)$, the respective multi-class classifier is activated and employed to predict the values of that property from Brick ontology. Alternatively, the point is identified as not having such property.

3.3 Single Step Approach

The second approach considers the task of predicting the values of properties as a single step process. Unlike the hierarchical approach, the single step approach doesn't employ the binary classifiers and aims to directly predict the values of the properties by applying a set of multi-class classifiers. We augment the values of each property to include 'NA' if a point doesn't have that property associated with it. Hence, in addition to the actual values of any properties, a classifier predicts 'NA' to represent the absence of a property without utilising any binary classifier. Moreover, like in hierarchical approach, we deployed the same set of ML algorithms (DT, RF, AdaBoost, SVM) as the classifiers and optimise their hyper-parameters by grid searching based on 10-fold cross-validation of the training data.

4   Experiments and Results

4.1   Evaluation Metric

The effectiveness of the proposed approaches for classifying point properties is assessed through the use of the F-score metric. Unlike accuracy, which simply measures the overall proportion of correct predictions across all test samples, the F-score offers a more robust evaluation by taking into account the model's perfor-mance on individual classes, particularly in datasets where there is a significant imbalance between the number of instances belonging to different classes. For a binary classification task, the F-score (1) characterizes the harmonic mean of precision and recall. Precision denotes the ratio of accurately predicted positive instances to all predicted positive instances, whereas recall measures the frac-tion of accurately predicted positive instances out of all actual positive instances. This means that the F-score provides a more comprehensive understanding of the model's ability to correctly classify instances across various classes, making it particularly useful in scenarios where class distribution is uneven. Conversely, in our multi-class classification task, we employ the weighted F-score, which computes the average F-score adjusted by the count of instances in each class. The F-score spans between 0 and 1, where a score of 1 denotes perfect precision and recall, and a score of 0 signifies entirely incorrect predictions made by the model.

$$F score = 2 \times \frac{(precision \times recall)}{(precision + recall)} \quad (1)$$

4.2   Dataset

We use a proprietary dataset from Australian Data Clearing House[1]. The dataset comprises text point names from 129 buildings (including corporate offices, libraries, schools, research labs, etc.) distributed across 30 different sites in Aus-tralia. Altogether, there are 28580 instances (name strings) of heterogeneous points in the dataset. To assess the performance of the proposed approaches, we split entire dataset into two non-overlapping subsets: *training set* and *testing set*. The *training set* comprises 80% of the points in the entire dataset. It is used to find the optimal hyper-parameters of the ML algorithms and train the classifiers. On the other hand, the *testing set* contains remaining 20% of the point names and is used as the *hold-out* set to compute the accuracy (F-score, *in subsequent part of this paper, we use accuracy and F-score interchangeably*) of the classifiers.

---

[1] https://research.csiro.au/dch/.

It is important to note that the data in the *testing set* was not exposed to the model during the training phase or model development. The split of the data between two subsets (i.e., *training set* and *testing set*) has been done randomly in a stratified manner to maintain the ratio of instances across classes in both subsets. To exclude any possible bias in random split of data between *training set* and *testing set*, we repeat data split 10 times and re-run the experiments. In the subsequent discussion, we report the mean value of the assessment metric obtained for 10 runs of the experiments unless otherwise stated.

4.3   Classification Accuracy

Table 1 presents the predictive skill in terms of F-score of both binary and multi-class classifiers (employing the embedding from BERT model as inputs) in the first approach. The results using the embedding from other language models like RoBERTa and MPNET are very similar and hence are not shown here for brevity. The binary classifiers in the hierarchical approach apply 4 different ML models (DT, AdaBoost, RF, and SVM) to decide if any point have certain properties or not. All these 4 models achieves over 80% classification accuracy: F-score varies between 0.83 ± 0.06 to 0.96 ± 0.03 regardless of the properties we consider. However, a comparison of the performance for the set of binary classifiers with the 4 ML models indicates that the classifiers with SVM show the highest accuracy followed by RF and AdaBoost. The classifiers with DT show the lowest accuracy which is not surprising as DT is more susceptible to over-fitting and affected by the noise in the data originated by the inconsistencies in the naming conventions of points. Besides, the performance of the binary classifier for different properties show that the existence of *Electrical Phase Count, Power Flow*, and *Power Complexity* can be predicted with relatively higher accuracy: F-score is 0.96 for these 3 properties vs 0.91 for *Electrical Phases*.

Like the binary classifiers, the final multi-class classifiers also show the best performance with SVM and RF for all the properties. In addition, the results indicate that prediction of the values for *Electrical Phases* is least accurate with a F-score 0.68 ± 0.21 with the best performing SVM model. The values for remaining three properties can be predicted with a F-score of 0.84 to 0.89. The main reason of lower accuracy for *Electrical Phases* compared to other properties is similarity among possible values. For example, the *Power Flow* can have 3 different values {*net, export, and import*}. The classifiers can easily distinguish the values of this properties from their dissimilar embedding generated by BERT from the point names. In contrast, there are 7 possible values for *Electrical Phases* that include {*A, B, C, AB, AC, BC, ABC*}. The embeddings for the point names with these values especially {A, AB, AC}, {AB, ABC}, {B, BC} show similarity which makes the prediction challenging. Utilisation of time series data as a complementary input to the model in such cases could be useful to improve the accuracy of predictions.

Moreover, the multi-class classifiers in hierarchical approach using the embedding generated based on MPNET and RoBERTa also exhibits similar performance. Like the results with BERT, SVM achieves the best F-score using embed-

Table 1. Performance of binary and multi-class classifiers (utilising embedding from BERT model) in the hierarchical approach. $x \pm y$ indicates the average (x) $\pm$ std. dev. (y) of the assessment metric (F-score) across 10 runs of the experiments based on different random 80:20 splits of the instances into training and testing sets.

ML Model	Electrical Phase Count	Electrical Phases	Power Flow	Power Complexity
F-score (binary classifiers)				
DT	0.84±0.05	0.83±0.06	0.88±0.03	0.89±0.04
AdaBoost	0.87±0.05	0.87±0.06	0.90±0.05	0.90±0.03
RF	0.93±0.03	0.91±0.06	0.94±0.04	0.95±0.03
SVM	0.96±0.04	0.91±0.07	0.96±0.03	0.96±0.04
F-score (multi-class classifiers)				
DT	0.76±0.11	0.43±0.07	0.73±0.11	0.68±0.12
AdaBoost	0.76±0.08	0.24±0.04	0.62±0.19	0.50±0.16
RF	0.84±0.08	0.52±0.13	0.76±0.22	0.86±0.13
SVM	0.88±0.08	0.68±0.21	0.84±0.15	0.89±0.15

ding from either NLP model. However, a comparison of the performance of the multi-class classifiers using embedding from 3 NLP models indicate that, there is no significant variation in their prediction accuracy (F-score) regardless of the NLP model considered. For instance, the F-score of muti-label classifier with best performing SVM model are 0.68–0.89, 0.68–0.87, 0.72–0.88 using the embedding from BERT, MPNET, and RoBERTa, respectively. This highlights that similar strength of the 3 NLP models to generated useful embedding from the point name strings in our dataset and the ability of the ML classifiers to generalise the relationship between the inputs (embedding) and the outputs (values of properties) in the dataset.

Table 2. Performance of the single step approach for predicting point properties.

ML Model	F-score			
	Electrical Phase Count	Electrical Phases	Power Flow	Power Complexity
DT	0.79±0.06	0.68±0.12	0.85±0.05	0.85±0.05
AdaBoost	0.71±0.11	0.65±0.08	0.69±0.16	0.80±0.13
RF	0.86±0.07	0.76±0.09	0.92±0.09	0.91±0.05
SVM	0.91±0.09	0.87±0.07	0.95±0.04	0.94±0.05

Table 2 shows the performance of the single step approach utilising embedding from BERT NLP model. Similar to the hierarchical approach, the single step approach also achieves best performance with SVM followed by RF regardless of the properties. The F-score of the classifiers with best performing ML model (i.e.,

SVM) are 0.91, 0.87, 0.95, and 0.94 for *Electrical Phase Count*, *Electrical Phases*, *Power Flow*, and *Power Complexity*, respectively. The F-score for these 4 properties when utilising embedding from MPNET are in the range of 0.85–0.95 and 0.88–0.96, respectively. Figure 2 presents a comparison of performance of both approaches utilising BERT embedding for classification of 4 types of properties. As we can see, the single step approach outperforms the hierarchical approach using all ML models and for all properties. A comparison of both approaches with the best performing ML model (SVM) indicates that the single step approach achieves 3.4%, 14.7%, 13.09%, and 5.62% improvement of performance in terms of F-score over the hierarchical approach for *Electrical Phase Count*, *Electrical Phases*, *Power Flow*, and *Power Complexity*, respectively. These improvements are found statistically significant at $p \leq 0.05$. In addition, the better performance of the single step approach is achieved with less computational requirements – the hierarchical approach requires double training time compared to the single step approach as it requires to train both a binary and a multi-class classifier for each properties.

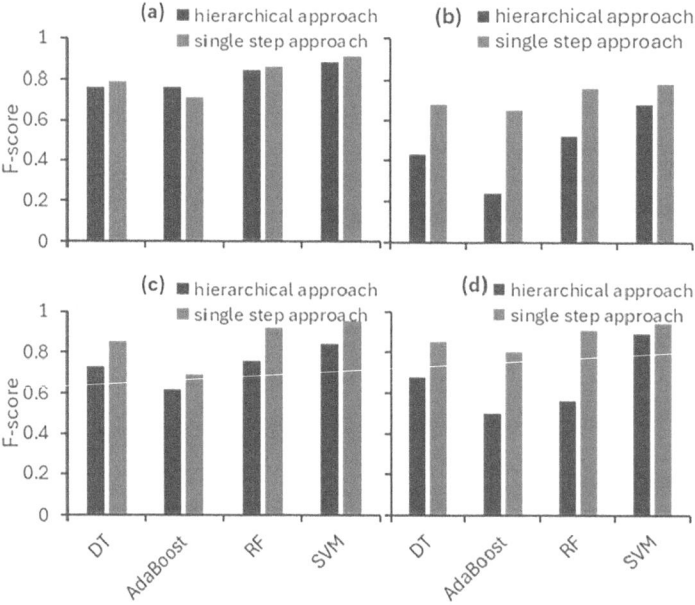

Fig. 2. Comparison of accuracy for hierarchical and single step approaches (using BERT embedding): (a) Electrical Phase Count, (b) Electrical Phases, (c) Power Flow, and (d) Power Complexity.

5   Conclusions

We introduced two machine learning based approaches (a hierarchical and a single step) for predicting point properties in buildings. The hierarchical approach employs binary classifiers to determine if a point possesses certain properties and multi-class classifiers to predict the values of these properties. On the other hand, the single-step approach directly predicts the values of point properties using multi-class classifiers. Both approaches utilise embeddings generated from point names by a NLP model which consider the semantic relationships among the words within the name strings.

Experimental results using a dataset from 129 buildings demonstrated that the single-step approach outperforms the hierarchical approach, achieving higher accuracy across all properties. For instance, the single-step approach achieves an improvement of up to 14.7% in F-score compared to the hierarchical approach. Additionally, the single-step approach requires less computational resources, making it more efficient.

The presented results support widespread applications of energy analytics tools by standardizing point data interpretation, ensuring inter-operability, and streamlining development. Accurately predicting point properties facilitates consistent data understanding, enabling seamless tool integration across buildings. This standardization enhances adaptability, reducing customization needs and optimizing tool development and deployment efficiency. Consequently, energy analytics tools become easily transferable and applicable across buildings, leading to improved energy efficiency and sustainability in smart building environments.

References

1. Project haystack (2014). https://project-haystack.org/
2. Almashor, M., Rana, M., McCulloch, J., Rahman, A., Sethuvenkatraman, S.: What's The Point: autoencoding building point names. In: Proceedings of the 10th ACM International Conference on Systems for Energy-Efficient Buildings, Cities, and Transportation (BuildSys'23), pp. 256–260 (Nov 2023)
3. Balaji, B., et al.: Brick: towards a unified metadata schema for buildings. In: Proceedings of the 3rd ACM International Conference on Systems for Energy-Efficient Built Environments, pp. 41–50 (2016)
4. Balaji, B., Verma, C., Narayanaswamy, B., Agarwal, Y.: Zodiac: organizing large deployment of sensors to create reusable applications for buildings. In: Proceedings of the 2nd ACM International Conference on Embedded Systems for Energy-Efficient Built Environments, pp. 13–22 (2015)
5. Breiman, L.: Random forests. Mach. Learn. **45**, 5–32 (2001)
6. Cortes, C., Vapnik, V.: Support-vector networks. Mach. Learn. **20**, 273–297 (1995)
7. Department of Climate Change, Energy, the Environment and Water, Australia: Commercial Buildings Energy Consumption Baseline Study 2022. Tech. rep. (2022). https://www.energy.gov.au/publications/commercial-buildings-energy-consumption-baseline-study-2022

8. Devlin, J., Chang, M.W., Lee, K., Toutanova, K.: BERT: pre-training of deep bidirectional transformers for language understanding. arXiv preprint arXiv:1810.04805 (2018)
9. Eskin, E., Weston, J., Noble, W., Leslie, C.: Mismatch string kernels for SVM protein classification. In: Advances in Neural Information Processing Systems, vol. 15 (2002)
10. EU Directorate-General for Climate Action: 2030 Climate Target Plan (2022). https://ec.europa.eu/clima/eu-action/european-green-deal/2030-climate-target-plan_en, European Commission
11. Freund, Y., Schapire, R.E.: A desicion-theoretic generalization of on-line learning and an application to boosting. In: Vitányi, P. (ed.) EuroCOLT 1995. LNCS, vol. 904, pp. 23–37. Springer, Heidelberg (1995). https://doi.org/10.1007/3-540-59119-2_166
12. Gao, J., Ploennigs, J., Berges, M.: A data-driven meta-data inference framework for building automation systems. In: Proceedings of the 2nd ACM International Conference on Embedded Systems for Energy-Efficient Built Environments, pp. 23–32 (2015)
13. He, F., Wang, D.: Cloze: a building metadata model generation system based on information extraction. In: Proceedings of the 9th ACM International Conference on Systems for Energy-Efficient Buildings, Cities, and Transportation (BuildSys'22), pp. 109–118 (2022)
14. Hong, D., Gu, Q., Whitehouse, K.: High-dimensional time series clustering via cross-predictability. In: Proceedings of the 20th International Conference on Artificial Intelligence and Statistics, pp. 642–651 (2017), ISSN: 2640-3498
15. Hong, D., Wang, H., Ortiz, J., Whitehouse, K.: The building adapter: towards quickly applying building analytics at scale. In: Proceedings of the 2nd ACM International Conference on Embedded Systems for Energy-Efficient Built Environments, pp. 123–132 (2015)
16. Jiao, Y., Li, J., Wu, J., Hong, D., Gupta, R., Shang, J.: SeNsER: learning cross-building sensor metadata tagger. In: Findings of the Association for Computational Linguistics: EMNLP 2020, pp. 950–960. Association for Computational Linguistics, Online (2020)
17. Koh, J., Balaji, B., Sengupta, D., McAuley, J., Gupta, R., Agarwal, Y.: Scrabble: transferrable semi-automated semantic metadata normalization using intermediate representation. In: Proceedings of the 5th Conference on Systems for Built Environments, pp. 11–20 (2018)
18. Liu, Y., et al.: RoBERTa: a robustly optimized BERT pretraining approach. arXiv preprint arXiv:1907.11692 (2019)
19. Mishra, S., et al.: Unified architecture for data-driven metadata tagging of building automation systems. Autom. Constr. **120**, 103411 (2020)
20. Quinlan, J.R.: C4. 5: programs for machine learning. Elsevier (2014)
21. Rana, M., Rahman, A., Almashor, M., McCulloch, J., Sethuvenkatraman, S.: Automatic classification of sensors in buildings: learning from time series data. In: Australasian Joint Conference on Artificial Intelligence, pp. 367–378. Springer (2023)
22. Song, K., Tan, X., Qin, T., Lu, J., Liu, T.Y.: MPNet: masked and permuted pre-training for language understanding. Adv. Neural. Inf. Process. Syst. **33**, 16857–16867 (2020)
23. Waterworth, D., Sethuvenkatraman, S., Sheng, Q.Z.: Advancing smart building readiness: automated metadata extraction using neural language processing methods. Adv. Appl. Energy **3**, 100041 (2021)

Integrating Explainable AI with Multisource Time-Series Data for Apple Yield Prediction

Boyuan Zheng[1]($\boxtimes$) , Zhitan Wu[2], and Victor W. Chu[1]

[1] University of Technology Sydney, Sydney, NSW 2007, Australia
{Boyuan.Zheng,WingYan.Chu}@uts.edu.au
[2] The University of New South Wales, Sydney, NSW 2052, Australia
zhitan.wu@student.unsw.edu.au

Abstract. Accurate apple yield prediction is critical for sustainable orchard management and food security, especially under the growing threat of climate variability. This study presents a comprehensive framework that integrates multisource time-series data with explainable artificial intelligence techniques to enhance apple yield forecasting. We collect and align daily weather data, extreme weather records, and annual apple yield data, focusing on major apple-producing regions in the United States. Sequential machine learning models are employed to capture temporal dependencies in climate and yield patterns. To address the interpretability challenges of complex models, we apply SHapley Additive exPlanations (SHAP) to provide post-hoc insights into feature contributions, further aligned with key apple phenological stages. Importantly, the generated explanations exhibit strong consistency with established domain knowledge, confirming the biological relevance of key climate-yield interactions. Our results demonstrate that LSTM with an attention mechanism achieves the highest predictive accuracy (86.64%) across all regions, while SHAP-based interpretations reveal the dynamic influence of climate factors at different growth stages. This study highlights the importance of integrating XAI into agricultural modeling, enabling stakeholders to make informed decisions based on both accurate predictions and transparent, domain-aligned explanations.

Keywords: Yield prediction · Explainable AI · Time-series modeling

1 Introduction

Urban development is dependent on agriculture for essential resources, including food security, economic stability, and environmental management. Among various agricultural sectors, apple production stands out for its substantial economic contributions. However, apple yields have become increasingly vulnerable to climate-related stresses. Extreme weather events, including frost, prolonged droughts, and heatwaves, have caused severe damage on apple crops. A notable example is the catastrophic 90% decline in Michigan's apple production in 2012, caused by early-season heat accumulation followed by severe frost [7]. Such events underscore the need for advanced apple yield forecasting systems. To mitigate the impact of climate change on apple production, more robust and high-performing artificial intelligence (AI) models are required to unravel the complex dependencies between climate and crop performance, providing farmers, policymakers, and industry stakeholders with essential tools for decision-making.

Modeling apple yields under such dynamic conditions presents numerous practical challenges. Traditional statistical approaches often fail to account for complex, non-linear relationships between climate variables and crop yield. Recent advances in machine learning (ML) have improved yield prediction accuracy but have introduced new challenges related to model interpretability. Many ML models operate as "black boxes", preventing stakeholders from understanding, trusting, and acting upon their predictions. In this case, eXplainable AI (XAI) offers a bottleneck to address these concerns by providing transparency into how models generate predictions. For apple farmers, agronomists, and policymakers, interpretability is not just a technical consideration but a practical necessity. For instance, orchard managers could use feature explanations to optimize irrigation and fertilizer schedules [10]. In regions prone to extreme events, clear model explanations can also inform crop insurance decisions and disaster preparedness.

To address the abovementioned problems, this paper contributes to the field by presenting a comprehensive framework for explainable apple yield modeling, which integrates multiple data sources and preprocessing steps to capture the complex relationships between climate factors and apple yield. We perform a detailed statistical analysis and association mining to uncover significant feature interactions. Various sequential models are then trained on the processed data to account for the temporal dependencies. To enhance interpretability, we explain feature contributions with respect to key stages of the apple lifecycle, such as dormancy, bud break, fruit set, and harvest. Finally, we evaluate the alignment between the model-generated explanations and expert domain knowledge from literature, identifying areas where model insights support or diverge from established agricultural practices.

2   Related Work

2.1   Apple Yield Modeling

Accurate modeling of apple yield is crucial for optimizing orchard management, ensuring food security, and mitigating the impacts of climate variability. Over the years, various methodologies have been explored for yield prediction, ranging from traditional statistical approaches to advanced machine learning techniques. Early studies, such as that by Aggelopoulou et al. [1], investigated the spatial variation in apple yield and quality within a small orchard in northern Greece over two growing seasons using statistical methods. Their findings revealed a strong correlation between flowering intensity and final yield, while certain fruit quality attributes, including soluble solids content and acidity, were found to impact overall yield. Subsequent research leveraged more sophisticated techniques for early apple yield prediction. Cheng et al. [8] introduced an approach integrating image processing with Support Vector Machine (SVM) modeling, enabling yield estimation based on canopy fruit and foliage characteristics. This method demonstrated improved predictive accuracy compared to statistical models, marking a transition toward data-driven modeling in precision agriculture. With the rapid advancement of neural network-based techniques, such models have been increasingly adopted in apple yield prediction. Sun et al. [14] proposed a UAV-based multimodal measurement approach to monitor orchard canopy. Other existing work such as [6,13] underscoring the potential of advanced computational methodologies in precision agriculture.

2.2   XAI in Agriculture

The increasing use of ML models in agriculture has intensified the need for XAI to ensure that predictions and recommendations are interpretable for domain experts, farmers, and policymakers. XAI plays a crucial role in bridging the gap between complex black-box models and practical decision-making by revealing model behavior and feature importance. Generally, XAI methods in agriculture can be grouped into two main categories: (i) inherently interpretable models and (ii) post-hoc explainability techniques. Inherently interpretable models, such as linear regression, decision trees, and K-nearest neighbors, are transparent by design and often serve as baselines in agricultural studies. For instance, Hu et al. [9] used linear regression to predict crop yield, highlighting its transparency and ease of interpretation. In these models, the magnitude and direction of feature contributions can be inferred from model coefficients. However, despite their interpretability, such models often struggle to capture the complex, non-linear relationships present in agricultural data [5]. Post-hoc explainability techniques, on the other hand, focus on interpreting trained models, offering insights into how predictions are generated, particularly for complex deep learning models that lack intrinsic interpretability. Local Interpretable Model-agnostic Explanations (LIME) were applied existing work such as [2,12] to assess feature importance in yield prediction, identifying key predictors like meteorological, agrochemical, and soil physiographic factors. Similarly, Paudel et al. [11] engaged domain experts in feature selection for non-neural network structured models and conducted a survey to rank the influential features before model training. SHAP explanations were then compared with expert empirical predictions to assess explanation quality. In this study, we adopt post-hoc explainability techniques to enhance the interpretability of sequential models, enabling a deeper understanding of how climate variables influence yield predictions over time.

3   Methodology

Predicting apple yield under changing climate conditions presents a significant challenge due to the intricate interplay between environmental factors, extreme weather events, and crop physiology. ML techniques provide a powerful means of modeling these dependencies by leveraging historical data to identify patterns and enhance yield forecasts. However, existing frameworks often lack a comprehensive approach that integrates multiple data sources while ensuring interpretability for stakeholders in the apple industry. Figure 1 contrasts conventional apple yield modeling workflows with the proposed framework, which addresses these limitations. Our methodology follows a structured pipeline comprising data collection, preprocessing and alignment, data mining and modeling, and explanation generation. As illustrated in Fig. 1b, this framework enables the integration of multisource data to improve the predictive accuracy of apple yield models, particularly under the influence of extreme weather conditions.

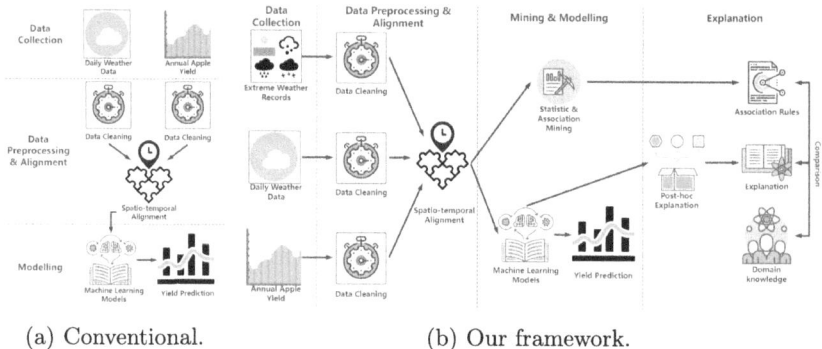

(a) Conventional.   (b) Our framework.

Fig. 1. Comparison between traditional workflow and proposed framework.

The first stage, data collection, consists of gathering three primary types of data: daily weather data, annual apple yield records, and extreme weather records. Daily weather data includes key meteorological variables such as temperature, precipitation, solar radiation, and atmospheric conditions, which are essential for capturing seasonal patterns that influence yield variability. Annual apple yield data, serving as the target variable, is aggregated at the regional level over a period exceeding ten years to account for long-term trends and interannual variability. Although climate change has been reported to have a limited direct impact on global food supply [4], extreme weather events can result in substantial agricultural losses. In this context, we incorporate extreme weather records to capture historical climate shocks, including heatwaves, droughts, and frosts, which have pronounced implications for apple yield in the target regions.

Following data collection, the data preprocessing and alignment phase ensures that the dataset is cleaned and structured for modeling. Data cleaning involves handling missing values and removing redundant features. Since both spatial and temporal factors play a critical role in agricultural modeling, a spatio-temporal alignment process is applied. Spatial alignment ensures that weather and yield data are mapped to specific regions, while temporal alignment synchronizes the data into the same granularity.

In the data mining and modeling stage, statistical and association mining techniques are applied to identify key feature correlations and dependencies, ensuring that relevant climate variables are effectively incorporated into predictive models. Explicit feature selection step is omitted, as the selected machine learning models inherently possess mechanisms for automatic feature selection. Subsequently, a set of machine learning models is employed to forecast apple yield, with a particular emphasis on sequential pattern learning. The selected models include Long Short-Term Memory (LSTM), Gated Recurrent Units (GRU), and one-dimensional Convolutional Neural Networks (1D-CNN), all of which are well-suited for capturing temporal dependencies in yield patterns. Additionally, attention mechanisms are integrated to enhance the performance of deep learning models and facilitate temporal analysis by visualizing attention weights.

The final stage, explanation generation, focuses on improving the interpretability of yield predictions. Post-hoc explainability techniques, SHapley Additive exPlanations (SHAP), is applied to quantify the contribution of individual climate features to yield predictions. Association rules extracted in data mining step is also employed to uncover the relationships between statistical results and post-hoc explanations. The generated explanations are also compared with domain knowledge extracted from existing literature to ensure alignment with real-world agronomic insights. These validation steps are critical for enhancing stakeholder trust in AI-driven predictions and ensuring that the model's outputs can inform practical decision-making in orchard management.

Overall, our proposed workflow establishes a comprehensive framework that integrates data-driven yield prediction with domain-specific explainability, enabling a deeper understanding of the impact of climate variability on apple production. This methodology not only improves predictive accuracy but also ensures that the decision-support system remains interpretable, transparent, and reliable for agricultural stakeholders.

4 Overview of Domain Data and Implementation Details

4.1 Data Scope and Granularity

This study leverages multisource data spanning from January 1, 2007, to December 31, 2023, integrating climate variables, extreme weather events, and annual apple yield records. The data sources include:

- **Daily Weather Data**: Obtained from NOAA Climate Data Online, providing meteorological variables recorded at the weather station level.
- **Extreme Weather Events**: Sourced from the NOAA Storm Events Database, originally available at an hourly resolution and recorded at the county level.
- **Apple Yield Data**: Collected from the USDA, available at a yearly resolution and aggregated at the state level.

To ensure consistency in temporal resolution, the extreme weather data is upsampled to a daily scale. For model input, all daily features are concatenated into a sequential tensor representation of size $[365 \times \text{num\_features}]$, ensuring alignment with annual yield records. However, the datasets exhibit inherent spatial misalignment due to differences in data granularity: daily weather data is recorded at the weather station level, extreme event data is county-level, and yield data is reported at the state level. To address this discrepancy, we focus on four key apple-producing counties from the top apple-producing states in the U.S.: Yakima (WA), Sonoma (CA), Kent (MI), and Adams (PA). These counties represent primary apple production regions within their respective states, ensuring a regionally representative dataset. For daily climate data, we selected airport-based weather stations within each county, as they provide a more meteorological variables compared to non-airport stations. The integration of these data sources enables a refined spatial representation, aligning county-level extreme events and station-level weather data with state-level yield records.

4.2   Implementation Details

Following spatio-temporal alignment, the aggregated dataset undergoes statistical preprocessing and exploratory analysis. To examine the relationships between climate factors, extreme weather events, and yield fluctuations, we compute the correlation matrix of key meteorological variables. Furthermore, association rule mining is applied to uncover frequent co-occurring patterns between climate conditions and yield outcomes. We set the minimum support threshold to 0.25 and the confidence threshold to 0.4, ensuring that only significant associations are considered in subsequent analyses.

The machine learning models employed for apple yield forecasting capture temporal dependencies in climate-driven yield patterns, utilizing 1D-CNN, GRU, LSTM, and LSTM with an attention mechanism. Each model is trained with the Adam optimizer (1×10^{-3} learning rate) and the Huber loss function ($\delta = 1.0$) for robustness against outliers. The experiments were conducted on an NVIDIA Tesla V100-PCIE-32GB GPU. The 1D-CNN model consists of two convolutional layers with ReLU activation, followed by max-pooling and dropout (0.2) for regularization, with extracted features processed through a fully connected layer using LeakyReLU and L_2 regularization. The GRU and LSTM architectures comprise three stacked layers (150, 100, 50 units) with LeakyReLU activation, dropout (0.2), and L_2-regularized dense output layers. To enhance temporal focus, an attention-augmented LSTM is employed, assigning dynamic importance to time steps by applying the attention mechanism to final LSTM outputs before concatenation with the last hidden state.

To enhance the interpretability of the proposed model, we employ SHAP as a post-hoc explainability method. The SHAP values are visualized using beeswarm and waterfall plots to illustrate the contributions of different climate variables at various stages of the prediction process. To ensure that the explanations remain biologically meaningful, the SHAP value matrix is restructured to align with the metabolic stages of apple growth. Specifically, feature importance analysis is segmented according to key phenological phases, including bud break, flowering, fruit set, and harvest, thereby providing an intuitive understanding of how climate variables influence apple yield across different developmental stages.

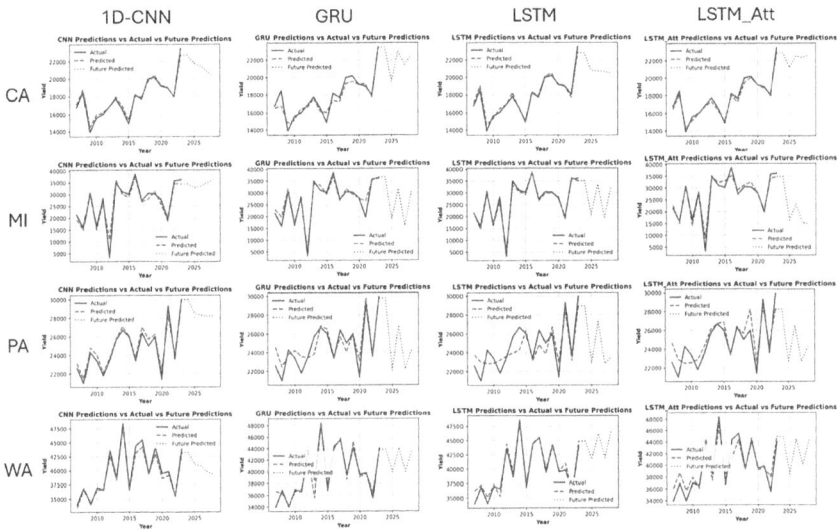

Fig. 2. Model Forecasting Performance: The blue line represents the ground truth yield values, the red line shows the predicted values, and the green dashed line indicates future values. (Color figure online)

5   Results

Among all the models evaluated, the LSTM with an attention mechanism achieved the highest average prediction accuracy (**86.64%**) across all states, demonstrating strong generalization performance. However, model performance varied significantly across different states. Notably, LSTM_Att exhibited the highest accuracy in Washington (WA) (**94.55%**) and Michigan (MI) (**87.57%**), whereas the GRU model performed best in Pennsylvania (PA) (**89.25%**). The 1D-CNN model, despite its strong performance, was outperformed by LSTM_Att in most cases, except in California (CA), where it achieved the highest accuracy (**86.95%**). When assessing visual consistency, the CA dataset displayed relatively stable performance across models (see Fig. 2), with minimal variation in accuracy between different architectures. Conversely, the MI dataset showed the greatest fluctuation, with LSTM with attention layer performing significantly better than GRU and LSTM, indicating that sequential architectures with attention mechanisms are more effective in capturing temporal dependencies in this region. Additionally, the strong performance of LSTM with attention layer in WA and MI suggests that these datasets contain more structured temporal patterns that are effectively leveraged by attention-based architectures. The variability in model performance can be attributed to several factors, including feature availability, dataset balance, and the presence of seasonal trends. For instance, the absence of the "Snow" attribute in the PA dataset may have contributed to the instability observed in the 1D-CNN model. The GRU model was chosen for the PA dataset, where it achieved the best prediction accuracy.

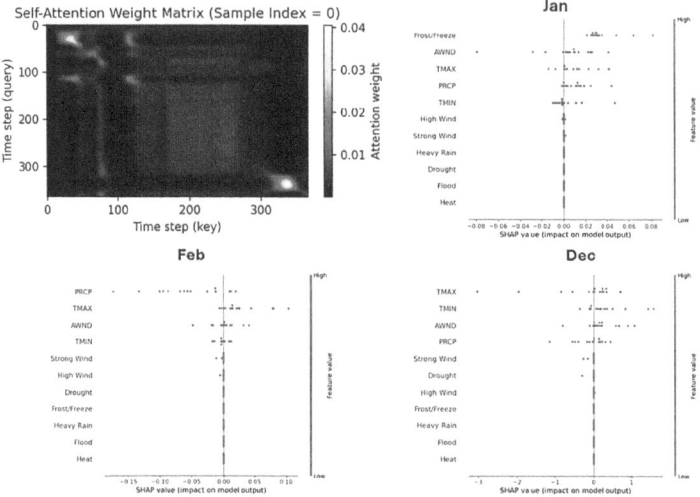

Fig. 3. Integrated visual explanations for apple yield in Sonoma, California.

Table 1. Prediction accuracy percentage with the best in bold.

	CA	WA	MI	PA	AVG
LSTM	86.21	87.86	71.25	88.38	83.43
LSTM Attention	79.50	**94.55**	**87.57**	84.95	**86.64**
1D-CNN	**86.95**	92.60	78.40	86.26	86.05
GRU	78.66	92.79	66.31	**89.25**	81.75

6   Discussion

Apple yield is strongly influenced by distinct phenological stages, each governed by specific climatic conditions. According to existing literature [3,6], bud break typically occurs in March in the Northern Hemisphere, requiring moderate temperatures and sufficient rainfall to transition into the vegetative growth stage. Flowering follows in April, favoring moderate temperatures with minimal rain and wind to ensure successful pollination. Fruit set occurs around May under warm temperatures and moderate rainfall, while fruit maturation progresses through summer, culminating in the harvest period between October and November, depending on fruit maturity, where cool temperatures and moderate rainfall are ideal. Dormancy, preceding these stages, is crucial for yield formation, as adequate chilling hours during winter are essential to synchronize bud break and subsequent growth. These climatic dependencies shape the productivity of apple orchards across different regions, influencing yield variability in response to local environmental conditions.

Figure 3 integrates temporal and feature-level explanations for apple yield predictions in Sonoma, California. Distinguished by its Mediterranean climate - mild, wet winters and dry, temperate summers moderated by maritime influences—Sonoma exhibits pronounced seasonal dynamics. By integrating self-attention maps and feature-level SHAP values, the visualization underscores how winter conditions and relevant meteorological variables jointly drive the apple yield outcomes in this fruit-growing region. The upper-left panel presents the model's self-attention weight matrix over 365 daily time steps, highlighting distinct "hot spots" during December–February. This temporal emphasis aligns with key orchard management and physiological processes, such as chill-hour accumulation, bud differentiation, and dormancy release, which are sensitive to Sonoma's cool, wet winters. The SHAP plots further emphasize the role of maximum/minimum temperatures (TMAX, TMIN), average wind speed (AWND), precipitation (PRCP), and frost/freeze events during this period, which aligns with the domain knowledge in existing literature. These climatic variables are crucial in Sonoma's microclimate, where moderated extremes still permit winter frosts and substantial precipitation, fostering optimal chilling conditions for uniform bud break. However, the occurrence of frost events during this vulnerable stage poses significant risks to sprout, impacting apple yield.

Figure 4 offers an integrated explanation of apple yield predictions in Kent County, Michigan, combining temporal dynamics with influential climatic factors. Kent's climate is shaped by lake-effect weather from the Great Lakes, resulting in heavier winter snowfall and moderate summers conducive to orchard health. The self-attention matrix highlights critical time windows in January–March, underscoring the importance of winter dormancy and chill-hour accumulation for synchronized bud development and flowering. The accompanying SHAP plots reveal that snow depth (SNWD), temperature extremes, and precipitation are dominant predictors during this phase, while wind speed highlights the orchard's exposure to cold air masses. These insights reflect Kent's climatic realities, where prolonged snowfall and sub-freezing temperatures influence bud survival, metabolic activity, and overall orchard management practices.

Figure 5 presents an integrated explanation of apple yield predictions in Adams County, Pennsylvania, which experiences a humid continental climate

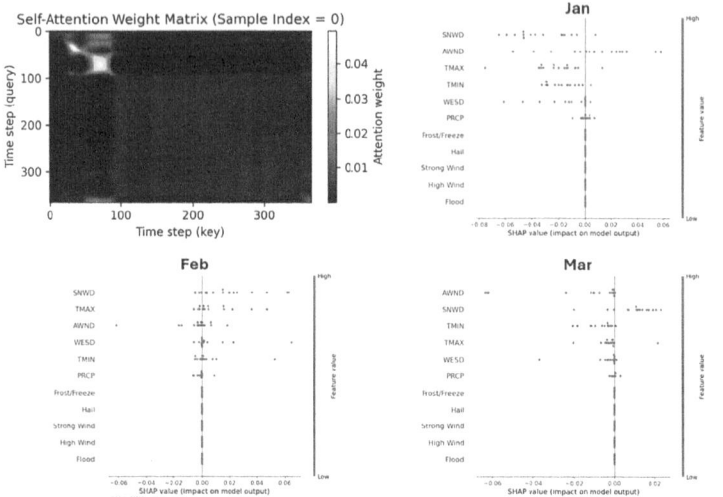

Fig. 4. Integrated visual explanations for apple yield in Kent County, Michigan.

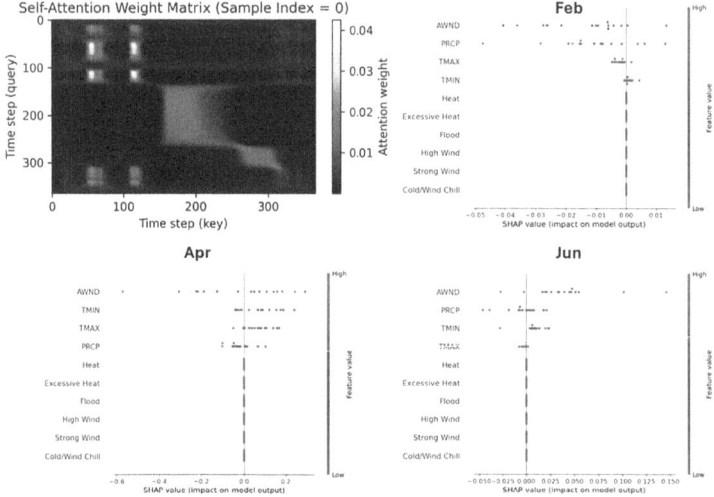

Fig. 5. Integrated visual explanations for apple yield in Adams, Pennsylvania.

characterized by consistent precipitation and seasonal variability. The attention matrix indicates heightened model focus during February, as well as April–June. The identified importance of February to June reflects key biological transitions from dormancy release to flowering and fruit set, aligning with known stages where moderate temperatures and rainfall support bud break (March) and flowering (April), followed by fruit set in May. SHAP plots identify wind speed, precipitation, and temperature extremes as key factors driving model

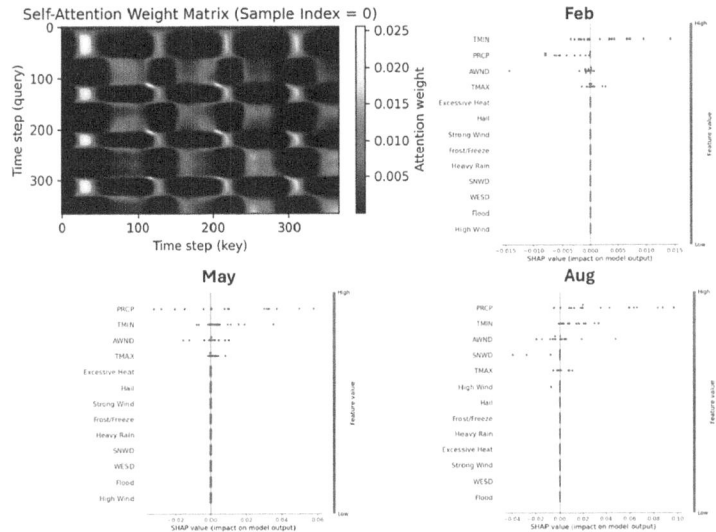

Fig. 6. Integrated visual explanations for apple yield in Yakima, Washington.

predictions during these critical growth stages. Consistent spring precipitation supports uniform bud emergence, while cooler June temperatures can delay fruit maturation. These climatic patterns, inherent to Adams County, interact with the apple tree's metabolic cycles, emphasizing the importance of timely chill accumulation, bud viability, and the smooth progression from dormancy to fruit development.

Figure 6 provides a combined temporal and feature-level interpretation of apple yield predictions in Yakima, Washington, a region defined by its semi-arid climate with cool winters and hot, dry summers necessitating extensive irrigation. The self-attention matrix reveals stripe-like patterns, indicating the model's recurrent focus on key phenological stages such as dormancy release, blossoming, and fruit maturation. The SHAP plots highlight February, May, and August as pivotal months, where temperature, precipitation, and wind conditions substantially influence yield outcomes. In late winter, chill-hour accumulation ensures uniform bud set, while strategic spring and summer irrigation, alongside favorable wind conditions, support successful pollination and fruit enlargement. These patterns reflect Yakima's reliance on managed water resources and its sensitivity to temperature fluctuations and wind exposure, underscoring the complex interplay between climate and orchard management in this semi-arid region.

7   Conclusion

This study introduces a robust and interpretable framework for apple yield prediction that integrates multisource time-series data with advanced machine learning models and explainable AI techniques. By leveraging sequential models alongside attention mechanisms, we capture complex temporal dependencies critical to accurate yield forecasting. The application of SHAP enables interpretations of climate-yield interactions, providing actionable insights for orchard managers, agronomists, and policymakers. This work demonstrates the consistency between model-generated explanations and expert domain knowledge, validating the biological plausibility of the identified climate influences on apple yield. This alignment not only enhances stakeholder trust but also supports practical decision-making in orchard management.

However, one notable limitation of this study lies in the encoding and representation of extreme weather events. While the current approach integrates extreme event records into the modeling pipeline, it may not fully capture the complex and episodic nature of such events and their nuanced impacts on apple yield. Future work could explore more sophisticated methods for representing extreme events, such as event-based temporal encoding or advanced data augmentation techniques, to better capture their influence on crop performance. In addition, future research could also focus on integrating additional data sources, such as soil conditions and management practices, and refining explainability techniques to further strengthen the coherence between model outputs and domain expertise, enhancing the model's practical applicability in real-world agricultural decision-making.

References

1. Aggelopoulou, K., Wulfsohn, D., Fountas, S., Gemtos, T., Nanos, G., Blackmore, S.: Spatial variation in yield and quality in a small apple orchard. Precision Agric. **11**, 538–556 (2010)
2. Ahmed, M.S., et al.: Yield response of different rice ecotypes to meteorological, agro-chemical, and soil physiographic factors for interpretable precision agriculture using extreme gradient boosting and support vector regression. Complexity **2022**, 1–20 (2022)
3. Bai, X., et al.: Comparison of machine-learning and casa models for predicting apple fruit yields from time-series planet imageries. Remote Sens. **13**(16), 3073 (2021)
4. Betts, R.A., et al.: Changes in climate extremes, fresh water availability and vulnerability to food insecurity projected at 1.5 c and 2 c global warming with a higher-resolution global climate model. Philos. Trans. Royal Soc. A: Math. Phys. Eng. Sci. **376**(2119), 20160452 (2018)
5. Cai, Y., et al.: Integrating satellite and climate data to predict wheat yield in Australia using machine learning approaches. Agric. For. Meteorol. **274**, 144–159 (2019)
6. Duan, Z., et al.: Cold climate during bud break and flowering and excessive nutrient inputs limit apple yields in Hebei province, China. Horticulturae **8**(12), 1131 (2022)
7. Gallardo, R.K., et al.: Perceptions of precision agriculture technologies in the us fresh apple industry. HortTechnology **29**(2), 151–162 (2019)
8. Hong, C., Damerow, L., Blanke, M., Yurui, S.: Early yield estimation of 'Gala' apple trees using image processing combined with support vector machine. Nongye Jixie Xuebao/Trans. Chin. Soc. Agric. Mach. **46**(3) (2015)

9. Hu, T., et al.: Crop yield prediction via explainable AI and interpretable machine learning: Dangers of black box models for evaluating climate change impacts on crop yield. Agric. For. Meteorol. **336**, 109458 (2023)
10. Huber, F., Engler, H., Kicherer, A., Herzog, K., Töpfer, R., Steinhage, V.: Grouping Shapley value feature importances of random forests for explainable yield prediction. In: Arai, K. (ed.) Intelligent Systems and Applications: Proceedings of the 2023 Intelligent Systems Conference (IntelliSys) Volume 3, pp. 210–228. Springer Nature Switzerland, Cham (2024). https://doi.org/10.1007/978-3-031-47715-7_15
11. Paudel, D., de Wit, A., Boogaard, H., Marcos, D., Osinga, S., Athanasiadis, I.N.: Interpretability of deep learning models for crop yield forecasting. Comput. Electron. Agric. **206**, 107663 (2023)
12. Ryo, M.: Explainable artificial intelligence and interpretable machine learning for agricultural data analysis. Artif. Intell. Agric. **6**, 257–265 (2022)
13. Singha, C., Gulzar, S., Swain, K.C., Pradhan, D.: Apple yield prediction mapping using machine learning techniques through the google earth engine cloud in Kashmir Valley, India. J. Appl. Remote Sens. **17**(1), 014505–014505 (2023)
14. Sun, G., Wang, X., Yang, H., Zhang, X.: A canopy information measurement method for modern standardized apple orchards based on UAV multimodal information. Sensors **20**(10), 2985 (2020)

Sparse Attention-Based Imputation Network for Time Series

Aman Atman$^{(\boxtimes)}$ ⓘ and Santosh Nannuru ⓘ

IIIT Hyderabad, Hyderabad, India
`aman.atman@research.iiit.ac.in`, `santosh.nannuru@iiit.ac.in`

Abstract. Time series data related to traffic and air quality are useful indicators for urban planning but they frequently have missing values. Imputation of multi-sensor time series data is thus a vital pre-processing step for forecasting, anomaly detection and other downstream tasks. We develop an efficient architecture – Sparse Attention-based Imputation Network for Time series (**SAINT**) – which outperforms the state-of-the-art imputation networks. Efficiency is achieved by separating the computations on the space-time product graph sequentially into channel independent temporal attention and sparse space-time transformer. This channel independent network can reduce overfitting to effectively represent general temporal patterns. Sparse space-time transformer performs message passing on the spatial graph conditioned on time. We consider real-world datasets for evaluation – PEMS-BAY, METR-LA and AQI – which are gathered from sensor networks in major cities. We demonstrate the effectiveness and robustness of SAINT across complex missing data scenarios. Additionally, SAINT generalizes well to short-term forecasting and is practical for long-term forecasting with limited resources.

Keywords: Time series · Imputation · Attention

1 Introduction

Missing values are ubiquitous in datasets. In the case of time series, where forecasting research has received prominent attention, it is often plagued by intermediate missing values in addition to the unknown future values. For example, sensor networks monitoring air quality or traffic may get disrupted momentarily due to connectivity issues or a sensor could stop working. Time series imputation is a vital pre-processing step for forecasting, anomaly detection and other downstream tasks. Additionally, forecasting can be modeled as an imputation problem where we consider the future values as missing.

While data analysis can also use visual features [4], we focus on works using time series features only. Methods based on auto-regressive models have been commonly applied for multi-variate time series imputation, for example, ARIMA [2] and RNN [3,5]. But these methods in addition to being slow also suffer from the problem of compounding errors. State-of-the-art methods like SPIN [8] and NRTSI [12] use attention [14] to overcome these limitations. GCASTN [11] uses

S. Yuan et al. (Eds.): PAKDD 2025 Workshops, LNAI 15835, pp. 44–55, 2025.
https://doi.org/10.1007/978-981-96-8197-6_4

missing-aware attention mechanism to adaptively learn representations. Trafformer [6] uses transformer on spatiotemporal product graph for traffic forecasting. SPIN uses graph attention network applied on space-time graph where it only considers edges between spatially connected nodes. SPIN however remains memory intensive and gives out-of-memory error under hardware constraints. SPIN-H, the smaller variant of SPIN, reduces temporal dimension to a few hierarchical dummy nodes. Despite this pooling, SPIN-H performs better than SPIN on a few benchmarks. Space-time product graphs can get huge and may even be unnecessary. Further, real-world time series are notoriously noisy and non-stationary. Complex networks often fail to capture generalizable patterns, and are prone to overfit [16].

In channel independent processing, for each variate (i.e., channel or spatial node) in a multi-variate time series, the outputs are generated independently without considering the other variates. It may seem limiting, but has been consistently used in state-of-the-art networks for multi-variate forecasting [9,16,17] and imputation [8,12]. These can be used for regularization and are effective as many of the real-world time series are noisy. Despite having a space-time attention module which jointly learns representations from the product graph, SPIN [8] requires a separate channel-independent temporal self-attention network to achieve the state-of-the-art results on imputation. Similarly [16] demonstrates improved results on forecasting by making Informer [20] channel-independent.

Transformers have also gained popularity on graph benchmarks. Sparse graph transformer like [13] operate on the edges of the graph instead of computing attention coefficient between every possible node pair. We are inspired by sparse graph transformers, as imputation can be formulated as a node regression problem on a product graph. We propose a novel adaptation of sparse graph attention to the multi-variate time-series domain. The vertices of the graph lie across both space and time.

1.1   Our Contribution

We present a novel architecture – Sparse Attention-based Imputation Network for Time series (**SAINT**). To reduce the time complexity, we propose separating computations on the product graph sequentially into—channel independent temporal attention and sparse space-time transformer. The channel independent layer can prevent overfitting and effectively represent general temporal pattens. Sparse space-time transformer performs spatial message passing for each time step independently, where 'time' refers to the temporal dimension of the space-time series. By leveraging channel-independent networks and sparse architectures, we develop an efficient network that outperforms the state-of-the-art baselines in many time-series imputation tasks. SAINT can easily fit in memory without requiring pooling unlike many other product graph based networks. Further, we show that SAINT can make long-term forecasting and imputation practical even with limited computing resources.

2   Problem Setup

We describe the multi-variate imputation problem. Let $Y \in \mathbb{R}^{N \times T}$ denote the spatiotemporal series, without any missing values. Here N denotes the number of spatial nodes and T the number of temporal nodes. $M \in \{0,1\}^{N \times T}$ is a binary mask of the same shape, denoting element wise missing (0) or present (1) values. Let the multi-variate input with missing values be $X \in \mathbb{R}^{N \times T}$ where $X \odot M = Y \odot M$. We want to learn a network f_θ which minimizes the mean absolute error (MAE) defined as,

$$\text{MAE} = \frac{1}{NT} \sum_{i=1}^{NT} |y_i - \hat{y}_i| \,, \tag{1}$$

where y is the flattened Y and $\hat{Y} = f_\theta(X, M)$.

The overall time-series T can be very long ($\sim 10^4$) and it can get too expensive to process it directly. As done by conventional forecasting and imputation baselines, we divide T into batches using a short window with a stride of 1. Therefore, from now onward T will be ~ 10.

3   Architecture

In this section, we will describe the architectural details of the Sparse Attention-based Imputation Network for Time series (SAINT). It reduces memory complexity to enable training on limited hardware, while still being able to perform computations on the space-time product graph. We achieve this by separating the computations into channel independent Time Transformer Encoder (TTE) and Sparse Transformer Encoder (STE). The components of our architecture are built using attention [14] and the importance of the individual components is empirically justified through ablation experiments.

3.1   Network Overview

In Fig. 1, we present the proposed multi-layered imputation network. Using the time series X and mask M as input, the network outputs point-wise imputations $\hat{Y}$. We borrow the initial embedding procedure to generate $H^{(0)} \in \mathbb{R}^{N \times T \times c}$ from [8]. For encoding time, we use relative positional encodings [14] and embedded day and week information. For encoding spatial information, we use learnable node embeddings ($\mathbb{R}^c$).

The network uses a sequence of several SAINT blocks to learn layer-wise representations $H^{(l)} \in \mathbb{R}^{N \times T \times c}$. Residual connections are used as proven to be effective for deep networks, mitigating vanishing gradients and slow training. Therefore, in the lth layer, we use $\text{MLP}_2(X) \times M + H^{(l-1)}$ as input to the attention backbone. The multi-layer perceptron (MLP) has one hidden layer, $\text{MLP}_2 : \mathbb{R}^1 \to \mathbb{R}^c \to \mathbb{R}^c$. We use ReLU activation across all MLP, unless otherwise stated. Additionally, not explicitly shown in the figure, we also introduce

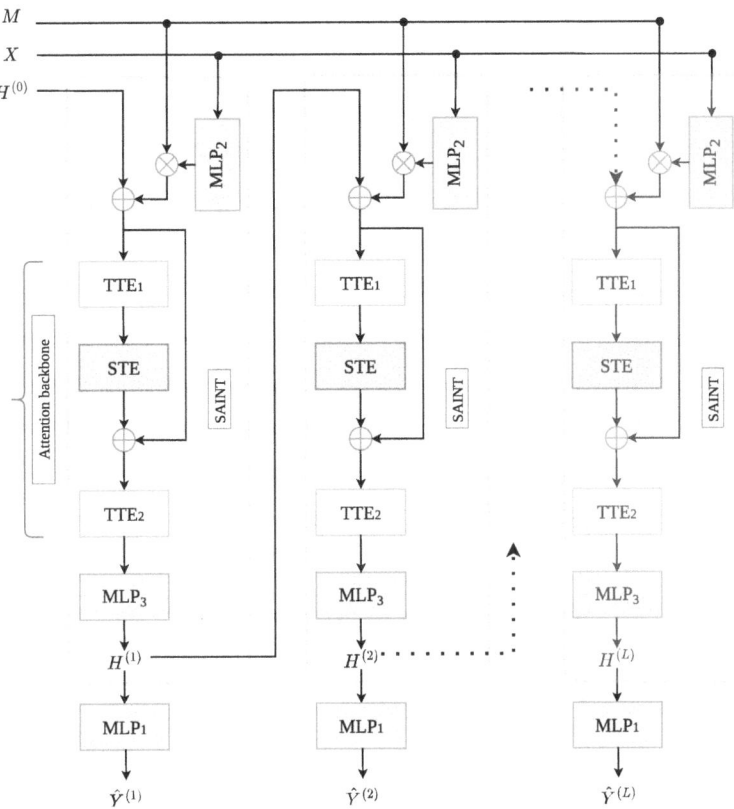

Fig. 1. Overall imputation network with multiple SAINT blocks. In each block, previous layer embeddings, raw time series X and mask M are passed as inputs to obtain next representations.

positional mask embeddings having shape $\mathbb{R}^c$ to the indicate missing and present values. We add these mask embeddings immediately before TTE_1. In the Fig. 1, $\oplus$ denotes positional addition and $\otimes$ denotes positional multiplication.

We use $\text{MLP}_1 : \mathbb{R}^c \to \mathbb{R}^c \to \mathbb{R}^1$ as the imputation head, which converts the outputs from SAINT into imputations $\hat{Y}^{(l)}$. We aggregate layer-wise losses while training, but only consider the final layer imputation $\hat{Y}^{(L)}$ for testing.

3.2 SAINT

The main components of the SAINT block are Sparse Transformer Encoder (STE) and Time Transformer Encoder (TTE). These consist of multi-head self-attention layers which process the space-time embedded tensors. There are two time transformer encoders – TTE_1 and TTE_2. Taken together, TTE_1 and STE can be considered as attention on a product graph where computations have

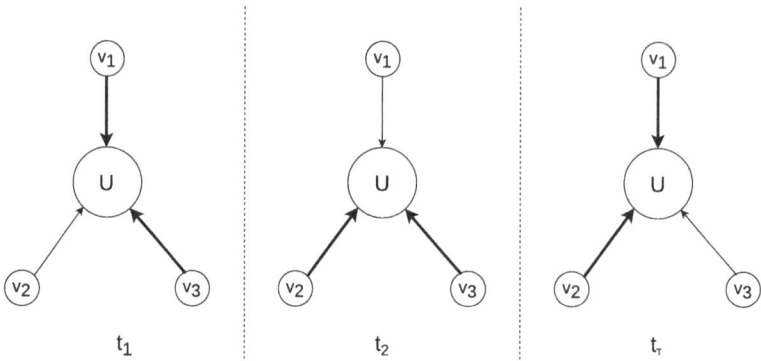

Fig. 2. Spatial aggregation that can change with time – a node can flexibly adjust its neighborhood attention weights (represented by edge thickness) at each time step.

been split and performed sequentially instead. We don't use dropout as the inputs themselves are sparse with randomly missing values.

Time Transformer Encoder is a channel independent temporal self-attention network. We denote its input as $H_{TTE} \in \mathbb{R}^{N \times T \times c}$. TTE processes these inputs as N batches of time series with T time steps and c features. Now, we will describe the multi-head attention equations represented collectively by Att.

Denoting $H_{TTE}[u] \in \mathbb{R}^{T \times c}$ by H_u for each spatial node u, we learn representations $\hat{H}_u$ parallely across the spatial dimension. We use $n_h = 4$ heads, and for each head i we do the following,

$$\mathbf{Q}_u = H_u \times W_Q^i, \quad \mathbf{K}_u = H_u \times W_K^i, \quad \mathbf{V}_u = H_u \times W_V^i$$
$$\hat{H}_u^i = \text{softmax}\left(\frac{\mathbf{Q}_u \mathbf{K}_u^\top}{\sqrt{c}}\right) \mathbf{V}_u. \tag{2}$$

Each of these network parameters have the same shape $W_Q^i, W_Q^i, W_V^i \in \mathbb{R}^{c \times \frac{c}{n_h}}$. We finally stack these outputs to obtain Att outputs.

$$\text{Att}(H_u) = [\hat{H}_u^1 \hat{H}_u^2 \ldots \hat{H}_u^{n_h}] W^{out} \tag{3}$$

where $W^{out} \in \mathbb{R}^{c \times c}$. The overall encoder network with no dropout is,

$$H'_{TTE} = H_{TTE} + \text{Att}(\text{LayerNorm}(H_{TTE}))$$
$$H''_{TTE} = H'_{TTE} + \text{MLP}_0(\text{LayerNorm}(H'_{TTE})), \tag{4}$$

where $\text{MLP}_0 : \mathbb{R}^c \to \mathbb{R}^{c_h} \to \mathbb{R}^c$ with ELU activation. The output of TTE is denoted by H''_{TTE}. We share the network weights across the space (channel) dimension.

This same network is used for both TTE_1 and TTE_2. We add time positional encodings replicated across the spatial dimension before applying TTE_1 to introduce the relative ordering information. STE output and the residual connections from the input are added positionally and is consumed by TTE_2.

Sparse Transformer Encoder is a sparse space-time attention-based encoder where we keep all the space-time vertices in memory but selectively consider the edges. Sparsity comes from using sparse space-time attention, where STE doesn't attend over all possible combinations space-time vertices with $\mathcal{O}(N^2 T^2)$ complexity. Instead, for each time-step we restrict the attention computation of vertex u to its spatial neighbourhood. We may consider the space-time graph as a time series of spatial graphs (see Fig. 2). From the given spatial edges E_S in the dataset, we use graph-attention on these temporally disconnected components in parallel to aggregate node wise representations. We replicate E_S to obtain these edges across time $E_{ST} := \{E_S^{t_1}, E_S^{t_2}, \ldots E_S^{t_T}\}$, which will be used for attention weight calculation resulting in $\mathcal{O}(E_s T)$ complexity. The remaining edges from space-time product graph have already been considered by temporal self-attention in TTE.

We use attention block Att in the encoder which is described next. For each node u at time step t, from the input $h_u^t = H[u, t] \in \mathbb{R}^c$ we use the following attention mechanism to update its representations to $\hat{h}_u^t$

$$\mathbf{Q}_u^t = h_u^t \times W_Q, \quad \mathbf{K}_u^t = h_u^t \times W_K, \quad \mathbf{V}_u^t = h_u^t \times W_V,$$

$$\hat{h}_u^t = \frac{\sum_{v \in \mathcal{N}(u)} \exp\left(\frac{(\mathbf{K}_v^t)^\top \mathbf{Q}_u^t}{\sqrt{c}}\right) \mathbf{V}_v^t}{\sum_{v \in \mathcal{N}(u)} \exp\left(\frac{(\mathbf{K}_v^t)^\top \mathbf{Q}_u^t}{\sqrt{c}}\right)}. \tag{5}$$

We share the network parameters across time, but restrict the computations to the same time step for each node. We use multi-head attention similar to TTE with weight parameters having same shape as before (not explicitly presented in the equations to avoid redundancy). The overall transformer encoder STE is,

$$H'_{\text{STE}} = H_{\text{STE}} + \text{Att}(\text{LayerNorm}(H_{\text{STE}}), E_{ST})$$
$$H'_{\text{STE}} = H'_{\text{STE}} + \text{MLP}_0(\text{LayerNorm}(H'_{\text{STE}})) \tag{6}$$

We add positional encoding of the flattened space-time sequence before processing the STE inputs.

Final Output of the SAINT block, $H^{(l+1)}$, is obtained using a two-layer MLP, $\text{MLP}_3 : \mathbb{R}^c \to \mathbb{R}^c \to \mathbb{R}^c$ which transforms inputs from attention backbone further increasing complexity of the network.

4   Simulation Setup and Dataset

We use RTX 2080 Ti with 11 GB memory to train the models. We implement in PyTorch [10] using open-source code from the baselines wherever available. We next describe the common datasets, baselines and training configuration we will use for evaluation on the imputation and forecasting tasks. We also further test the robustness of the network to increasing missing rates. We follow the exact train-test split methods as described in the baselines for each benchmark.

4.1 Imputation

We use real-world datasets for evaluation – AQI, AQI36 [19] which consist of air quality measurements recorded over cities in China; PEMS BAY, METR LA [7] which record traffic speed using road sensors in the Bay Area and Los Angeles respectively. For AQI and AQI36, evaluation mask is generated following the distribution of truly missing values.

The evaluation mask tests on two scenarios – uniform random point missing and block missing. For POINT missing, values are uniformly missing with 25% probability. BLOCK missing introduces contiguous blocks of missing values, simulating sensor failure for consecutive timestamps. We report performance statistics on 5 random initializations. Mean Absolute Error (MAE) is the evaluation metric.

Across all benchmarks, we use the training configuration and common hyperparameters from the transformer baseline in [8] to skip the exhaustive hyperparameter search. We use 8 as batch size, 8×10^{-4} as learning rate, cosine scheduler with restarts, Adam optimizer, 5.0 as gradient clipping value, 300 as number of epochs and 40 epochs as patience. The imputation is done in sliding windows of size 24 usually, but for AIR36 it is 36. We have 5 layers of SAINT blocks, with each transformer encoder having 4 heads, $c = 64$ as the hidden dimension.

4.2 Forecasting

PEMS BAY: Spatio-temporal traffic prediction benchmark from MIDM [15] uses PEMS BAY dataset with 12 steps ($12 \times 5 = 60$ minutes) forecasting horizon and same length look-back window. As this dataset was also used in imputation, we use the same training and model configuration described previously in the imputation section.

ETTm1: Electricity Transformer Temperature (ETTm1) is a conventional forecasting dataset with 7 channels, consisting of 7 oil and load features of electricity transformers readings. There are total $69,680$ time steps. It is recorded at 5 minute granularity. As there is no given graph, we use assume a fully connected spatial graph. Following SSSD [1], we evaluate on various forecasting horizons from short to long – $\{24, 48, 96, 288, 672\}$ with the look-back windows as $\{96, 48, 284, 288, 384\}$ respectively.

We share the training configuration with DLinear [16] and evaluate on 2021 as the random seed. We reduce the number of layer to 2 and use 0.2 as dropout. Transformers are prone to overfit on such datasets [16] therefore we reduce complexity and regularize the network following [9,17].

Table 1. Performance (in terms of MAE) averaged over 5 independent runs.

	Block missing		Point missing		Simulated failures	
	PEMS-BAY	METR-LA	PEMS-BAY	METR-LA	AQI-36	AQI
KNN	4.30 ± 0.00	7.79 ± 0.00	4.30 ± 0.00	7.88 ± 0.00	30.21 ± 0.00	34.10 ± 0.00
Transformer	1.70 ± 0.02	3.54 ± 0.00	0.74 ± 0.00	2.16 ± 0.00	11.98 ± 0.53	18.11 ± 0.25
GRIN	1.14 ± 0.01	2.03 ± 0.00	0.67 ± 0.00	1.91 ± 0.00	12.08 ± 0.47	14.73 ± 0.15
SPIN	1.06 ± 0.02	1.98 ± 0.01	0.70 ± 0.01	1.90 ± 0.01	11.77 ± 0.54	13.92 ± 0.15
SPIN-H	1.05 ± 0.01	2.05 ± 0.02	0.73 ± 0.01	1.96 ± 0.03	10.89 ± 0.27	14.41 ± 0.13
SAINT	**1.00** ± 0.01	**1.92** ± 0.01	**0.65** ± 0.00	**1.82** ± 0.01	**10.82** ± 0.21	**13.67** ± 0.17

Table 2. Performance on PEMS-BAY for 1 h Forecasting averaged over 5 independent runs.

	GMAN	MIDM	SAINT
MAE	1.86 ± 0.02	1.83 ± 0.04	**1.57** ± 0.01
MRE (%)	4.31 ± 0.02	4.21 ± 0.06	**2.51** ± 0.01

4.3 Robustness

Imputation: We evaluate on the benchmark presented in [8]. We test both on the POINT and BLOCK masks. In POINT mask, values are uniformly missing with rates – $50\%, 75\%, 95\%$. BLOCK masks simulate sensor failure. The failure probability p_f of $5\%, 10\%$ and 15% corresponds to an overall missing rate ranging between ≈ 70-75%, ≈ 90-92%, and ≈ 96-97% respectively.

Forecasting: We evaluate the sensitivity of forecasting networks to increasing missing values in the look-back window. We simulate BLOCK missing values. Usual forecasting networks need pre-processing as they commonly don't account for missing values. We use k-Nearest Neighbour (kNN) for look-back window imputation as it is efficient and can process the entire dataset in a single step. We use correlation matrix for obtaining the spatial graph and select $k = 1$ as it gave least error. We present a comparison of,

1. DLinear (forecasting) + kNN (look-back window imputation)
2. SAINT (forecasting) + kNN (look-back window imputation)
3. SAINT for forecasting as well as imputation simultaneously.

5 Results

In this section, we provide empirical evidence for the usefulness of SAINT on multiple tasks and various datasets described previously.

Table 3. Performance (MAE) on Time series forecasting results on the ETTm1 data set. Random seed 2021

Model	24	48	96	288	672
CSDI	0.370	0.546	0.756	0.530	0.891
SSSD	0.361	0.479	0.547	0.648	0.783
DLinear	0.312	0.501	**0.345**	**0.390**	**0.419**
SAINT	**0.308**	**0.473**	0.373	0.462	0.549

Table 4. Performance (MAE) with increasing data sparsity in the *Point missing* setting (averaged over 5 evaluation masks).

	METR-LA			PEMS-BAY			AQI		
	Missing rate			Missing rate			Missing rate		
	50 %	75 %	95 %	50 %	75 %	95 %	50 %	75 %	95 %
Transformer	2.31 ± 0.00	2.71 ± 0.00	5.13 ± 0.01	0.85 ± 0.00	1.13 ± 0.00	2.70 ± 0.01	9.11 ± 0.02	12.56 ± 0.05	25.65 ± 0.11
GRIN	2.05 ± 0.00	2.39 ± 0.00	4.08 ± 0.02	0.79 ± 0.00	1.09 ± 0.00	2.70 ± 0.01	8.43 ± 0.01	10.97 ± 0.02	20.38 ± 0.10
SPIN	2.02 ± 0.00	2.24 ± 0.00	2.89 ± 0.01	0.79 ± 0.00	1.00 ± 0.00	1.71 ± 0.00	8.15 ± 0.01	9.96 ± 0.02	15.51 ± 0.08
SPIN-H	2.01 ± 0.00	2.20 ± 0.00	2.82 ± 0.00	0.79 ± 0.00	**0.97 ± 0.00**	1.68 ± 0.00	8.67 ± 0.02	10.27 ± 0.02	15.75 ± 0.07
SAINT	**1.93 ± 0.01**	**2.16 ± 0.01**	**2.79 ± 0.01**	**0.75 ± 0.00**	**0.96 ± 0.00**	1.69 ± 0.01	**7.94 ± 0.02**	**9.64 ± 0.02**	**14.55 ± 0.03**

5.1   Imputation

In Table 1, we compare performance of SAINT against the state-of-the-art networks. SAINT is consistently better or comparable than all the baselines. Additionally, while evaluating on our machine, SPIN gives out-of-memory (OOM) error on PEMS BAY dataset for batch size as small as 1. SPIN has a time complexity of $\mathcal{O}(NT^2 + E_s T^2)$ in comparison to SAINT with $\mathcal{O}(NT^2 + E_s T)$. The performance and efficiency of SAINT can be attributed to two key factors – regularization through channel independent layers and preservation of space-time inductive biases using sparse graph transformer. Although not included in the table due to limited data, MIDM [15] is a recent diffusion based model, reporting the following MAE values – BAY POINT (0.60) and BAY BLOCK (1.03). In comparison, SAINT achieves 0.65 and 1.00 respectively.

5.2   Forecasting

PEMS BAY: In Table 2 we evaluate on PEMS-BAY forecasting. We also report Mean Relative Error (MRE) metric along with MAE. We observe that SAINT outperforms the baselines by a significant margin. MIDM [15] doesn't use space inductive biases. GMAN [18] considers space and time independently.

ETTm1: In Table 3, we observe that SAINT is better than CSDI and SSSD in all cases. The state-of-the-art forecasting specific model DLinear is more suitable for long-term forecasting ($L \geq 96$). SAINT being a general imputation model doesn't impose forecasting biases, failing to model for trend and seasonality.

Table 5. Performance (MAE) with an increasing number of simulated failures in the *Block missing* setting (averaged over 5 evaluation masks).

	METR-LA			PEMS-BAY			AQI		
	Failure probability			Failure probability			Failure probability		
	5 %	10 %	15 %	5 %	10 %	15 %	5 %	10 %	15 %
Transformer	6.03 ± 0.04	7.19 ± 0.05	8.06 ± 0.05	3.69 ± 0.06	5.09 ± 0.05	6.02 ± 0.04	29.21 ± 0.33	33.62 ± 0.16	37.31 ± 0.14
GRIN	3.05 ± 0.02	4.52 ± 0.05	5.82 ± 0.06	2.26 ± 0.03	3.45 ± 0.06	4.35 ± 0.04	15.62 ± 0.24	22.08 ± 0.39	29.03 ± 0.42
SPIN	2.71 ± 0.02	3.32 ± 0.02	3.87 ± 0.05	1.78 ± 0.03	**2.15 ± 0.03**	**2.41 ± 0.02**	14.29 ± 0.24	18.71 ± 0.34	24.34 ± 0.46
SPIN-H	2.64 ± 0.02	3.17 ± 0.02	3.61 ± 0.04	**1.75 ± 0.04**	2.16 ± 0.03	2.48 ± 0.02	14.55 ± 0.26	19.37 ± 0.36	25.38 ± 0.37
SAINT	**2.58 ± 0.04**	**3.14 ± 0.04**	**3.57 ± 0.05**	1.77 ± 0.01	2.22 ± 0.05	2.64 ± 0.07	**13.57 ± 0.11**	**17.00 ± 0.32**	**21.75 ± 0.65**

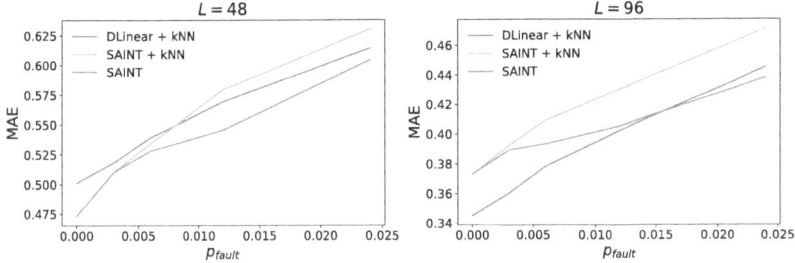

Fig. 3. Forecasting with increasing data sparsity in the look-back window.

5.3 Robustness

Imputation: In Tables 4 and 5, we report robustness of the models to increasing missing rates. Table 4 tests the POINT missing case. SAINT outperforms the baselines in many cases and can be used reliably for imputation even under very high missing rates. In Table 5 we evaluate on BLOCK missing mask. SAINT gives better performance generally, and by a significant margin for the AIR dataset. SAINT has higher error on the PEMS-BAY dataset for the extremely high missing rate scenario.

Forecasting: In Fig. 3 we plot the sensitivity of forecasting networks to increasing missing rates in the look-back window. We observe that for $L = 48$, SAINT consistently outperforms DLinear. For $L = 96$, SAINT gives lower error when missing rate is high ($\approx 46.93\%$). It is more effective to use SAINT for both imputation and forecasting instead of only for forecasting, i.e., SAINT + kNN.

6   Ablation

In Table 6 we provide empirical justification for various components of SAINT. This is done by removing components (one at a time) and analyzing the performance of the resulting network. We evaluate on AQI36 and LA BLOCK datasets with 5 independent initializations.

Table 6. Ablation study: Performance (in terms of MAE) averaged over 5 independent runs. SAINT is compared with modified versions of itself.

Model	AQI 36	LA BLOCK
SAINT	10.82, 0.21	1.92, 0.01
Remove STE	14.67, 0.10	2.69, 0.02
Remove TTE_1 and STE	14.81, 0.16	2.69, 0.02
Remove TTE_1	11.30, 0.18	1.98, 0.01
Remove TTE_2	10.85, 0.26	1.96, 0.01
Replace E_{ST} with E_{PG}	10.86, 0.13	1.89, 0.02
Remove residual connection	10.84, 0.33	1.93, 0.01

1. **Attention components:** STE plays a key role in the overall performance as removing it increases the error significantly. Similarly, TTE_1 is also important as it is being used in conjunction with STE for space-time representation learning. Removing TTE_2 results in a slight drop in performance.
2. **Product graph:** We replace E_{ST} with product graph E_{PG}, i.e., all possible space-time connections. Despite using the product graph, which can be huge, results are marginally better only for LA BLOCK. It suggests that many of the edges in the product graph may not be useful and E_{ST} is able to capture most of the information.
3. **Residual connections:** We remove the residual connection (on the right of TTE_1 and STE in Fig. 1) in the attention backbone. It causes a small decrease in performance across both datasets.

7   Discussion

We presented SAINT, a novel sparse attention based imputation network which consistently outperforms the state-of-the-art imputation baselines on various real-world datasets. We justify the architecture choices using ablations. It is memory efficient and robust against higher missing rates. While SAINT is practical for even long term forecasting, it shows better performance compared to state-of-the-art forecasting networks only for short-term forecasting. Future work can introduce forecasting biases to make a more general imputation model which is also effective for long term forecasting.

References

1. Alcaraz, J.L., Strodthoff, N.: Diffusion-based time series imputation and forecasting with structured state space models. Trans. Mach. Learn. Res. (2022)
2. Bashir, F., Wei, H.L.: Handling missing data in multivariate time series using a vector autoregressive model-imputation (VAR-IM) algorithm. Neurocomputing **276**, 23–30 (2018)

3. Cao, W., Wang, D., Li, J., Zhou, H., Li, L., Li, Y.: BRITS: Bidirectional recurrent imputation for time series. In: Advances in Neural Information Processing Systems, vol. 31 (2018)
4. Chen, J., Tan, E., Li, Z.: A machine learning framework for real-time traffic density detection. Int. J. Pattern Recogn. Artif. Intell. **23**(07), 1265–1284 (2009)
5. Cini, A., Marisca, I., Alippi, C.: Filling the G_ap_s: multivariate time series imputation by graph neural networks. In: International Conference on Learning Representations (2022)
6. Jin, D., Shi, J., Wang, R., Li, Y., Huang, Y., Yang, Y.B.: Trafformer: unify time and space in traffic prediction. Proc. AAAI Conf. Artif. Intell. **37**(7), 8114–8122 (2023)
7. Li, Y., Yu, R., Shahabi, C., Liu, Y.: Diffusion convolutional recurrent neural network: data-driven traffic forecasting. In: International Conference on Learning Representations (2018)
8. Marisca, I., Cini, A., Alippi, C.: Learning to reconstruct missing data from spatiotemporal graphs with sparse observations. Adv. Neural. Inf. Process. Syst. **35**, 32069–32082 (2022)
9. Nie, Y., Nguyen, H., Sinthong, P., Kalagnanam, J.: A time series is worth 64 words: long-term forecasting with transformers. In: International Conference on Learning Representations (2023)
10. Paszke, A., et al.: PyTorch: an imperative style, high-performance deep learning library (2019). http://arxiv.org/abs/1912.01703
11. Peng, W., Lin, Y., Guo, S., Tang, W., Liu, L., Wan, H.: Generative-contrastive-attentive spatial-temporal network for traffic data imputation. In: Kashima, H., Ide, T., Peng, W.C. (eds.) Advances in Knowledge Discovery and Data Mining (2023)
12. Shan, S., Li, Y., Oliva, J.B.: NRTSI: non-recurrent time series imputation. In: International Conference on Acoustics, Speech and Signal Processing (ICASSP) (2023)
13. Shirzad, H., Velingker, A., Venkatachalam, B., Sutherland, D.J., Sinop, A.K.: Exphormer: Sparse transformers for graphs (2023). https://arxiv.org/abs/2303.06147
14. Vaswani, A.: Attention is all you need. In: Advances in Neural Information Processing Systems (2017)
15. Wang, X., et al.: An observed value consistent diffusion model for imputing missing values in multivariate time series. In: Proceedings of the 29th ACM SIGKDD Conference on Knowledge Discovery and Data Mining (2023)
16. Zeng, A., Chen, M., Zhang, L., Xu, Q.: Are transformers effective for time series forecasting? In: Proceedings of the AAAI Conference on Artificial Intelligence, vol. 37, pp. 11121–11128, June 2023
17. Zhang, Y., Ma, L., Pal, S., Zhang, Y., Coates, M.: Multi-resolution time-series transformer for long-term forecasting. In: International Conference on Artificial Intelligence and Statistics. PMLR (2024)
18. Zheng, C., Fan, X., Wang, C., Qi, J.: GMAN: a graph multi-attention network for traffic prediction (2019). https://arxiv.org/abs/1911.08415
19. Zheng, Y., et al.: Forecasting fine-grained air quality based on big data. In: Proceedings of the 21st ACM SIGKDD International Conference on Knowledge Discovery and Data Mining (2015)
20. Zhou, H., et al.: Informer: beyond efficient transformer for long sequence time-series forecasting (2020). https://arxiv.org/abs/2012.07436

Federated Learning for Smart and Sustainable Aquaponics: A Decentralized AI Approach for Urban Resilience

Ahmet Kasif[1] and Cagatay Catal[2(✉)]

[1] Computer Engineering, Bursa Technical University, Bursa, Türkiye
`ahmet.kasif@btu.edu.tr`
[2] Computer Science and Engineering, Qatar University, Doha, Qatar
`ccatal@qu.edu.qa`

Abstract. The majority of machine learning models rely on centralized methods, which require large data transfers to central repositories. Federated learning, a decentralized machine learning approach, offers a solution by enabling local computations and aggregating local models to build a global model. While federated learning has been applied in some agricultural domains, its use in aquaponics remains unexplored. Aquaponics, a sustainable agricultural method combining fish farming and soil-less plant production, presents unique opportunities for federated learning applications, particularly in urban farming environments. By using edge-based federated learning, we improve scalability, data privacy, and sustainability and also, reduce data transmission needs in smart urban agriculture. This research, a collaboration between research teams from three countries, highlights how federated learning and deep learning can enhance environmental monitoring and sustainability in urban resilience strategies, particularly in smart agriculture. The Flower framework was used to implement federated learning, and ResNet-18 was employed for fish disease detection. This paper introduces novel contributions in federated learning and deep learning techniques for the management of aquaponics systems, highlighting the potential of these technologies to optimize aquaponics systems' efficiency.

Keywords: Federated learning · aquaponics · agriculture · transfer learning

1 Introduction

The majority of deployed machine learning models rely on centralized methods and utilize data stored in centralized repositories [1]. However, centralized approaches require the transfer of large amounts of data to central repositories, as well as significant bandwidth and reliable connections. For small-scale and resource-constrained farmers, these limitations require a different machine learning strategy that minimizes data transfers and reduces dependency on high-bandwith infrastructure [2].

© The Author(s), under exclusive license to Springer Nature Singapore Pte Ltd. 2025
S. Yuan et al. (Eds.): PAKDD 2025 Workshops, LNAI 15835, pp. 56–66, 2025.
https://doi.org/10.1007/978-981-96-8197-6_5

Federated learning, a state-of-the-art decentralized machine learning approach, enables local computations and builds a global machine learning model by aggregating the parameters of local models [3]. A recent review study discusses several use cases of federated learning in agriculture, including crop yield estimation, food disease diagnosis, securing agricultural IoT infrastructures, pest classification, animal activity recognition, and crop classification [1].

While federated learning has been recently applied to some areas in agriculture, to the best of our knowledge, no study has investigated its use in the field of aquaponics. Aquaponics, a sustainable agricultural approach that combines fish farming (aquaculture) with soil-less plant production (hydroponics) creates an ecosystem where fish waste is used to feed plants [4], presenting unique challenges and opportunities for the application of federated learning.

We recently received funding for a research project aimed at addressing these challenges by integrating machine learning technologies in aquaponics, which we refer to as Precision Aquaponics. While the term Precision Aquaculture has been used in fish farming [5], to the best of our knowledge, Precision Aquaponics has not been used in the literature to describe the integration of advanced Artificial Intelligence technologies in aquaponics. Our definition of Precision Aquaponics is the integration of advanced Machine Learning (ML), Deep Learning (DL), Computer Vision (CV), and Internet of Things (IoT) solutions, along with other data-driven technologies, to automate and optimize aquaponics systems for the efficient management of fish, plants, and the environment. This research project involves collaboration between research teams from three different countries, bringing together diverse expertise to advance the field. This paper presents our initial results in applying edge-based federated learning, a technique that combines local data processing on edge devices with federated learning to improve the efficiency of machine learning models for precision aquaponics.

In an aquaponics farm with multiple areas, such as fish ponds and plant cultivation zones, each area plays an important role in maintaining an efficient ecosystem. First, federated learning helps aggregate insights from different parts of the farm (e.g., fish ponds and plant areas), resulting in a more robust model. Second, from a scalability perspective, as the aquaponics farm grows and more ponds or plant cultivation areas are added, federated learning allows the system to scale easily. Third, since each area is processed locally on edge devices, the need for extensive data transmission is reduced. Finally, real-time data collected from the different ponds and plant areas can help improve the overall system's performance.

In this study, we adopted the Flower framework [6] for federated learning due to its flexibility and support for several federated learning algorithms, such as Federated Averaging. Flower utilizes the gRPC protocol to facilitate efficient communication between server and client devices. For the machine learning model, we selected ResNet-18, an efficient model for resource-constrained devices such as Raspberry Pi.

The main contributions of this paper are as follows:

– This is the first study to apply federated learning in Precision Aquaponics, a term we define and propose for the integration of advanced Artificial Intelligence techniques in aquaponics systems.
– We provide a proof of concept to demonstrate the feasibility and effectiveness of federated learning in optimizing aquaponics systems.
– We introduce a novel edge-based federated learning framework tailored for precision aquaponics, which utilizes edge devices to improve scalability and efficiency.
– We present a deep learning approach using ResNet-18 for fish disease detection, showcasing the potential of AI models in enhancing the management of aquaponics ecosystems.

The following section presents the Related Work. Section 3 details the methodology. Section 4 discusses the experimental results. Section 5 concludes the paper and outlines future work.

2   Related Work

Aquaponics is a sustainable farming system that integrates aquaculture (i.e., the cultivation of aquatic animals like fish or shrimp) with hydroponics (i.e., the cultivation of plants in water without soil) [7]. In aquaponics, fish waste provides nutrients for plant growth, while plants help filter and clean the water, creating a closed-loop, symbiotic ecosystem. This approach not only conserves water and space but also reduces the need for chemical fertilizers and improves efficiency in food production. Aquaponics inherently relies on its aquaculture component, where the health and productivity of aquatic species play a central role. Aquaculture itself is a rapidly growing sector, addressing the increasing global demand for seafood. However, it faces several challenges, such as disease outbreaks, water quality management, and the need for sustainable farming practices. These challenges become even more critical in aquaponics, where the health of fish directly impacts the nutrient supply for plants and the overall system balance. Currently, there are many open research questions in the intersection of aquaponics and artificial intelligence, such as water quality management [8], fish disease monitoring and prevention and cost and resource constraints [9,10].

Federated Learning (FL) is a relatively recent machine learning approach that enables decentralized model training across multiple devices or nodes. Unlike traditional centralized methods, FL trains models directly on clients (e.g., edge devices) where the data is generated. McMahan et al. introduced the foundational framework for FL through the widely adopted Federated Averaging (FedAvg) algorithm [11]. In fields like aquaponics and aquaculture, FL offers significant potential by using data collected from IoT devices deployed across fish farms. This approach not only enhances decision-making without requiring centralized data storage but also reduces the need for data transmission [2].

Recent studies have highlighted the application of FL in IoT systems, focusing on its ability to handle heterogeneous data and work in resource-constrained environments. FL enables edge devices to collaboratively train shared models without transferring raw data, improving communication efficiency and ensuring privacy. Wang et al. investigated the applicability of FL for IoT devices, proposing gradient compression techniques to mitigate communication overhead. This approach is particularly valuable in aquaponics and aquaculture systems, where edge devices often encounter bandwidth limitations [12,13].

FL has also proven effective in domains with fragmented data, such as healthcare, finance, and agriculture. In agriculture, FL is still emerging, but it shows promise by enabling distributed data analysis across farms or aquaponics systems. By using FL, these systems can develop predictive models that are better suited to local variations in environmental conditions and farming practices.

In aquaponics and aquaculture, where IoT devices and edge computing are commonly used, FL has immense potential. These systems naturally align with FL's decentralized framework, enabling tasks like anomaly detection, yield prediction, and system optimization. Additionally, the integration of FL with edge-cloud architectures, as demonstrated by Cheng et al., supports scalable and privacy-preserving learning for precision aquaculture [2].

IoT and Cloud integration are also key aspects of aquaponics and aquaculture environments. Cheng et al. proposed a cloud environment powered by Amazon Web Services (AWS) alongside a federated learning platform serving as a server prototype [2]. On the edge side, a Raspberry Pi device is employed, while AWS IoT Greengrass facilitates the local execution of pre-trained machine learning models directly on IoT devices. Edge devices perform two primary functions. First, they operate with a rule-based alert system, and second, they serve as input providers for the federated learning client. This setup highlights the integration of edge computing capabilities with federated learning, allowing IoT devices to contribute to decentralized model training while maintaining local functionality. The solution focuses on connecting multiple local nodes to a centralized cloud provider. This raises a critical question: why would farmers share their models? This observation leads to a key research question: How can federated learning frameworks be designed to encourage collaboration among decentralized entities while addressing the practical and privacy-related concerns of individual participants?

This study focuses on creating a proof-of-concept environment to utilize image processing and AI for efficient and accurate fish disease detection in aquaponics systems. By integrating IoT-enabled imaging devices with AI models, the system aims to address the limitations of traditional detection methods. The study also explores how decentralized frameworks, like federated learning, can be used to train AI models on distributed datasets while preserving privacy and reducing data transmission requirements. This research contributes to creating sustainable and efficient aquaponics systems by ensuring fish health through advanced technological solutions.

3   Methodology

Machine learning (ML) is a subset of Artificial Intelligence (AI) that focuses on creating algorithms capable of learning patterns from data and making predictions or decisions without being explicitly programmed. This centralized approach works well in scenarios where data can be aggregated at one location. However, as data privacy regulations (e.g., GDPR) become stricter and the volume of data increases with the rise of IoT devices, the traditional ML approach faces several challenges, such as privacy concerns, communication overhead, and scalability issues. FL is a decentralized paradigm in machine learning designed to address these challenges. Unlike traditional ML, FL trains models across multiple devices or nodes (e.g., smartphones, IoT devices) without requiring the data to be centralized. Instead, the data remains on local devices, and only model updates (e.g., gradients or model parameters) are shared with a central server for aggregation.

Figure 1 shows the proof-of-concept FL architecture developed for fish disease detection in aquaponics, based on fish images. The following five steps are designed to implement the federated learning approach:

1. Define a global model and push the global model to all edge devices.
2. Train local models with local device data on edge devices.
3. Pull model updates from edge devices back to the server.
4. Aggregate model updates into a new global model.
5. Repeat the previous steps until the global model converges.

To implement FL architecture, the Flower framework has been utilized. Flower is a high-level federated framework that provides tools to both simulate and experiment with real-world federated learning platforms [6]. The framework is compatible with many pre-trained models as well as custom models. While it comes with its own data provider, Flower is also compatible with HuggingFace datasets, making it an even more practical choice in terms of compatibility and scalability [14].

The dataset used in this study is a fish disease classification dataset [15]. The FishLens dataset is a publicly available dataset hosted on Roboflow Universe, designed for tasks related to fish detection and classification. Dataset model v2 is used in this study, as it contains more than 7,000 images, each with 640×640 pixels, and includes pre-processing and augmentation already implemented for the task. The fish disease classes used in the study are "Aeromonas Septicemia", "Columnaris Disease", "Edwardsiella Ictaluri (Bacterial Red Disease)", "Epizootic Ulcerative Syndrome (EUS)", "Flavobacterium (Bacterial Gill Disease)", "Fungal Disease (Saprolegniasis)", "Healthy Fish", "Ichthyophthirius (White Spots)" , "Parasitic Disease", "Streptococcus", "Tilapia Lake Virus (TiLV)". Fish images depicting several health conditions, including parasitic disease, are shown in Figure 2.

To use this dataset with Flower seamlessly, a HuggingFace dataset has been created. The dataset consists of two columns, "img" and "label". While "img"

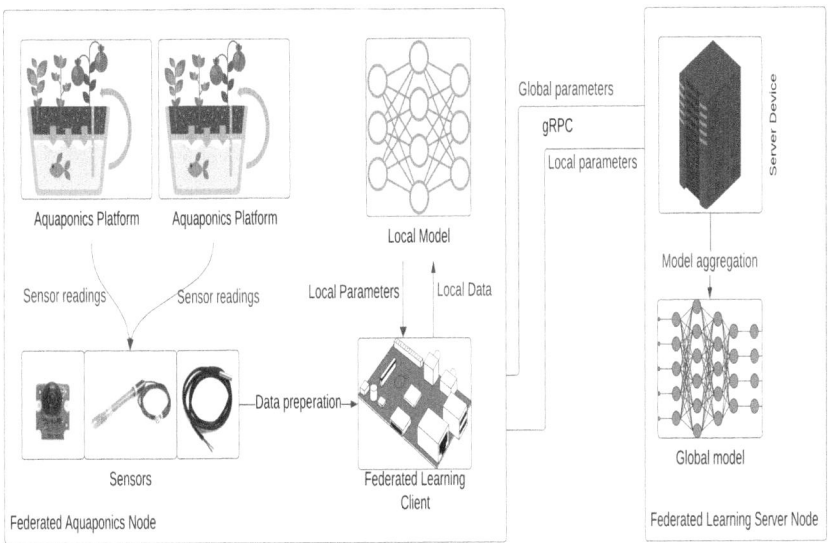

Fig. 1. Proof-of-concept smart aquaponics setup environment powered by federated learning

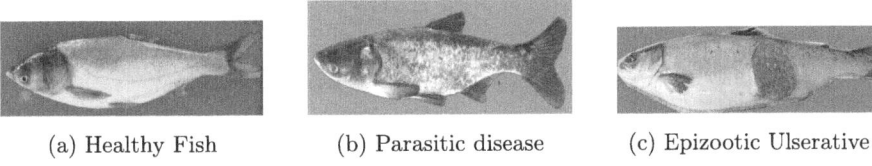

(a) Healthy Fish   (b) Parasitic disease   (c) Epizootic Ulserative

Fig. 2. Different fish health conditions

refers to the image, "label" refers to one of 11 different classes. The final dataset is then converted to Parquet format [16]. The .parquet extension refers to files in the Apache Parquet format, a widely used columnar storage file format designed for efficient storage and processing of large datasets.

ResNet-18 has been used as a pre-trained model for this study [17]. ResNet-18 is a deep convolutional neural network architecture introduced by He et al. It is part of the ResNet family, which revolutionized deep learning by introducing residual connections that help mitigate the vanishing gradient problem during backpropagation in very deep networks (Fig. 3). This architecture not only provides srong classification performance over images but also has low computational overhead and memory requirements compared to other pre-trained models (Table 1). In the ResNet-18 architecture, only the final layer is trained to accommodate the custom dataset, while the pre-trained layers are frozen to utilize their feature extraction capabilities.

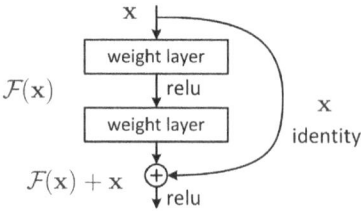

Fig. 3. Residual learning: a building block [17]

Table 1. Parameter size comparison of popular pre-trained models

Model	Parameters (Millions)	Depth	Key Characteristics
ResNet-18	11M	18 layers	Lightweight and efficient baseline model.
ResNet-50	25M	50 layers	Introduces bottleneck blocks for better accuracy.
VGG-16	138M	16 layers	Large parameter size, older but still effective.
Inception-V3	23M	48 layers	Efficient design with Inception modules.

4   Experiments

4.1   Environment

The experimental infrastructure for this study was built using two Raspberry Pi 4 Model B+ devices (client nodes), each equipped with a quad-core ARM Cortex-A72 processor and 4GB of RAM and one Raspberry Pi 3 Model B device (server node) with 1GB of RAM.

This setup reflects a typical federated learning architecture, where a central server coordinates and aggregates updates from distributed clients. A local WiFi network was established to connect the devices, ensuring low-latency communication and simulating the decentralized nature of federated learning systems. The hardware setup is depicted in Fig. 4. The figure shows the Raspberry Pi devices, with the server positioned centrally and the two clients on either side, connected via a local wireless network. The neat arrangement highlights the simplicity and reproducibility of the experimental setup.

Python version 3.11 was used as the programming language for this study. Its modern features, enhanced performance, and compatibility with both the Flower federated learning framework and PyTorch machine learning library made it a suitable choice for system implementation. The Flower framework was selected for its flexibility and comprehensive support for federated learning algorithms, including Federated Averaging (FedAvg). Also, Flower facilitates seamless communication between the server and client devices using its built-in gRPC protocol, ensuring efficient parameter exchanges.

Fig. 4. Experimental environment setup for federating learning with Raspberry PI devices

The ResNet-18 model was used as the machine learning model for this study. Its compact architecture and computational efficiency made it ideal for the resource-constrained Raspberry Pi devices. This model's design also supports robust performance, ensuring that it could meet the demands of federated learning while accommodating hardware limitations.

The rationale for this setup lies in its balance of practicality, cost-effectiveness, and reproducibility. The Raspberry Pi devices provide an accessible and standardized platform for edge computing research, while the local WiFi network offers a simple and reliable communication infrastructure. Python 3.11, the Flower framework, and ResNet-18 together collectively form a robust software stack that supports efficient implementation and evaluation of federated learning methodologies.

4.2 Results

Figure 5 depicts the training loss and accuracy for centralized learning across the training phase. The training loss demonstrates a rapid and smooth decline. The accuracy curve rises sharply in the initial epochs and surpasses 75% by epoch 10. It continues to improve steadily, reaching close to 80% accuracy by the end of training.

Figure 6 shows that the training loss and accuracy over 40 epochs in a federated learning setup. The training loss pictures a consistent and rapid decline in the early epochs, stabilizing at a minimal value beyond epoch 15. However, compared to centralized learning, the loss remains slightly higher overall, reflecting the inherent challenges in federated optimization, such as variability in client updates and aggregation noise. The accuracy curve exhibits a sharp rise in the initial epochs, stabilizing near 76% towards the end of training. These results suggest that the federated model successfully learned from distributed client data, achieving good generalization despite the distributed setup.

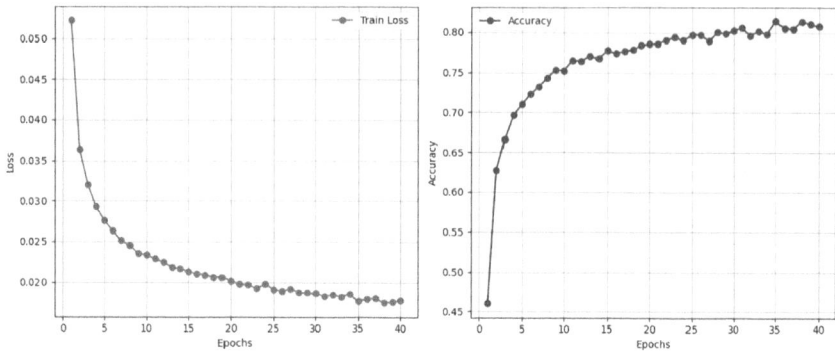

Fig. 5. Centralized learning training performance

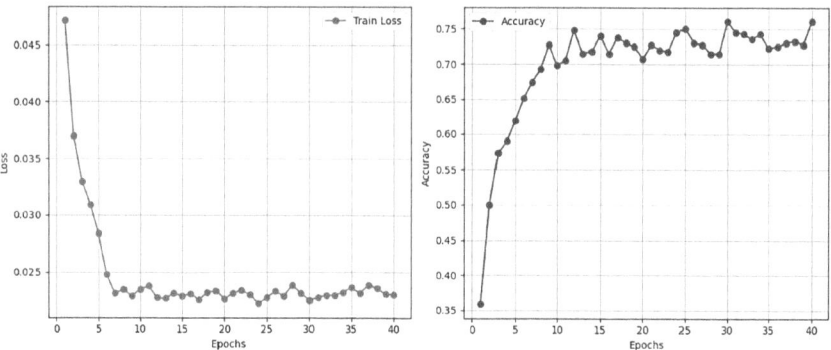

Fig. 6. Federated learning training performance

5 Conclusion

In this paper, we have introduced the concept of Precision Aquaponics, a novel integration of deep learning, computer vision, and IoT technologies to optimize and automate aquaponics systems for efficient management of fish, plants, and the environment. We have demonstrated the feasibility of applying federated learning in Precision Aquaponics, highlighting its potential to address challenges such as communication overhead and scalability. Our approach utilizes edge-based federated learning to process data locally on resource-constrained devices, reduces the need for extensive data transmission, and improves system efficiency. Additionally, we presented a deep learning solution for fish disease detection, showcasing the effectiveness of deep learning in Precision Aquaponics. This work provides a proof of concept for federated learning in the aquaponics field, offering practical applications for improving the management of these systems. Future research will explore further enhancements to the performance of federated learning in aquaponics. The research team will investigate the applicability of a Reinforcement Learning approach, along with the use of edge-based federated learning, and deploy the system on the research platform to be built near one of the campuses of the research teams.

Acknowledgments. This research was supported by the Qatar Research, Development, and Innovation (QRDI) Council under Grant [FSC05-0327-240008]. This publication was made possible by the support of Qatar University. The findings achieved herein are solely the responsibility of the authors.

References

1. Žalik, K.R., Žalik, M.: A review of federated learning in agriculture. Sensors **23**(23), 9566 (2023)
2. Cheng, W.K., Khor, J.C., Liew, W.Z., Bea, K.T., Chen, Y.L.: Integration of federated learning and edge-cloud platform for precision aquaculture. IEEE Access (2024)
3. Moshawrab, M., Adda, M., Bouzouane, A., Ibrahim, H., Raad, A.: Reviewing federated machine learning and its use in diseases prediction. Sensors **23**(4), 2112 (2023)
4. Chandramenon, P., Aggoun, A., Tchuenbou-Magaia, F.: Smart approaches to Aquaponics 4.0 with focus on water quality * Comprehensive review. Comput. Electron. Agric. **225**, 109256 (2024)
5. O'Donncha, F., Grant, J.: Precision aquaculture. IEEE Internet Things Mag. **2**(4), 26–30 (2019)
6. Beutel, D.J., Topal, T., Mathur, A., Qiu, X., Fernandez-Marques, J., Gao, Y., et al.: Flower: a friendly federated learning research framework, arXiv preprint arXiv:2007.14390 (2020)
7. Arvind, C.S., Jyothi, R., Kaushal, K., Girish, G., Saurav, R., Chetankumar, G.: Edge computing based smart aquaponics monitoring system using deep learning in IoT environment. In: Proceedings of the 2020 IEEE Symposium Series on Computational Intelligence (SSCI), pp. 1485–1491 (2020)
8. Metin, A., Kasif, A., Catal, C.: Temporal fusion transformer-based prediction in aquaponics. J. Supercomput. **79**(17), 19934–19958 (2023)
9. Imai, T., Arai, K., Kobayashi, T.: Smart aquaculture system: a remote feeding system with smartphones. In: Proceedings of the 2019 IEEE 23rd International Symposium on Consumer Technologies (ISCT), pp. 93–96 (2019)
10. Wu, B., Wang, C., Du, K.-L.: A smart aquaculture system exploiting IoT, AI and cloud computing. In: Proceedings of the 2023 IEEE International Conference on Smart Internet of Things (SmartIoT), pp. 251–256 (2023)
11. McMahan, B., Moore, E., Ramage, D., Hampson, S., y Arcas, B.A.: Communication-efficient learning of deep networks from decentralized data. In: Artificial Intelligence and Statistics, pp. 1273–1282. PMLR (2017)
12. Wang, T., Liu, Y., Zheng, X., Dai, H.-N., Jia, W., Xie, M.: Edge-based communication optimization for distributed federated learning. IEEE Trans. Netw. Sci. Eng. **9**(4), 2015–2024 (2021)
13. Chen, M., Shlezinger, N., Poor, H.V., Eldar, Y.C., Cui, S.: Communication-efficient federated learning. Proc. Natl. Acad. Sci. **118**(17), e2024789118 (2021)
14. Face, H.: Hugging face datasets, Hugging Face, https://huggingface.co/datasets, Accessed 14 Dec 2024

15. fishlens, "fishlens-modelv1 Dataset," *Roboflow Universe*, Open Source Dataset, https://universe.roboflow.com/fishlens/fishlens-modelv1, Publisher: Roboflow, Sep. 2024, Visited: 14 Dec 2024

16. Vohra, D.: Apache parquet. In: Practical Hadoop Ecosystem, pp. 325–335. Apress, Berkeley, CA (2016). https://doi.org/10.1007/978-1-4842-2199-0_8

17. He, K., Zhang, X., Ren, S., Sun, J.: Deep residual learning for image recognition. In: Proceedings of the IEEE Conference on Computer Vision and Pattern Recognition (CVPR), pp. 770–778 (2016)

18. Dash, S., Ojha, S., Muduli, R.K., Patra, S.P., Barik, R.C.: Fish type and disease classification using deep learning model based customized cnn with resnet 50 technique. J. Adv. Zoology **45**(3) (2024)

19. Huang, Y.-P., Khabusi, S.P.: A cnn-oselm multi-layer fusion network with attention mechanism for fish disease recognition in aquaculture. IEEE Access **11**, 58729–58744 (2023)

20. Suhana, R., Mahmudy, W.F., Budi, A.S.: Fish image classification using adaptive learning rate in transfer learning method. Knowl. Eng. Data Sci. **5**(1), 67–77 (2022)

Multi-Task Learning Model for Mobile Threat Detection and Cyber Resilience in Urban Systems

Shimaa Ibrahim[1], Cagatay Catal[1(✉)], and Thabet Kacem[2]

[1] Department of Computer Science and Engineering, Qatar University, Doha, Qatar
{st1103366,ccatal}@qu.edu.qa
[2] Department of Computer Science and Information Technology,
University of the District of Columbia, Washington DC, USA
thabet.kacem@udc.edu

Abstract. The rapid expansion of Android devices in urban environments has led to an increase in sophisticated cyber threats, including zero-day malware, which pose risks to public safety and critical infrastructure. Traditional signature-based detection methods are often ineffective against these evolving threats. This paper presents a multi-task learning (MTL) framework designed to enhance Android malware detection and classification, contributing to the resilience of urban digital ecosystems. The model simultaneously performs binary classification (malware detection) and multi-class classification (malware family identification), using shared representations to improve efficiency and accuracy compared to single-task learning (STL) models. The framework is trained and evaluated on the CCCS-CIC-AndMal-2020 dataset, utilizing API-based static features of Android applications. Feature dimensionality is reduced through Principal Component Analysis (PCA), and class imbalance is addressed with a weighted loss function. Hyperparameter tuning with Optuna further optimizes configurations, including layer sizes, learning rate, and task-specific weights. Experimental results demonstrate that the MTL framework outperforms STL models, offering a promising approach for strengthening cybersecurity. Future work will explore real-time deployment and dynamic analysis to enhance urban resilience against emerging cyber threats.

Keywords: Android Malware · Multi-task learning · Hyperparameter tuning · CCCS-CIC-AndMal-2020

1 Introduction

Traditional malware detection techniques, based on signatures or behavioral analysis, often fail to detect advanced threats such as zero-day attacks and obfuscated malware [6]. These challenges underscore the need for adaptive and scalable detection systems. Deep Learning (DL) has shown promise in analyzing complex patterns and detecting sophisticated malware [14,16]. However,

S. Yuan et al. (Eds.): PAKDD 2025 Workshops, LNAI 15835, pp. 67–78, 2025.
https://doi.org/10.1007/978-981-96-8197-6_6

Single-Task Learning (STL) models face limitations due to class imbalance and high-dimensional feature spaces, restricting their generalization capabilities and reliability, especially for rare malware families [15,17].

To address these issues, this study proposes a Multi-Task Learning (MTL) model that concurrently performs malware detection (binary classification) and malware family identification (multiclass classification). By leveraging shared representations, the MTL approach enhances detection accuracy and generalization across diverse malware types, overcoming STL limitations. While MTL has shown success in other cybersecurity applications, its application in Android malware detection remains underexplored as presented in the SLR conducted by [8] which motivates the present study. The key contributions of this work are:

1. Development of an MTL framework for Android malware detection and family classification.
2. Implementation of shared and task-specific layers for improved efficiency and performance.
3. Use of PCA for dimensionality reduction, weighted loss for class imbalance handling, and Optuna for hyperparameter optimization.
4. Demonstration of superior results compared to STL models.

The MTL model is trained on the CCCS-CIC-AndMal-2020 dataset, utilizing API-based static features. PCA reduces computational complexity, and a weighted loss function addresses class imbalance. Hyperparameter tuning optimizes key parameters, achieving 97% accuracy in malware detection and 91% in family classification. These results demonstrate the model's scalability and effectiveness in addressing complex malware patterns. The remainder of this paper is organized as follows: Sect. 2 reviews related work, Sect. 3 describes the proposed framework, Sect. 4 presents experimental results, and Sect. 5 concludes with implications and future directions.

2   Related Work

Recent advancements in malware detection underscore the effectiveness of machine learning and deep learning models, particularly with customized feature selection. Approaches like Mahindru and Sangal's PermDroid and Almakayeel et al.'s DLBITM-AMD utilize deep learning with feature prioritization and hybrid optimization techniques, achieving high detection accuracy [3,11]. Wang et al.'s FGL_Droid integrates API call sequence graphs with permission features, demonstrating competitive results [18].

Deep learning architectures such as CNNs and RNNs are foundational in Android malware detection, providing high accuracy and rapid predictions [5]. Transfer learning further enhances detection by reducing training time while maintaining robustness [13]. Models like PermDroid and DLBITM-AMD highlight the role of feature selection in improving accuracy and computational efficiency.

Table 1. Comprehensive Comparison of Models Using the CCCS-CIC-AndMal-2020 Dataset for Android Malware Detection and Classification

Study	Model Approach	Feature Type	Feature Selection	Year	Samples	Task	Accuracy
[2]	CNN on APK Images	Static	CNN	2020	21K	Detection	96.4%
[9]	Ensemble Model	Dynamic	Information Gain & Backward Feature Elimination	2022	50K	Detection	96.8%
[19]	WGAN-GP and Auto-Encoder	Static	RBM-based Subspace	2022	227K	Classification	83.17%
[4]	SVM	Static & Dynamic	–	2023	13K	Detection	96.64%
Our MTL Model	DNN (MTL)	Static	PCA	2024	190K	Detection, Classification	97%, 91%

The CCCS-CIC-AndMal-2020 dataset is a standard benchmark for Android malware detection, primarily in STL frameworks. Table 1 compares STL models, their feature selection techniques, and performance. However, applying MTL for simultaneous malware detection and classification using this dataset remains unexplored.

MTL has shown significant benefits across various cybersecurity applications, including network intrusion detection, cyberbullying classification, and infrastructure protection. Drawing from a comprehensive SLR conducted in our previous study [8], it is illustrating MTL's effectiveness in handling challenges such as data imbalance and task complexity. These insights underscore MTL's value in cybersecurity by improving generalization and robustness through shared task-specific knowledge.

3 Methodology

This section describes the development of the MTL framework for Android malware detection and family classification. The methodology illustrated in Fig. 1 has key stages, including data preparation, feature selection, and the design of an architecture with shared and task-specific layers. Additionally, hyperparameter tuning is employed for optimization. The following subsections provide further details on each component of the methodology.

3.1 Experimental Setup

The MTL model was trained on a HP Pavilion Gaming Laptop with an Intel Core i7 processor (6 cores), NVIDIA GTX 1050 Ti GPU, and 16 GB of RAM

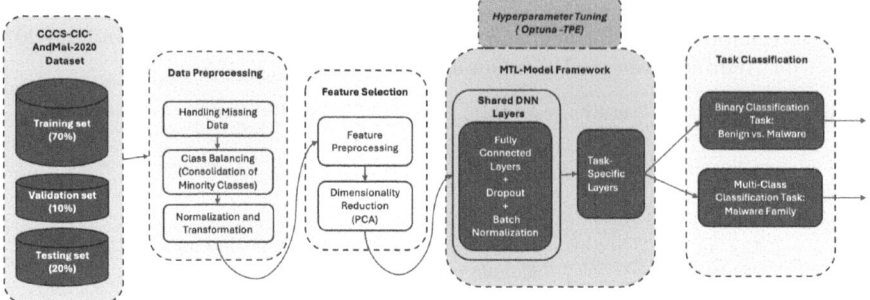

Fig. 1. High-Level Architecture of the Proposed methodology.

with SSD storage. While sufficient for binary and multi-class classification tasks, the consumer-grade GPU and moderate processing capabilities resulted in a training time of approximately two hours, longer than server-grade hardware with high-performance GPUs.

3.2 Data Preparation

The CCCS-CIC-AndMal-2020 dataset, consisting of 400k Android application samples with static features such as API calls and permissions [12], was used. A subset of 190,165 samples was selected to address class imbalance:

- **Balanced sample size**: Prioritized prominent malware types with significant behavior.
- **Consolidation of minority variants**: Combined Trojan variants (e.g., Banker, Dropper, SMS, Spy) under a single class to ensure consistent model performance.

The dataset was balanced for binary classification but exhibited imbalance in multi-class tasks, as shown in Table 2.

Table 2. Number of Samples for Each Class

Class	Number of Samples
Benign	98,340
Adware	47,210
Trojan (Consolidated)	23,413
Zero-day Malware	15,000
Ransomware	6,202
Total	**190,165**

3.3   Feature Selection

The dataset initially included 9,000 static features, increasing computational cost and introducing noise. A multi-step feature selection process was applied:

– **Feature Preprocessing**: Removed features with zero values across all samples, reducing the feature space to approximately 7,000.
– **Dimensionality Reduction via PCA**: PCA reduced the feature space further to 2,000 components, capturing 95% of the variance.

This reduction enhanced model efficiency and accuracy by focusing on the most relevant patterns for malware detection and classification.

3.4   Model Architecture

The proposed MTL model utilizes DNNs to process high-dimensional, non-sequential feature spaces efficiently [7].

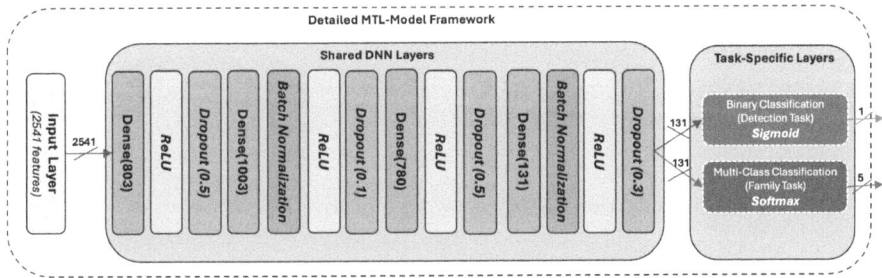

Fig. 2. Detailed Architecture of the MTL Model Framework.

As shown in Fig. 2, the architecture consists of three main components:

– **Shared Layers**: Extract generalized features across tasks, reducing redundancy.
– **Task-Specific Layers**: Refine features for binary and multi-class classification.
– **Regularization Techniques**: Use dropout to prevent overfitting and batch normalization to stabilize training.

The MTL model processes the input feature vector $x \in \mathbb{R}^d$ through shared layers using ReLU activation to extract generalized features, followed by task-specific branches for binary and multi-class classification. Outputs are generated using sigmoid and softmax activations and optimized with cross-entropy loss functions. Dropout and batch normalization are applied to enhance generalization and training stability. The total loss balances detection and classification tasks using a weighted sum:

$$\mathcal{L}_{\text{total}} = \alpha \mathcal{L}_{\text{binary}} + (1 - \alpha)\mathcal{L}_{\text{multi}} \tag{1}$$

This architecture employs efficient feature sharing and simultaneous optimization across tasks.

3.5   Hyperparameter Tuning

Hyperparameter tuning was conducted using the Optuna framework, which employs the Tree-structured Parzen Estimator (TPE) algorithm for efficient exploration of key parameters such as the number of layers, hidden layer size, dropout rates, learning rate, and binary task weight [1]. Optuna's pruning mechanism terminated underperforming trials early, reducing computation time while ensuring optimal performance.

The hyperparameter search spanned 200 trials, with the best configuration stabilizing after approximately 150 trials, as shown in Fig. 3. Binary task weight was identified as the most influential hyperparameter, alongside parameters such as epochs and dropout rates as presented in Fig. 4.

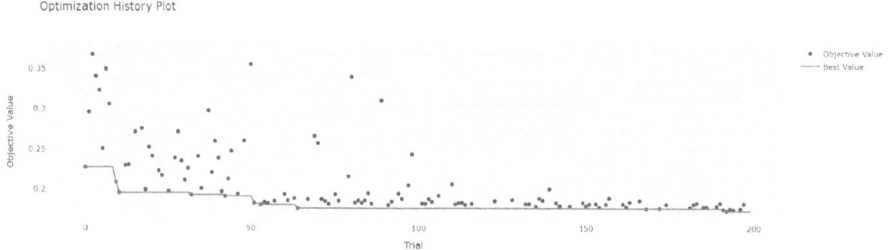

Fig. 3. Optimization history showing objective loss across trials.

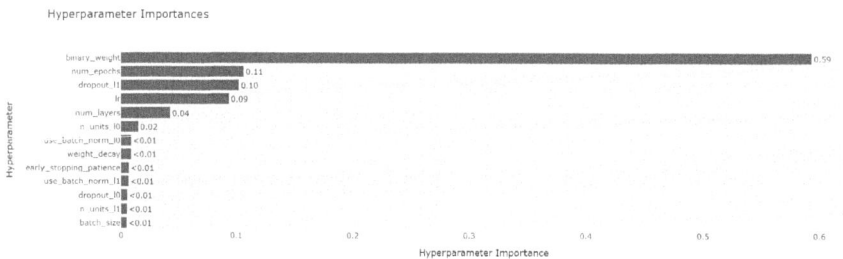

Fig. 4. Hyperparameter importance showing relative impact on model performance.

For the training process, The model was trained to minimize total loss using the Adam optimizer [10], which adapts learning rates dynamically. Each training epoch includes a forward pass, backward pass, and parameter update based on:

$$\theta_{t+1} = \theta_t - \eta \cdot \frac{m_t}{\sqrt{v_t} + \epsilon} \tag{2}$$

where m_t and v_t are moment estimates, η is the learning rate, and ϵ ensures numerical stability. A learning rate scheduler adjusted rates for higher convergence.

Evaluation Process. The model's performance was evaluated using task-specific metrics for both binary and multi-class classification tasks.

For malware detection, metrics such as accuracy, precision, recall, and F1 score were used to assess detection quality and balance. These metrics provide insights into the model's ability to correctly classify malicious and benign samples, focusing on relevance, sensitivity, and overall performance.

For the multi-class classification task, macro-averaged metrics, including macro-precision, macro-recall, and macro-F1 score, were employed. These metrics ensure balanced evaluation across all classes, particularly addressing class imbalance, and validate the model's performance in minority classes.

4 Results

This section evaluates the MTL model for Android malware detection and classification. It compares the baseline and optimized MTL models and highlights improvements in performance achieved through hyperparameter tuning. Additionally, comparisons with STL models underscore MTL's advantages in efficiency and robustness.

4.1 Hyperparameter Tuning

Hyperparameter tuning was conducted using Optuna's TPE algorithm to optimize the MTL model's performance. The baseline model utilized default hyperparameters, while the optimized model employed parameters derived from tuning, as summarized in Tables 3 and 4. The baseline model used a simpler architecture with three layers, fixed dropout rates, no batch normalization, and a learning rate of 0.0001, as shown in Table 3. After tuning, the optimized model employed a four-layer architecture, variable dropout rates, selective batch normalization, and a learning rate of 2.87×10^{-5}, as shown in Table 4.

Table 3. Baseline Model Hyperparameters

Hyperparameter	Value
Number of Layers	3
Units per Layer	1024, 512,256
Dropout Rate	0.3–0.5
Batch Size	64
Learning Rate	0.0001
Binary Task Weight (α)	0.5

Table 4. Optimized Model Hyperparameters

Hyperparameter	Value
Number of Layers	4
Units per Layer	803,1003,780,131
Dropout Rate	0.1–0.5
Batch Size	64
Learning Rate	2.87×10^{-5}
Binary Task Weight (α)	0.69

4.2 Training and Validation Performance

The training and validation performance of the baseline and optimized models was evaluated using 5-fold cross-validation. The optimized model exhibited smoother convergence, lower training and validation losses, and improved generalization compared to the baseline model as shown in both plots Figs. 5 and 6.

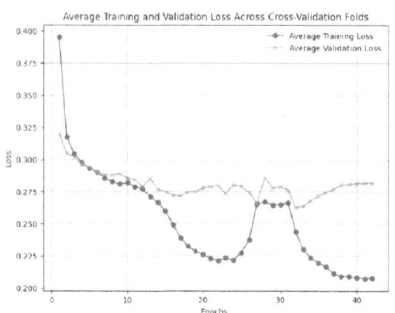 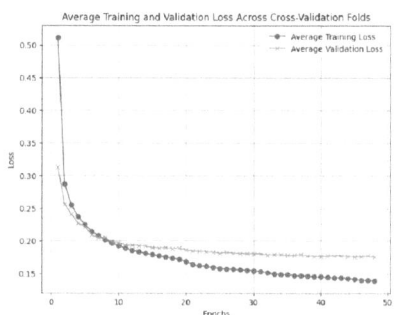

Fig. 5. Training and Validation Loss (Baseline).

Fig. 6. Training and Validation Loss (Optimized).

4.3 Performance Metrics

The optimized model achieved higher stability and accuracy across binary and multi-class tasks compared to the baseline. Binary classification metrics such as accuracy, precision, recall, and F1-score showed reduced variability which is presented in Figs. 7 and 8. Similarly, multi-class classification performance improved significantly, particularly in precision and recall (Figs. 9 and 10).

4.4 Test Evaluation

The MTL model achieved good performance in both binary and multi-class classification tasks. For binary classification, it reached 97% accuracy with high precision and recall, while for multi-class classification, it attained 91% accuracy,

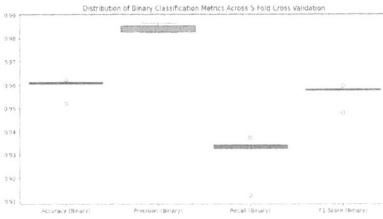

Fig. 7. Binary Metrics Across Folds (Baseline).

Fig. 8. Binary Metrics Across Folds (Optimized).

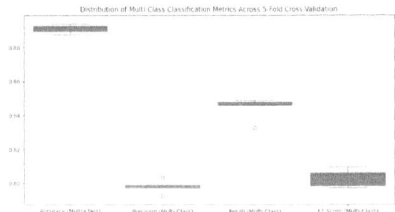

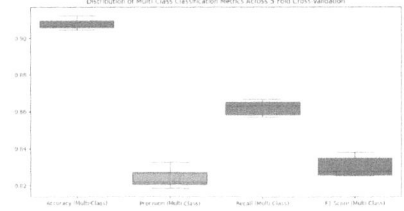

Fig. 9. Multi-Class Metrics Across Folds (Baseline).

Fig. 10. Multi-Class Metrics Across Folds (Optimized).

excelling in major classes but with moderate performance for some minor classes like Trojan and Zeroday the detailed evaluation in Table 5.

The model's compact size of 14.33 MB makes it suitable for deployment on resource-constrained devices such as mobile and embedded systems. This combination of high accuracy and small size highlights the model's efficiency and practicality for real-world malware detection applications.

Table 5. Classification Report for Binary and Multi-Class Tasks

Class	Precision	Recall	F1-Score	Support
Benign (Binary)	0.96	0.99	0.97	19620
Malware (Binary)	0.99	0.95	0.97	18080
Benign (Multi-Class)	0.96	0.98	0.97	19620
Adware	0.97	0.91	0.94	9548
Ransomware	0.73	0.95	0.82	1255
Trojan	0.92	0.66	0.76	4671
Zeroday	0.54	0.81	0.65	2606
Binary Accuracy	0.97			
Multi-Class Accuracy	0.91			

4.5   Comparison with Single-Task Learning (STL)

The MTL model was compared with STL models designed for binary and multi-class tasks. Both STL models were built with identical architectures and training settings for fair comparison. The MTL model demonstrated superior performance due to shared feature learning, which reduces redundancy and enhances generalization.

As shown in Fig. 11, the MTL model achieved faster convergence and lower training and validation losses than STL models. This aligns with observations in the literature [2,12], where STL models exhibited slower convergence and higher loss due to independent task handling.

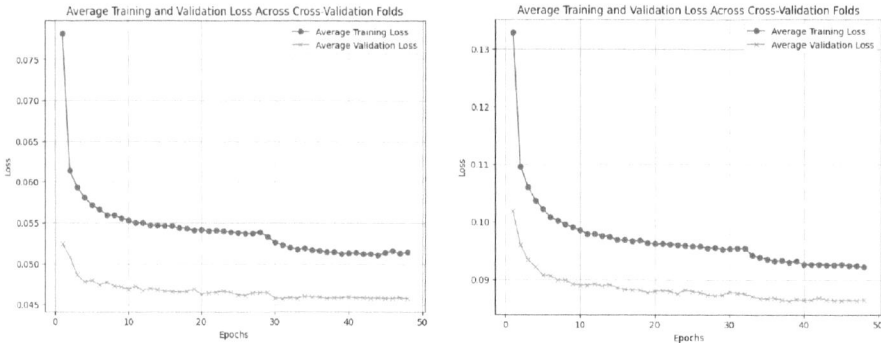

Fig. 11. Comparison of Training and Validation Loss for STL Binary-Class and Multi-Class Models

In terms of execution time, the MTL model trained in 94 min, significantly faster than the combined 351 min required for STL models. The MTL model outperformed STL models in computational efficiency, proving effective for resource-limited environments.

5   Conclusion

This study presented an MTL model for Android malware detection and classification, demonstrating its effectiveness in computational efficiency, generalization, and scalability compared to conventional STL models. The shared feature extraction layers enabled simultaneous task processing, reducing training time without sacrificing accuracy, making it ideal for deployment in resource-constrained environments such as mobile devices, IoT systems, and enterprise networks. The model's ability to classify rare malware families shows its potential for applications in healthcare, finance, and critical infrastructure, contributing to the resilience of urban digital ecosystems. By addressing challenges such as class imbalance and computational overhead, it can be integrated into mobile device management systems, cloud-based security platforms, and IoT monitoring frameworks, enhancing cybersecurity in smart cities and critical urban infrastructures. Despite its strong performance, challenges remain in detecting minority malware classes.

Future work will focus on enhancing minority class detection through synthetic data generation, data augmentation, and advanced architectures including transformers and attention mechanisms. Additional efforts will explore real-time deployment and integrating dynamic analysis features, aligning the model with urban resilience frameworks to strengthen its resilience and adaptability to evolving cybersecurity threats.

Acknowledgments. This publication was supported by Qatar University Internal Grant No. QUUG-CENG-CSE-2022. The findings achieved herein are solely the responsibility of the authors.

References

1. Akiba, T., Sano, S., Yanase, T., Ohta, T., Koyama, M.: Optuna: a next-generation hyperparameter optimization framework. In: Proceedings of the 25th ACM SIGKDD International Conference on Knowledge Discovery & Data Mining, KDD 2019, pp. 2623–2631. Association for Computing Machinery, New York, NY, USA (2019). https://doi.org/10.1145/3292500.3330701
2. Al-Fawa'reh, M., Saif, A., Jafar, M.T., Elhassan, A.: Malware detection by eating a whole apk. In: 2020 15th International Conference for Internet Technology and Secured Transactions (ICITST), pp. 1–7 (2020). https://doi.org/10.23919/ICITST51030.2020.9351333
3. Almakayeel, N.: Deep learning-based improved transformer model on android malware detection and classification in internet of vehicles. Sci. Rep. **14**(1), 25175 (2024). https://doi.org/10.1038/s41598-024-74017-z
4. Awais, M., Tariq, M.A., Iqbal, J., Masood, Y.: Anti-ant framework for android malware detection and prevention using supervised learning. In: 2023 4th International Conference on Advancements in Computational Sciences (ICACS), pp. 1–5 (2023). https://doi.org/10.1109/ICACS55311.2023.10089629
5. Bilal, A., Alzahrani, A., Almuhaimeed, A., Khan, A.H., Ahmad, Z., Long, H.: Advanced ckd detection through optimized metaheuristic modeling in healthcare informatics. Sci. Rep. **14**(1), 12601 (2024)
6. Catal, C., Giray, G., Tekinerdogan, B.: Applications of deep learning for mobile malware detection: a systematic literature review. Neural Comput. Appl. **34**(2), 1007–1032 (2022). https://doi.org/10.1007/s00521-021-06597-0
7. Goodfellow, I.J., Bengio, Y., Courville, A.: Deep Learning. MIT Press, Cambridge, MA, USA (2016). http://www.deeplearningbook.org
8. Ibrahim, S., Catal, C., Kacem, T.: The use of multi-task learning in cybersecurity applications: a systematic literature review. Neural Comput. Appl. (2024). https://doi.org/10.1007/s00521-024-10436-3
9. Khalid, S., Hussain, F.B.: Evaluating dynamic analysis features for android malware categorization. In: 2022 International Wireless Communications and Mobile Computing (IWCMC), pp. 401–406 (2022). https://doi.org/10.1109/IWCMC55113.2022.9824225

10. Kingma, D., Ba, J.: Adam: a method for stochastic optimization. In: International Conference on Learning Representations (ICLR), San Diega, CA, USA (2015)
11. Mahindru, A., et al.: Permdroid a framework developed using proposed feature selection approach and machine learning techniques for android malware detection. Sci. Rep. **14**(1), 10724 (2024). https://doi.org/10.1038/s41598-024-60982-y
12. Rahali, A., Lashkari, A.H., Kaur, G., Taheri, L., Gagnon, F., Massicotte, F.: Didroid: android malware classification and characterization using deep image learning. In: ACM International Conference Proceeding Series, pp. 70–82, November 2020. https://doi.org/10.1145/3442520.3442522, https://www.unb.ca/cic/datasets/andmal2020.html
13. Raza, A., Qaisar, Z.H., Aslam, N., Faheem, M., Ashraf, M.W., Chaudhry, M.N.: Tl-gnn: android malware detection using transfer learning. Appl. AI Lett. **5**(3), e94 (2024). https://doi.org/10.1002/ail2.94, https://onlinelibrary.wiley.com/doi/abs/10.1002/ail2.94
14. Sagar, R., Jhaveri, R., Borrego, C.: Applications in security and evasions in machine learning: a survey. Electronics **9**(1) (2020). https://doi.org/10.3390/electronics9010097, https://www.mdpi.com/2079-9292/9/1/97
15. Seltzer, M.L., Droppo, J.: Multi-task learning in deep neural networks for improved phoneme recognition. In: 2013 IEEE International Conference on Acoustics, Speech and Signal Processing, pp. 6965–6969 (2013). https://doi.org/10.1109/ICASSP.2013.6639012
16. Taddeo, M.: Three ethical challenges of applications of artificial intelligence in cybersecurity. Minds Mach. **29**, 187–191 (2019). https://doi.org/10.1007/S11023-019-09504-8/METRICS, https://link.springer.com/article/10.1007/s11023-019-09504-8
17. Vandenhende, S., Georgoulis, S., Van Gansbeke, W., Proesmans, M., Dai, D., Van Gool, L.: Multi-task learning for dense prediction tasks: a survey. IEEE Trans. Pattern Anal. Mach. Intell. **44**(7), 3614–3633 (2022). https://doi.org/10.1109/TPAMI.2021.3054719
18. Wang, W., Ren, C., Song, H., Zhang, S., Liu, P.: Fgl_droid: an efficient android malware detection method based on hybrid analysis. Secur. Commun. Netw. 8398591 (2022). https://doi.org/10.1155/2022/8398591, https://onlinelibrary.wiley.com/doi/10.1155/2022/8398591
19. Xie, J., Li, S., Zhang, Y., Sun, P., Xu, H.: Analysis and detection against network attacks in the overlapping phenomenon of behavior attribute. Comput. Secur. **121**, 102867 (2022). https://doi.org/10.1016/j.cose.2022.102867, https://www.sciencedirect.com/science/article/pii/S0167404822002619

Workshop of Foundational AI for Pervasive Computing (FairPC 2025)

The Workshop of Foundational Artificial Intelligence for Pervasive Computing (FairPC): Theory, Algorithms, and Applications

Data collected from sensors and embedded devices are becoming pervasive in numerous real-world applications, e.g., IoT devices, healthcare, wearable devices, smart vehicles, environmental sciences, etc. With massive amounts of data collected everywhere, we have entered an era promising a much greater understanding of complex and fundamental challenges in diverse applications through Foundational AI for Pervasive Computing.

Through this workshop, we aimed to provide a platform for researchers and AI practitioners from both academia and industry to discuss potential research directions, especially in foundational models, and key technical issues, and present solutions to tackle related challenges in practical applications. We discussed foundational models and large language models (LLMs) and their potential impacts on pervasive computing applications. This workshop was well-aligned with the main conference focusing on data science, data mining, knowledge discovery, and topics in generative AI and LLMs.

April 2025

Beiyu Lin
Dongjin Song
Tianlong Chen
Pan He
Sihong He
Jing Ma

Organization

Organizers

Beiyu Lin	University of Oklahoma, USA
Dongjin Song	University of Connecticut, USA
Tianlong Chen	University of North Carolina at Chapel Hill, USA
Pan He	Auburn University, USA
Sihong He	University of Texas at Arlington, USA
Jing Ma	Case Western Reserve University, USA

Program Committee

Nam Huynh	University of Oklahoma, USA
Ahsan Bilal	University of Oklahoma, USA
Mehdi Zaeifi	University of Oklahoma, USA
Saeed Tajik H.	University of Oklahoma, USA
Ziying Jia	University of Texas at Arlington, USA
Ningkang Chang	University of Texas at Arlington, USA
Yiran Qiao	Case Western Reserve University, USA
Xinyu Zhao	University of North Carolina at Chapel Hill, USA
Pingzhi Li	University of North Carolina at Chapel Hill, USA
Sukwon Yun	University of North Carolina at Chapel Hill, USA
Mufan Qiu	University of North Carolina at Chapel Hill, USA
Huaizhi Qu	University of North Carolina at Chapel Hill, USA
Darryl Jacob	Auburn University, USA
Stephen Cortez	Auburn University, USA
Darahaas Nallagatla	Auburn University, USA
Owen Strength	Auburn University, USA
Matthew Rogers	Auburn University, USA
Xuyang Shen	University of Connecticut, USA
Zijie Pan	University of Connecticut, USA
Yushan Jiang	University of Connecticut, USA
Binghao Lu	University of Connecticut, USA
Qianying Ren	University of Connecticut, USA

A Comprehensive Survey on Bias and Fairness in Large Language Models

Farsheed Haque[ID], Depeng Xu[✉][ID], and Xi Niu[ID]

University of North Carolina at Charlotte, Charlotte, USA
{fhaque,dxu7,xniu2}@charlotte.edu

Abstract. Large Language Models (LLMs) excel in various applications, from conversational agents to medical diagnostics. However, they often inherit societal biases from training data, risking discriminatory outcomes. This survey provides a structured review of fairness in LLMs, analyzing the origins of bias in data and model design. We summarize key bias evaluation metrics, including embedding-based and probability-based approaches, and examine mitigation techniques in pre-processing, in-processing, and post-processing strategies. We also highlight essential datasets and tools for fairness research and discuss current challenges, offering future directions to guide the development of fairer LLMs.

Keywords: Fairness and Bias · Large Language Model

1 Introduction

Large Language Models (LLMs) have significantly advanced the field of Natural Language Processing (NLP) by showcasing exceptional performance across many applications, such as medical diagnostics [77], financial consulting [67], and legal assistance [85]. Their influence extends beyond traditional language tasks like translation and sentiment analysis [83], highlighting their versatility and impact across diverse industries.

However, alongside these advancements, LLMs have been found to inherit and sometimes amplify biases present in the extensive datasets used for their training [27,88]. These biases often reflect societal prejudices related to sensitivie attributes such as gender, race, age, nationality, occupation, and religion [1,35]. When embedded within LLMs, these biases can lead to unfair or discriminatory outcomes in downstream applications [6]. For instance, an analysis revealed that ChatGPT produced letters of recommendation that depicted individuals with typically female names as warm and amiable, while those with typically male names were described as leaders and innovators [76]. In healthcare, racial biases have resulted in race-based medicine suggestions [58].

Addressing the fairness concerns associated with LLMs has become an essential area of research. The community has developed various methods for evaluating and mitigating bias within these models [21]. However, the rapid evolution of LLMs, particularly with the emergence of different LLM techniques like fine-tuning and prompting, introduces new complexities. Moreover, discrepancies in fairness definitions and evaluation methodologies across studies have created confusion and impeded progress. The absence of a standardized framework that aligns fairness notions with appropriate mitigation techniques complicates the development of algorithms for future fair LLMs. While existing surveys have provided insights into fairness in machine learning and NLP [8,16], they often do not specifically address the unique challenges posed by LLMs, especially those arising from different LLM techniques.

This survey aims to provide a structured overview of fairness in LLMs, bridging gaps in the current literature. Our key contributions include:

- In-Depth Exploration of LLM Fundamentals: We delve into the foundational aspects of LLMs, explaining what they are, how they are trained, and the sources of bias that can arise from large-scale data collection and model architectures.
- Comprehensive Review of Bias Evaluation Metrics: We systematically review the methods used to quantify bias in LLMs, including embedding-based approaches, probability-based approaches, and distribution-based techniques.
- Analysis of Bias Mitigation Strategies: We examine a range of strategies for mitigating bias in LLMs, categorizing them into pre-processing methods (modifying data before training), in-processing techniques (altering the training process itself), and post-processing approaches (adjusting outputs after generation). We discuss the effectiveness and limitations of methods.
- Discussion of Challenges and Future Directions: We highlight current challenges in achieving fairness in LLMs (e.g., interpretability, evolving biases, fairness-performance trade-offs). We also propose directions for future research, emphasizing the need for standardized evaluation frameworks and the development of models that are both fair and efficient.

2   Comparison with Existing Surveys

LLMs are relatively new, yet numerous studies note their susceptibility to social biases. In response, several surveys address fairness and bias, often with a broader or more conceptual focus. For instance, [8] discusses various bias types, fairness metrics, and debiasing approaches but primarily targets general NLP rather than LLMs. Meanwhile, [25,32] propose frameworks to formalize social biases, fairness definitions, and taxonomies of bias mitigation methods, yet these remain largely conceptual. [28] lists prior work on bias detection and mitigation without accounting for LLM-specific operational nuances like zero-shot, few shot learning, and [26] narrows its discussion of LLM bias to the education domain. Collectively, these surveys emphasize metrics and mitigation strategies but seldom explore how bias arises from training paradigms, prompting strategies, or adaptations (e.g., zero-shot, few-shot, parameter-efficient fine-tuning). Our survey addresses these gaps by providing an LLM-focused perspective: we detail how biases embed and manifest in foundational architectures, review bias measurement and mitigation techniques, catalog relevant datasets, and highlight persistent challenges. This unified treatment offers a more nuanced and comprehensive view of fairness and bias within the rapidly evolving sphere of large language models.

3   Survey Methodology

Our study followed a three-step process:

- **Scope Definition:** We targeted works focusing on *bias* or *fairness* in large-scale transformer-based LLMs.
- **Literature Search:** We searched ACL Anthology, IEEE Xplore, ACM, and arXiv (2018–present) with keywords such as **fairness, bias, debiasing, LLM**, then removed duplicates and irrelevant entries.
- **Screening:** We reviewed 89 full papers, emphasizing those that analyze or mitigate bias in LLMs, propose bias metrics, or provide domain-specific insights.

4   Large Language Models (LLMs)

Large Language Models (LLMs) are advanced computational systems designed to understand and generate human language. They are built using deep learning architectures, primarily decoder-based transformers [75], which capture complex patterns and relationships in language data to process and generate text effectively. Transformers utilize self-attention mechanisms, enabling models to consider the entire input sequence when generating the output. This capability allows them to capture long-term dependencies, thus enabling the models to grasp the nuances of language.

LLMs are typically pre-trained on massive text data, including books, articles, websites, and other textual sources [73]. This extensive training allows them to develop a broad understanding of language, grammar, facts, and even some reasoning abilities [13]. As a result, LLMs can perform a wide array of tasks. Such as translation, summarization, question answering, and creative writing without requiring task-specific training. A pre-trained LLM employs various techniques to adapt to different tasks:

Zero-Shot Prompting refers to the ability of a model to perform tasks that it has not been explicitly trained on by leveraging its generalized language understanding [61]. In the context of LLMs, zero-shot learning allows the model to generate appropriate responses to new tasks based solely on the instructions provided in the input prompt.

Few-Shot Prompting involves providing the model with a small number of examples or demonstrations of a task within the input prompt [11]. This technique helps the model understand the desired output format and task requirements based on the given examples. For instance, by including a few examples of question-answer pairs in the prompt, the model can infer how to answer a new question following the same pattern.

Fine-Tuning adapts a pre-trained LLM to a specific task or domain by updating its parameters with task-specific data, which can greatly improve performance. However, fine-tuning very large models with billions of parameters can be computationally expensive [38]. To overcome this, **Parameter-Efficient Training** methods have emerged. For example, [38] proposed Low-Rank Adaptation (LoRA), which reduces the number of trainable parameters by inserting small rank-decomposition matrices into each transformer layer rather than updating all parameters. Building on LoRA, [18] introduced Quantized LoRA (QLoRA), employing quantization (e.g., 8-bit or 4-bit weights) to further lower memory usage while preserving performance.

5   Fairness in LLMs

5.1   Definition of Bias in LLMs

Bias in LLMs typically emerges when models associate specific traits or behaviors with certain social groups, often because of stereotypes embedded in the training data. This kind of social bias arises when a model learns that certain occupations or attributes are more likely to be associated with specific demographics, such as associating "nurse" with women or "programmer" with men. This happens due to models being trained on data containing imbalances and societal stereotypes, leading to biased behavior that impacts downstream tasks [28]. In LLMs, biases manifest as both representational bias and output bias. Representational bias appears in the model's learned embeddings, embedding societal stereotypes and skewed group representations into the model's understanding of language. Output bias becomes visible in downstream tasks where the model's predictions or generated texts reflect these internalized stereotypes. Both forms of bias influence the model's performance across various applications, from content generation to classification tasks [30].

5.2   Key Sources of Bias

Bias in LLMs stems from multiple interconnected sources, each affecting the model's predictions in unique ways. Here are some major contributors:

Training Data: The data used to train LLMs is often reflective of historical societal biases. If data used for model training is disproportionately representative of particular social views or demographics, models are likely to echo these biases in their responses [68]. For instance, when datasets overrepresent certain gender roles, the model may learn to associate particular occupations with specific genders, reinforcing occupational stereotypes [14].

Embedding: Embeddings, fundamental to model architecture, can inadvertently carry biases that subtly influence downstream tasks. For example, clustering male-associated words closer to professions like "doctor" while positioning female-associated words near "nurse" embeds stereotypes into the model's language representation. These biases are then perpetuated in responses, affecting the model's fairness and inclusivity [25].

Labeling and Human Annotations: LLMs often rely on labeled data to learn specific tasks. This labeling process is vulnerable to human subjectivity, as annotators may unconsciously introduce their own beliefs or stereotypes into the labels they assign [66]. Additionally, the practice of using reinforcement learning with human feedback (RLHF) to align model behavior with human values may also incorporate subjective notions into model responses [45].

Amplification: Models not only learn from biases in their data but also amplify them throughout the training process. During pre-training, models may internalize subtle biases, which become increasingly pronounced when fine-tuned on specific downstream tasks. This amplification effect can result in the over-representation of stereotypes, especially in high-stakes scenarios [53].

5.3   Bias Manifestation

Pre-training and Fine-tuning Paradigm. In the pre-training and fine-tuning, LLMs undergo an initial unsupervised pre-training phase on extensive textual data, followed by a supervised fine-tuning phase tailored to specific downstream tasks. This traditional training strategy has been predominant for medium-sized models such as BERT, which typically contain less than a billion parameters. Although some medium-sized LLMs like GPT-2 (1.5B parameters) and T5 (3B parameters) are larger, they still primarily rely on the fine-tuning paradigm. During the pre-training phase, the model learns from vast and diverse datasets that inevitably contain historical and societal biases. These intrinsic biases are encoded in the model's internal representations and remain task-independent. For instance, associations between gender and professions learned during pre-training can influence the model's behavior across various applications, leading to biased outputs regardless of the specific task [36]. Fine-tuning introduces task-specific biases as the model adapts its parameters to optimize performance on particular tasks. If the fine-tuning data is biased or unbalanced, it can exacerbate existing or introduce new biases specific to the task [9]. For example, fine-tuning a model on a sentiment analysis dataset with biased labels can result in unfair sentiment predictions for certain demographic groups [34].

Prompt-Based Paradigm. With the emergence of large models like GPT-3 and GPT-4, the prompt-based paradigm has become increasingly popular. In this approach, models generate responses to task-specific prompts with the help of zero-shot [61] and few-shot [11] prompting. Prompting offers flexibility, as models can be adapted to new tasks through carefully crafted input prompts. However, prompt-based models still retain biases from the pre-trained embeddings, which influence their generated outputs [76].

6 Bias Quantification in LLMs

6.1 Embedding-Based Metrics

Embedding-based metrics assess bias by analyzing the relationships between word or sentence embeddings. These methods extend techniques developed for static word embeddings to contextual embeddings used in LLMs.

Word Embedding Association Test (WEAT) [15] measures the association between two sets of target words and two sets of attribute words by computing the differences in cosine similarities between their embeddings.

Let X and Y be two sets of target words (e.g., occupations), and A and B be two sets of attribute words (e.g., male and female terms). The association of a word w with the attribute sets A and B is defined as:

$$s(w, A, B) = \frac{1}{|A|} \sum_{a \in A} \cos(w, a) - \frac{1}{|B|} \sum_{b \in B} \cos(w, b), \tag{1}$$

where $\cos(w, a)$ represents the cosine similarity between the embeddings of words w and a. The test statistic $s(X, Y, A, B)$ is computed as:

$$s(X, Y, A, B) = \sum_{x \in X} s(x, A, B) - \sum_{y \in Y} s(y, A, B). \tag{2}$$

The effect size quantifies the bias and is calculated as:

$$\text{WEAT}(X, Y, A, B) = \frac{\mu_X - \mu_Y}{\sigma_{X \cup Y}}, \tag{3}$$

where $\mu_X = \mathbb{E}_{x \in X} s(x, A, B)$, $\mu Y = \mathbb{E}_{y \in Y} s(y, A, B)$, and $\sigma_{X \cup Y}$ is the pooled standard deviation of $s(w, A, B)$ for all $w \in X \cup Y$.

WEAT assesses biases in embeddings by comparing the association strengths between target words and attribute words, effectively capturing how closely certain concepts are linked in the embedding space.

Sentence Embedding Association Test (SEAT) [52] adapts WEAT to contextual embeddings by using sentence templates to generate context-independent sentence embeddings. Templates such as "This is a [BLANK]" or "[BLANK] is here" are filled with target and attribute words to create sentences whose embeddings are then used in the WEAT framework.

SEAT computes the effect size using the same equations as WEAT, but with embeddings obtained from sentences rather than individual words. Adjustments in the embedding selection, such as using different layers of the model or specific token representations (e.g., [CLS] token), can improve the reliability of bias measurements [44, 46]. However, SEAT has limitations. Different choices in embedding selection can yield significantly different results, and SEAT may fail to reliably indicate the presence of stereotypes in the model [44].

Contextualized Embedding Association Test (CEAT) [31] extends WEAT to dynamic settings by quantifying the distribution of effect sizes across different contexts in contextualized word embeddings. CEAT measures how the association between target groups and attributes varies depending on the context, providing a more nuanced understanding of bias in LLMs. Given a set of target groups and two polarity attribute sets, CEAT computes the effect size for each context and assesses the bias distribution. Lower effect sizes indicate that the target group is closer to the negative polarity of the attribute.

6.2   Probability-Based Metrics

Probability-based metrics formalize bias by analyzing the probabilities assigned by LLMs to different words or sentences.

Discovery of Correlations (DisCo) [80] uses templates with slots filled by occupation-related nouns and lets the model predict the most probable words for the remaining slots. For example, using a template like "[PERSON] often likes to [MASK]", the [PERSON] slot is filled with gendered terms, and the model predicts the top candidate words for the [MASK] slot. DisCo assesses bias by comparing the model's predictions across different demographic terms. The average score of the model's predictions serves as the measurement of bias, highlighting any associations between demographics and certain attributes.

Log Probability Bias Score (LPBS) [44] calculates bias by comparing the log probabilities of sentences differing in target attributes while correcting for prior probability inconsistencies. Given a sentence S, LPBS is defined as:

$$\text{LPBS}(S) = \log\left(\frac{P_{\text{target}_a}}{P_{\text{prior}_a}}\right) - \log\left(\frac{P_{\text{target}b}}{\text{prior}_b}\right), \tag{4}$$

where P_{target_a} is the probability of the sentence with attribute a, and $P\text{prior}_a$ is the prior probability of attribute a (e.g., the probability of "he" or "she" in isolation). This correction ensures that the measurement difference is due to the attribute and not inherent model biases.

Categorical Bias (CB) Score [3] extends LPBS to multi-class targets and uses a set of sentence templates to quantify bias across categories like race. The CB score is defined as the variance of the normalized probabilities between target and attribute words:

$$\text{CB Score} = \frac{1}{|T||A|} \sum_{t \in T} \sum_{a \in A} \text{Var}_{n \in N}\left(\log P'\right), \tag{5}$$

where T is the set of sentence templates, A is the set of attribute words, N is the set of target words, $P' = \frac{P_{\text{target}}}{P_{\text{prior}}}$ is the normalized probability. The CB score captures the degree of bias by measuring how the normalized probabilities vary across different targets and attributes.

Context Association Tests (CATs) [55] measure the association between target groups and stereotypes from both intra-sentence and inter-sentence perspectives. CATs utilize evaluation datasets such as StereoSet [55] and CrowS-Pairs [56] that contain sentences reflecting stereotyped and anti-stereotyped associations. The evaluation uses the pseudo-log-likelihood (PLL) to compute the model's preference for one sentence over the other. PLL is defined as:

$$\mathrm{PLL}(S) = \sum_{i=1}^{|S|} \log P(s_i | S \setminus s_i, M_\theta), \tag{6}$$

where S is the sentence, s_i is the i-th token, $S \setminus s_i$ represents the sentence with the i-th token masked, M_θ is the language model with parameters θ. The model's bias is indicated by a consistent preference for stereotyped sentences over anti-stereotyped ones.

All Unmasked Likelihood (AUL) [42] modifies the PLL approach by considering multiple correct predictions instead of testing only whether the target token is predicted. It avoids using the [MASK] token, which may not appear in downstream tasks, to provide a more accurate assessment of the model's bias in natural language understanding.

6.3   Generation-Based Metrics

Generation-based metrics evaluate bias in the text generated by LLMs, which is particularly useful for models where internal probabilities or embeddings are inaccessible (e.g., closed-source models).

Classifier-Based Metrics involve using an auxiliary classifier to detect bias, toxicity, or sentiment in the generated text [7]. For instance, a classifier might be trained to identify toxic language or negative sentiment. By applying the classifier to texts generated from similar prompts featuring different social groups, disparities in classification outcomes can reveal biases in the model. For example, gender bias can be quantified by analyzing the difference in true positive rates for male and female-associated texts in the classification outcomes [31].

Distribution-Based Metrics assess bias by comparing the distribution of tokens associated with different social groups in the generated text. The Co-Occurrence Bias score [10] is one such metric.

For any token w and two sets of gendered words G_{female} and G_{male}, the bias score is defined as:

$$\mathrm{Bias}(w) = \log \left(\frac{P(w|G_{\text{female}})}{P(w|G_{\text{male}})} \right), \tag{7}$$

where $P(w|G)$ is the probability of encountering word w in the context of gendered terms G, calculated as:

$$P(w|G) = \frac{d(w,G)/\sum_i d(w_i, G)}{d(G)/\sum_i d(w_i)}, \tag{8}$$

with $d(w,G)$ representing the count of w occurring within a certain contextual window around words in G, and $d(G)$ being the total count of words in G.

A positive bias score indicates that w is more commonly associated with female words than male words. In an infinite context, words like "doctor" and "nurse" would occur equally with both female and male words, resulting in bias scores of zero. Deviations from zero indicate bias in the model's generated text.

6.4 Benchmark Datasets

As shown in Table 1, we categorize the datasets into two groups based on the type of bias metrics they support Natural Language Understanding (NLU)-based and generation(Gen)-based datasets.

Table 1. Summary of Datasets for Bias Evaluation

Type	Dataset	Description	Citation
NLU-Based	WinoBias	Evaluates gender bias in coreference resolution tasks.	[89]
	WinoBias+	Neutralized version of WinoBias using rewriters.	[74]
	Winogender	Tests gender bias in pronoun resolution.	[65]
	WinoQueer	Assesses biases related to LGBTQ+ identities.	[23]
	BUG	Evaluates gender bias in machine translation.	[47]
	GAP	Gender-balanced ambiguous pronoun-name pairs.	[79]
	CrowS-Pairs	Tests stereotypes across nine bias categories.	[56]
	StereoSet	Evaluates stereotypical biases in context.	[55]
Gen-Based	RealToxicityPrompts	Prompts to measure propensity for toxic generation.	[29]
	BOLD	Assesses biases in open-ended generation tasks.	[20]
	Jigsaw Datasets	Annotated comments for toxicity classification.	[17]
	BBQ Dataset	Questions designed to reveal model biases.	[60]

Natural Language Understanding (NLU)-Based Datasets help to measure bias in LLMs by assessing their performance on specific NLU tasks. These datasets typically consist of template-based sentences or pairs of counterfactual sentences. In template-based datasets, sentences with placeholders are completed by the language model selecting from predefined demographic terms. The model's biases are then inferred from the probabilities assigned to different terms.

WinoBias [89] is a benchmark dataset for intra-clause coreference resolution that evaluates a model's ability to associate gender pronouns with occupations in contexts reflecting stereotypes and anti-stereotypes. It includes two types of sentence templates: (1) Sentences with syntactic ambiguity, requiring semantic understanding. (2) Sentences with clearer syntactic structures, making coreference resolution easier. The bias score is defined as the difference in the model's performance on stereotyped versus anti-stereotyped sentences.

WinoBias+ [74] extends WinoBias by converting gendered sentences to neutral using rule-based and neural neutral rewriters.

Winogender [65] is an English coreference resolution dataset based on the Winograd format. It includes neutral gender and considers one occupation in each instance, aiming to test the model's ability to resolve pronouns w/o gender bias.

WinoQueer [23] developed by members of the LGBTQ+ community, WinoQueer contains 45,540 sentence pairs designed to evaluate biases related to LGBTQ+ identities. The dataset includes sentences with LGBTQ+ identity descriptors and counterfactual versions without such markers.

BUG (Bias in Universal Genders) [47] focuses on evaluating gender bias in machine translation systems. It uses a large-scale, real-world English dataset to assess how translation models handle gendered language, aiming to uncover biases in translation outputs.

GAP (Gendered Ambiguous Pronouns) [79] is a gender-balanced dataset containing 8,908 ambiguous pronoun-name pairs extracted from Wikipedia. It is designed to evaluate coreference resolution systems and to identify gender bias by providing a balanced set of examples across genders.

CrowS-Pairs [56] dataset comprises 1,508 sentence pairs that test for stereotypes across nine categories, including race, gender, sexual orientation, religion, age, nationality, disability, physical appearance, and socioeconomic status. Each pair consists of a stereotyped and a non-stereotyped sentence, allowing for the assessment of model biases using metrics like pseudo-log-likelihood.

StereoSet [55] is a dataset comprising 4,200 sentences designed to evaluate stereotypical biases in language models across four domains: gender, profession, race, and religion. Each sentence provides a context followed by three types of continuations: a stereotyped continuation that aligns with common societal biases, an anti-stereotyped continuation that contradicts these biases, and an unrelated or meaningless continuation. This structure allows for the assessment of model biases by analyzing the probabilities assigned to each type of continuation.

Generation-Based Datasets provide initial prompts or sentence prefixes, requiring the language model to generate continuations. These datasets are used to evaluate biases in the model's generated text based on the content it produces when prompted. Key generation-based datasets include:

RealToxicityPrompts [29] is one of the largest datasets of its kind, RealToxicityPrompts offers 100,000 sentence prefixes sourced from web text, each annotated with a toxicity score using the Perspective API. The dataset is used to measure the propensity of language models to generate toxic content when given both neutral and potentially provocative prompts.

Bias in Open-Ended Language Generation (BOLD) [20] consists of 23,679 prompts aimed at assessing biases across various domains such as professions, gender, race, religion, and political ideology. The prompts are derived from English Wikipedia pages associated with specific demographic groups or topics, and sentences are truncated to create prompts that elicit continuations from the model.

Jigsaw Datasets [17] consist of large-scale collections of online comments that have been annotated for toxicity and various forms of abusive language, including hate speech, insults, and threats. They contain millions of comments sourced from platforms like Wikipedia talk pages and news sites, each labeled with indicators of toxic content.

BBQ Dataset [60] is a recently developed resource aimed at evaluating and revealing biases in Large Language Models, especially within generative tasks. It comprises a curated set of questions that are specifically designed to prompt responses from models that may reflect biases or stereotypes. The questions cover sensitive areas such as race, gender, religion, and political views.

7   Bias Mitigation in LLM

Mitigating bias in Large Language Models (LLMs) can be broadly classified into four categories: (1) **Pre-processing**, (2) **In-processing**, (3) **Post-processing**, (4) **Intra-processing**.

7.1   Pre-processing Methods

Pre-processing methods aim to reduce biases in the data provided to the model, including the training dataset and the prompts used during inference.

Counterfactual Data Augmentation (CDA) is a widely used technique that balances training data across different social groups by augmenting it with counterfactual examples [80]. This involves generating new data instances by swapping protected attributes (e.g., gender, race) with their counterparts. For example, a sentence like "He is a doctor" can be transformed into "She is a doctor" to balance gender representations in the dataset.

Subsequent works have enhanced CDA by introducing variations: **Counterfactual Data Substitution (CDS)** [80,81] randomly replaces gendered terms in the dataset with their counterparts at certain probabilities to alleviate gender bias. **Selective Augmentation** [87] identifies and removes augmented instances that could negatively impact fairness, optimizing the model's fairness by pruning less effective counterfactual samples.

Data Calibration involves modifying the training data to reduce harmful information and biases. This can be achieved by:

- **Removing Biased Texts**: Identifying and deleting subsets of data that contain biased or harmful content using differential analysis [12], programmatic identification [57], or token-level matching [62].
- **Data Intervention Strategies**: Creating an unbiased dataset using techniques like naive-masking, neutral-masking, and random-phrase-masking [71].

For languages with complex morphology, generating biased text from fair text using machine translation models and round-trip translation has been explored [5].

Prompt Tuning focuses on refining user-provided prompts to reduce biases in LLM outputs. Prompts can be categorized as: **Hard Prompts**: Predefined, static templates that guide the model's responses [51]. Less abstract prompts can encourage the use of gender-neutral language. **Soft Prompts**: Learnable embeddings that are dynamically updated during prompt tuning [22]. By freezing model parameters and updating biased word embeddings associated with occupations, soft prompts can effectively reduce bias. While hard prompts lack flexibility, soft prompts may suffer from a lack of interpretability.

7.2   In-processing Methods

In-processing methods integrate fairness constraints into the model's training process, requiring retraining or fine-tuning of the model.

Loss Function Modification to include fairness constraints guides the model towards learning fair representations. For example:

- **Causal Regularization** [78]: Identifies causal and spurious features using counterfactual frameworks. The loss function is adjusted to impose smaller penalties on causal features and larger penalties on spurious ones, encouraging the model to rely on causal relationships.
- **Embedding Regularization** [59]: Addresses the persistence of gender-related features by utilizing generated gender direction vectors during fine-tuning.

Auxiliary Modules to the model architecture can help reduce bias without altering the original model parameters.

- **Adapter Modules** [37,46]: Incorporate additional layers (adapters) into the model that are trained on counterfactual data while keeping the original model parameters fixed. For example, ADELE (Adapter-based DEbiasing of Language Models) [46] achieves debiasing by using adapters.
- **Iterative Null Space Projection (INLP)** [63]: Iteratively trains linear classifiers to predict protected attributes and projects the representations into their null space, making it difficult for the model to encode bias-related information.

Projection-Based Methods aim to separate biased information from useful representations [19,41,48]. By projecting representations into orthogonal subspaces, models can remove discriminatory correlations while preserving essential information. For example, Orthogonal Subspace Correction and Rectification (OSCAR) [19] disentangles biased concepts without aggressively removing all related information, striking a balance between debiasing and utility. Iterative Gradient-Based Projection (IGBP) [40] iteratively trains a probe classifier to predict sensitive attributes and uses gradient information to guide the projection of representations onto a hypersurface that minimizes bias.

Adversarial Learning involves training the model to be fair by hiding sensitive information from an adversary trying to predict protected attributes [33,64]. An adversarial network is set up where the *Encoder* aims to produce representations that make it difficult for the adversary to predict protected attributes. The *Adversary* tries to predict the protected attributes from the encoder's representations. The model is trained to minimize task loss while maximizing the adversary's loss, encouraging fair representations.

Contrastive Learning offers an efficient method to mitigate bias during classifier training by learning representations that cluster instances based on their main task labels rather than sensitive attributes [69]. The training objective combines contrastive loss with cross-entropy loss to maximize the similarity between instances sharing the same class label while minimizing the similarity between instances sharing the same sensitive attribute. This approach reduces the influence of sensitive attributes on the model's predictions. Research in [70] extends this concept by incorporating both target labels and sensitive attributes into the contrastive learning framework. They define optimization functions aimed at achieving representational fairness, ensuring that learned representations are equitable across different demographic groups.

7.3   Post-processing Methods

Post-processing techniques modify the model's outputs after training to mitigate biases. This is especially important for closed-source LLMs where internal modifications are not feasible.

Chain-of-Thought (CoT) Prompting enhances model fairness by guiding LLMs through step-by-step reasoning, which can reduce reliance on biases [43]. By encouraging models to articulate their reasoning process, CoT can mitigate default biases that might influence responses. For example, when tasked with determining the gender associated with certain occupations, models using CoT prompts were less likely to rely on societal biases.

Rewriting Techniques involve identifying and modifying biased or discriminatory language in the model's outputs [72]. Techniques include:

- **Text Style Transfer**: Automatically substituting biased content with neutral terms using style transfer models trained on non-parallel data.
- **Post-Processing Filters**: Applying rules or filters to generated text to replace or remove biased language before presenting it to the user.

7.4   Intra-processing Methods

Intra-processing methods mitigate bias during the inference stage without additional training, often by modifying the model's decoding process.

Prompt Editing allows for efficient adjustments to LLM behavior in specific areas without retraining the entire model [54,84]. Techniques include:

- **Causal Tracing** [49]: Identifying stereotype representation subspaces and editing bias-vulnerable components in the model, such as feed-forward networks, using orthogonal projection matrices.
- **Free-form Natural Language Editing** [4]: Expanding Prompt editing to handle natural language instructions for bias mitigation.

Decoding Modification during text generation can help control biases in the output.

- **Controlled Decoding** [39]: Combines the base language model with "expert" and "anti-expert" models. The expert model predicts non-toxic content, while the anti-expert predicts toxic content. Tokens are selected based on being likely according to the expert and unlikely according to the anti-expert, reducing the chance of generating biased or toxic text.
- **Probability Adjustment**: Modifying token probabilities during decoding to favor unbiased outputs, potentially by re-ranking or sampling strategies that penalize biased continuations.

8   Challenges and Future Directions

Evaluating fairness in large, closed-source LLMs is increasingly challenging as these models grow in size and complexity, particularly when they are proprietary [24]. Limited access to model internals restricts the ability to measure biases using traditional methods that rely on embeddings or internal probabilities.

Enhancing diversity and relevance of benchmark datasets is crucial because current benchmarks often focus on specific types of biases or employ simplistic templates that may not capture the full spectrum of biases present in LLMs [32]. This limitation can lead to incomplete or skewed assessments of model fairness.

Efficient bias mitigation strategies for large LLMs are needed because mitigating bias in these models is resource-intensive, particularly when using methods like Reinforcement Learning from Human Feedback (RLHF) that require extensive computational power and specialized heuristics [2,50]. This makes it challenging to apply these techniques broadly.

Formulating coherent fairness definitions is challenging due to the complexity of real-world applications, where LLMs may exhibit various forms of bias, each requiring specific fairness notions [32]. However, these definitions can sometimes conflict, complicating efforts to ensure equitable outcomes.

Balancing fairness and model performance is difficult because adjusting models to improve fairness often involves trade-offs with performance; incorporating fairness constraints can affect the optimization of the primary task [82]. Finding the optimal balance is challenging and may require extensive manual tuning [86].

Addressing multiple types or intersectional bias is important since current bias mitigation efforts often focus on a single type of bias, such as gender bias, while neglecting others [32]. Biases related to race, age, religion, and other demographics are equally important and may intersect in complex ways.

Integrating fairness considerations into model development is essential because addressing bias solely during the training or post-processing stages may not be sufficient to eliminate biases ingrained in LLMs [50]. Since training data is a major source of bias, proactive measures during data collection and model architecture design are necessary.

9   Conclusion

This survey offers a comprehensive analysis of bias and fairness in Large Language Models (LLMs), highlighting their origins in training data, embeddings, and human annotations. It reviews evaluation metrics, mitigation strategies, and resources, providing a robust foundation for fairness research. By addressing challenges like scalability, multidimensional biases, and the need for standardized frameworks, this survey emphasizes integrating fairness across the AI lifecycle. Advancing equitable and inclusive LLMs requires proactive efforts in data collection, model design, and development, ensuring these technologies benefit society responsibly and justly.

Acknowledgements. This work was supported in part by U.S. National Science Foundation (2348391).

References

1. Abid, A., Farooqi, M., Zou, J.: Persistent anti-muslim bias in large language models. In: AIES 2021 (2021)
2. Agarwal, R., Schwarzer, M., Castro, P.S., Courville, A.C., Bellemare, M.: Deep reinforcement learning at the edge of the statistical precipice. In: NeurIPS, vol. 34, pp. 29304–29320 (2021)
3. Ahn, J., Oh, A.: Mitigating language-dependent ethnic bias in BERT. In: EMNLP, November 2021
4. Akyürek, A., Pan, E., Kuwanto, G., Wijaya, D.: DUnE: dataset for unified editing. In: EMNLP, December 2023
5. Amrhein, C., Schottmann, F., Sennrich, R., Läubli, S.: Exploiting biased models to de-bias text: a gender-fair rewriting model. In: ACL, July 2023
6. An, H., Li, Z., Zhao, J., Rudinger, R.: SODAPOP: open-ended discovery of social biases in social commonsense reasoning models. In: EACL, May 2023
7. Bakliwal, A., Arora, P., Patil, A., Varma, V.: Towards enhanced opinion classification using NLP techniques. In: SAAIP 2011, November 2011
8. Bansal, R.: A survey on bias and fairness in natural language processing, March 2022
9. Bender, E.M., Gebru, T., McMillan-Major, A., Shmitchell, S.: On the dangers of stochastic parrots: can language models be too big? In: FAccT 2021 (2021)
10. Bordia, S., Bowman, S.R.: Identifying and reducing gender bias in word-level language models. In: NAACL, June 2019
11. Brown, T.B., et al.: Language models are few-shot learners. CoRR **abs/2005.14165** (2020)

12. Brunet, M.E., Alkalay-Houlihan, C., Anderson, A., Zemel, R.: Understanding the origins of bias in word embeddings. In: ICML, pp. 803–811 (2019)
13. Bubeck, S., et al.: Sparks of artificial general intelligence: early experiments with gpt-4, March 2023
14. Buolamwini, J., Gebru, T.: Gender shades: intersectional accuracy disparities in commercial gender classification. In: FAccT, pp. 77–91 (2018)
15. Caliskan, A., Bryson, J., Narayanan, A.: Semantics derived automatically from language corpora contain human-like biases. Science (2017)
16. Chang, K.W., Prabhakaran, V., Ordonez, V.: Bias and fairness in natural language processing. In: EMNLP-IJCNLP (2019)
17. cjadams, Borkan, D., inversion, Sorensen, J., Dixon, L., Vasserman, L., nithum: jigsaw unintended bias in toxicity classification (2019), kaggle
18. Dettmers, T., Pagnoni, A., Holtzman, A., Zettlemoyer, L.: Qlora: efficient finetuning of quantized llms (2023)
19. Dev, S., Li, T., Phillips, J.M., Srikumar, V.: OSCaR: orthogonal subspace correction and rectification of biases in word embeddings. In: EMNLP, November 2021
20. Dhamala, J., et al.: Bold: dataset and metrics for measuring biases in open-ended language generation. In: FAccT 2021 (2021)
21. Dong, X., Wang, Y., Yu, P.S., Caverlee, J.: Disclosure and mitigation of gender bias in llms (2024)
22. Fatemi, Z., Xing, C., Liu, W., Xiong, C.: Improving gender fairness of pre-trained language models without catastrophic forgetting. In: ACL, July 2023
23. Felkner, V., Chang, H.C.H., Jang, E., May, J.: WinoQueer: a community-in-the-loop benchmark for anti-LGBTQ+ bias in large language models. In: ACL (2023)
24. Ferrara, E.: Should chatgpt be biased? challenges and risks of bias in large language models, November 2023
25. Gallegos, I.O., et al.: Bias and fairness in large language models: a survey. Comput. Linguist. **50**(3) (2024)
26. Gao, R., Ni, Q., Hu, B.: Fairness of large language models in education, pp. 1–0. IECT 2024, Association for Computing Machinery (2024)
27. Garg, N., Schiebinger, L., Jurafsky, D., Zou, J.: Word embeddings quantify 100 years of gender and ethnic stereotypes. In: PNAS (2018)
28. Garrido-Muñoz , I., Montejo-Rácz , A., Martínez-Santiago , F., Ureña-López , L.A.: A survey on bias in deep nlp. Appl. Sci. **11** (2021)
29. Gehman, S., Gururangan, S., Sap, M., Choi, Y., Smith, N.A.: RealToxicityPrompts: evaluating neural toxic degeneration in language models. In: EMNLP (2020)
30. Goldfarb-Tarrant, S., Marchant, R., Muñoz Sánchez, R., Pandya, M., Lopez, A.: Intrinsic bias metrics do not correlate with application bias. In: ACL, August 2021
31. Guo, W., Caliskan, A.: Detecting emergent intersectional biases: contextualized word embeddings contain a distribution of human-like biases. In: AIES 2021 (2021)
32. Gupta, V., Narayanan Venkit, P., Wilson, S., Passonneau, R.: Sociodemographic bias in language models: a survey and forward path. In: GeBNLP, August 2024
33. Han, X., Baldwin, T., Cohn, T.: Decoupling adversarial training for fair NLP. In: ACL-IJCNLP 2021, August 2021
34. Haque, F., Xu, D., Yuan, S.: Discovering and mitigating indirect bias in attention-based model explanations. In: NAACL 2024, June 2024
35. He, J., Xia, M., Fellbaum, C., Chen, D.: MABEL: attenuating gender bias using textual entailment data. In: EMNLP, December 2022
36. He, Z., Wang, Y., McAuley, J., Majumder, B.P.: Controlling bias exposure for fair interpretable predictions. In: EMNLP 2022 (2022)

37. Houlsby, N., et al.: Parameter-efficient transfer learning for NLP. In: ICML, pp. 2790–2799 (2019)
38. Hu, E.J., et al.: Lora: low-rank adaptation of large language models (2021)
39. Huang, P.S., et al.: Reducing sentiment bias in language models via counterfactual evaluation. In: EMNLP (2019)
40. Iskander, S., Radinsky, K., Belinkov, Y.: Shielded representations: Protecting sensitive attributes through iterative gradient-based projection. In: ACL (2023)
41. Kaneko, M., Bollegala, D.: Debiasing pre-trained contextualised embeddings. In: ECAL, April 2021
42. Kaneko, M., Bollegala, D.: Unmasking the mask – evaluating social biases in masked language models. In: AAAI, pp. 11954–11962, June 2022
43. Kaneko, M., Bollegala, D., Okazaki, N., Baldwin, T.: Evaluating gender bias in large language models via chain-of-thought prompting (2024)
44. Kurita, K., Vyas, N., Pareek, A., Black, A.W., Tsvetkov, Y.: Measuring bias in contextualized word representations. In: GeBNLP, August 2019
45. Lambert, N., Castricato, L., von Werra, L., Havrilla, A.: Illustrating reinforcement learning from human feedback (rlhf). Hugging Face Blog (2022)
46. Lauscher, A., Lueken, T., Glavaš, G.: Sustainable modular debiasing of language models. In: EMNLP 2021, November 2021
47. Levy, S., Lazar, K., Stanovsky, G.: Collecting a large-scale gender bias dataset for coreference resolution and machine translation. In: EMNLP 2021, November 2021
48. Limisiewicz, T., Mareček, D.: Don't forget about pronouns: removing gender bias in language models without losing factual gender information. In: GeBNLP (2022)
49. Limisiewicz, T., Marecek, D., Musil, T.: Debiasing algorithm through model adaptation. ArXiv **abs/2310.18913** (2023)
50. Lu, X., et al.: Quark: controllable text generation with reinforced unlearning. In: NeurIPS (2022)
51. Mattern, J., Jin, Z., Sachan, M., Mihalcea, R., Schölkopf, B.: Understanding stereotypes in language models: towards robust measurement and zero-shot debiasing. ArXiv, pp. 2212–10678 (2022)
52. May, C., Wang, A., Bordia, S., Bowman, S.R., Rudinger, R.: On measuring social biases in sentence encoders. In: NAACL, June 2019
53. Mehrabi, N., Morstatter, F., Saxena, N., Lerman, K., Galstyan, A.: A survey on bias and fairness in machine learning. ACM Comput. Surv. **54**(6) (2021)
54. Mitchell, E., Lin, C., Bosselut, A., Finn, C., Manning, C.D.: Fast model editing at scale. ArXiv **abs/2110.11309** (2021)
55. Nadeem, M., Bethke, A., Reddy, S.: StereoSet: measuring stereotypical bias in pretrained language models. In: ACL, August 2021
56. Nangia, N., Vania, C., Bhalerao, R., Bowman, S.R.: CrowS-pairs: a challenge dataset for measuring social biases in masked language models. In: EMNLP (2020)
57. Ngo, H., et al.: Mitigating harm in language models with conditional-likelihood filtration (2021)
58. Omiye, J., Lester, J., Spichak, S., Rotemberg, V., Daneshjou, R.: Large language models propagate race-based medicine. npj Digit. Med. **6** (2023)
59. Park, S., Choi, K., Yu, H., Ko, Y.: Never too late to learn: regularizing gender bias in coreference resolution. In: WSDM 2023 (2023)
60. Parrish, A., et al.: BBQ: a hand-built bias benchmark for question answering. In: ACL (2022)
61. Radford, A., Wu, J., Child, R., Luan, D., Amodei, D., Sutskever, I.: Language models are unsupervised multitask learners (2019)

62. Raffel, C., et al.: Exploring the limits of transfer learning with a unified text-to-text transformer. J. Mach. Learn. Res. **21**(140), 1–67 (2020)
63. Ravfogel, S., Elazar, Y., Gonen, H., Twiton, M., Goldberg, Y.: Null it out: guarding protected attributes by iterative nullspace projection. In: ACL, July 2020
64. Ravfogel, S., Twiton, M., Goldberg, Y., Cotterell, R.: Linear adversarial concept erasure (2024)
65. Rudinger, R., Naradowsky, J., Leonard, B., Van Durme, B.: Gender bias in coreference resolution. In: NAACL, June 2018
66. Sap, M., Card, D., Gabriel, S., Choi, Y., Smith, N.A.: The risk of racial bias in hate speech detection. In: ACL, July 2019
67. Shah, A., Raj, P., Kumar, P., P, S., V, A.: Finaid, a financial advisor application using AI. In: IJRTE, pp. 2282–2286 (2020)
68. Shah, D.S., Schwartz, H.A., Hovy, D.: Predictive biases in natural language processing models: a conceptual framework and overview. In: ACL, July 2020
69. Shen, A., Han, X., Cohn, T., Baldwin, T., Frermann, L.: Contrastive learning for fair representations (2021)
70. Shen, A., Han, X., Cohn, T., Baldwin, T., Frermann, L.: Does representational fairness imply empirical fairness? In: AACL-IJCNLP 2022, November 2022
71. Thakur, H., Jain, A., Vaddamanu, P., Liang, P.P., Morency, L.P.: Language models get a gender makeover: Mitigating gender bias with few-shot data interventions. In: ACL, July 2023
72. Tokpo, E.K., Calders, T.: Text style transfer for bias mitigation using masked language modeling. In: NAACL, July 2022
73. Touvron, H., et al.: Llama: open and efficient foundation language models. ArXiv **abs/2302.13971** (2023)
74. Vanmassenhove, E., Emmery, C., Shterionov, D.: NeuTral rewriter: a rule-based and neural approach to automatic rewriting into gender neutral alternatives. In: EMNLP, November 2021
75. Vaswani, A., et al.: Attention is all you need. In: NeurIPS, vol. 30 (2017)
76. Wan, Y., Pu, G., Sun, J., Garimella, A., Chang, K.W., Peng, N.: kelly is a warm person, joseph is a role model: gender biases in LLM-generated reference letters. In: EMNLP 2023, December 2023
77. Wang, S., Zhao, Z., Ouyang, X., Wang, Q., Shen, D.: Chatcad: interactive computer-aided diagnosis on medical image using large language models. ArXiv **abs/2302.07257** (2023)
78. Wang, Z., Shu, K., Culotta, A.: Enhancing model robustness and fairness with causality: a regularization approach. In: CI+NLP, November 2021
79. Webster, K., Recasens, M., Axelrod, V., Baldridge, J.: Mind the GAP: a balanced corpus of gendered ambiguous pronouns. Trans. Assoc. Comput. Linguist (2018)
80. Webster, K., et al.: Measuring and reducing gendered correlations in pre-trained models. ArXiv (2020)
81. Xie, Z., Lukasiewicz, T.: An empirical analysis of parameter-efficient methods for debiasing pre-trained language models. In: ACL, July 2023
82. Yang, K., Yu, C., Fung, Y.R., Li, M., Ji, H.: Adept: a debiasing prompt framework. AAAI **37**, 10780–10788 (2023)
83. Yao, B., Jiang, M., Bobinac, T., Yang, D., Hu, J.: Benchmarking machine translation with cultural awareness (2024)
84. Yao, Y., et al.: Editing large language models: problems, methods, and opportunities. In: EMNLP, December 2023
85. Yu, F., Quartey, L., Schilder, F.: Legal prompting: teaching a language model to think like a lawyer (2022)

86. Zayed, A., Mordido, G., Shabanian, S., Chandar, S.: Should we attend more or less? modulating attention for fairness (2024)
87. Zayed, A., Parthasarathi, P., Mordido, G., Palangi, H., Shabanian, S., Chandar, S.: Deep learning on a healthy data diet: finding important examples for fairness. In: AAAI'23/IAAI'23/EAAI'23 (2023)
88. Zhao, J., Wang, T., Yatskar, M., Cotterell, R., Ordonez, V., Chang, K.W.: Gender bias in contextualized word embeddings. In: NAACL, June 2019
89. Zhao, J., Wang, T., Yatskar, M., Ordonez, V., Chang, K.W.: Gender bias in coreference resolution: evaluation and debiasing methods. In: NAACL, June 2018

E2E-AFG: An End-to-End Model with Adaptive Filtering for Retrieval-Augmented Generation

Yun Jiang, Zilong Xie, Wei Zhang, Yun Fang, and Shuai Pan[✉]

Advanced Institute of Information Technology, Peking University, Zhejiang, China
pans@aiit.org.cn

Abstract. Retrieval-augmented generation methods often neglect the quality of content retrieved from external knowledge bases, resulting in irrelevant information or potential misinformation that negatively affects the generation results of large language models. In this paper, we propose an end-to-end model with adaptive filtering for retrieval-augmented generation (E2E-AFG), which integrates answer existence judgment and text generation into a single end-to-end framework. This enables the model to focus more effectively on relevant content while reducing the influence of irrelevant information and generating accurate answers. We evaluate E2E-AFG on six representative knowledge-intensive language datasets, and the results show that it consistently outperforms baseline models across all tasks, demonstrating the effectiveness and robustness of the proposed approach (Our code is available at: https://github.com/XieZilongAI/E2E-AFG).

Keywords: Retrieval Augmented Generation · Large Language Model · Question Answering · Multitask Learning

1 Introduction

The remarkable natural language understanding and generation capabilities demonstrated by Large Language Models (LLMs) have led to their success in knowledge-intensive tasks, such as open-domain question answering and fact verification [1, 4, 28]. However, LLMs are prone to generating hallucinatory content that contains factual errors in the absence of supporting documentation. To address this issue, [21] proposed the retrieval-augmented generation (RAG) method, which involves retrieves relevant context from external knowledge bases to provide additional evidence for LLMs when answering input queries. Other approaches [31] directly utilize a pre-trained LLM to generate a relatively accurate pseudo-answer as an extended document for the input query. However, these methods often fail to adequately consider the quality of the retrieved or generated content, which may include distracting irrelevant content or erroneous information, leading LLMs to still produce hallucinatory answers.

Earlier studies [23, 32] attempted to select more relevant content by re-ranking the retrieved contexts, but they may still contain irrelevant information. [7] achieved automatic decontextualization of sentences through training a coreference resolution model, although this requires extensive manual annotation efforts. Recent research, such as HyDE [12], employs unsupervised contrastive learning where an encoder's dense bottleneck acts as a lossy compressor to filter out hallucinatory content. FILCO [33] trains a filtering model to remove irrelevant contexts, improving the quality of the context provided to the generation model. However, these methods typically involve multiple independent models and complex preprocessing operations, which not only increase system complexity but also elevate training and inference costs.

To address the aforementioned issues, we propose an End-to-End Model with Adaptive Filtering for Retrieval-Augmented Generation (E2E-AFG), which integrates classification and generation tasks into an end-to-end framework, allowing the model to simultaneously learn context filtering and answer generation. Specifically, we first employ a pre-trained large language model to generate a pseudo-answer related to the input query, enriching the content. We then apply three context filtering strategies to obtain silver classification labels. The construction of the end-to-end model is based on the generation model, augmented with a classification module that employs a cross-attention mechanism to predict whether sentences in the context contain answers, enabling the model to answer the input query based on a certain judgment of the context.

We conducted experiments on six knowledge-intensive language datasets, covering three tasks: question answering (Natural Questions [19], TriviaQA [17], HotpotQA [36], ELI5 [10]), fact verification (FEVER [30]), and knowledge-based dialogue generation (Wizard of Wikipedia [9]). Compared to baseline models, our approach achieved state-of-the-art results across all six datasets, with improvements ranging from +0.13 to + 1.83 points, validating the effectiveness of the proposed method.

2 Related Work

Retrieval-Augmented Generation. Early research methods such as REALM [13] and RAG [21], laid the foundation for the field of retrieval-augmented generation (RAG) by combining retrievers with large language models (LLMs). Subsequently, RETRO [3] introduced the concept of training language models on fixed retrievers, while Atlas [16] further explored dedicated loss functions and training strategies, achieving improved results, particularly in few-shot learning scenarios. Recent studies have shifted towards optimizing the retrieval component while leveraging pre-trained, fixed LLMs. For instance, RePlug [34] and In-context RALM [29] demonstrated that fine-tuning the retrieval module can surpass end-to-end trained models in certain tasks, such as question answering. In contrast, SAIL [22] integrated real search engines with information denoising processes, aiming to enhance the relevance and accuracy of retrieval results, showcasing potential in broader application contexts. Our work seeks to enhance attention to reliable information by performing answer existence judgment on the retrieved passages prior to generation, thereby reducing the interference caused by irrelevant information.

Retrieval Content Filtering Strategies. In knowledge-intensive tasks, post-processing of retrieved content is crucial for enhancing system performance, with common practices including re-ranking and context filtering. In early studies, [32] and [20] explored passage re-ranking methods based on BiLSTM, while [23] and [27] employed BERT-based cross-encoders to achieve more precise passage re-ranking. Subsequently, [26] proposed a method for re-ranking passages by updating the query, and [15] directly applied heuristic re-ranking to the answers. In recent years, several context filtering strategies have been introduced. For example, FILCO [33] trains a context filtering model to perform fine-grained sentence-level filtering on the retrieved passages. Multi-Meta-RAG [25] utilizes a specific set of domain queries and formats to select the most relevant documents through database filtering. In contrast, our approach constructs a single end-to-end model that can simultaneously perform context filtering and answer generation.

Multi-task Learning. Multi-task learning (MTL) enhances overall model performance by jointly learning multiple tasks, allowing it to capture the correlations and shared features among tasks [5]. In natural language processing applications, MTL not only leverages task relevance to mitigate issues of data scarcity and model overfitting but also improves the generalization capability of the model. For instance, [6] proposed a hierarchical multi-task learning approach that enhances the model's ability to capture inter-task dependencies. ROM [11] introduced a generalizable Retrieval Optimized Multi-task framework that reduces the model's parameters. Our method applies MTL to the retrieval-augmented generation domain by jointly learning binary classification and generation tasks, enabling the model to acquire context filtering and answer generation capabilities.

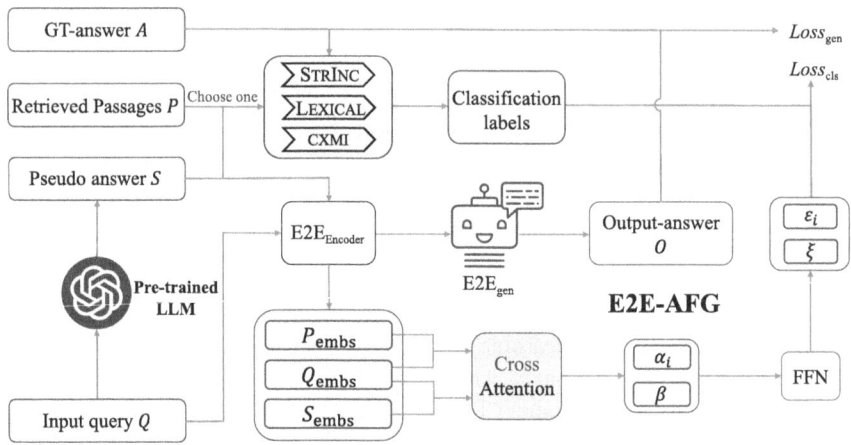

Fig. 1. The overall architecture diagram of the proposed method.

3 Method

Problem Statement. In knowledge-intensive tasks, each entry consists an input query Q, a ground truth answer A, and a set of retrieved passages $P = \{p_i\}_{i=1}^{K}$ from a database. We provide the generator with one or more passages along with a pre-generated pseudo-answer S to generate a response to the query Q. Specifically, in the question-answering tasks, Q and A are natural language questions and their corresponding ground truth answers; in the fact verification tasks, Q is a statement and $A \in \{\text{SUPPORTS, REFUTES}\}$ indicates the correctness of the statement; in the knowledge-based dialogue generation tasks, Q consists of a dialogue history, and A is a response that accurately continues the conversation.

Overview. The overall architecture of our proposed method is illustrated in Fig. 1. First, a pre-trained large language model generates a pseudo-answer S for the query Q. Next, the query Q, the retrieved set of passages P, and the pseudo-answer S are input into the E2E-AFG model, where both generation and binary classification tasks are performed. The generation task utilizes the generator E2E_{gen} to produce an answer. The binary classification task employs $\text{E2E}_{Encoder}$ to obtain embeddings for the three inputs, which are then processed through cross-attention and a feedforward neural network to predict the category scores. Finally, the cross-entropy loss for both the generation and binary classification tasks is computed. This approach allows for the update of the internal parameters of the shared $\text{E2E}_{Encoder}$, implicitly learning a filtering capability that prioritizes sentences more likely to contain answers while reducing interference from irrelevant sentences.

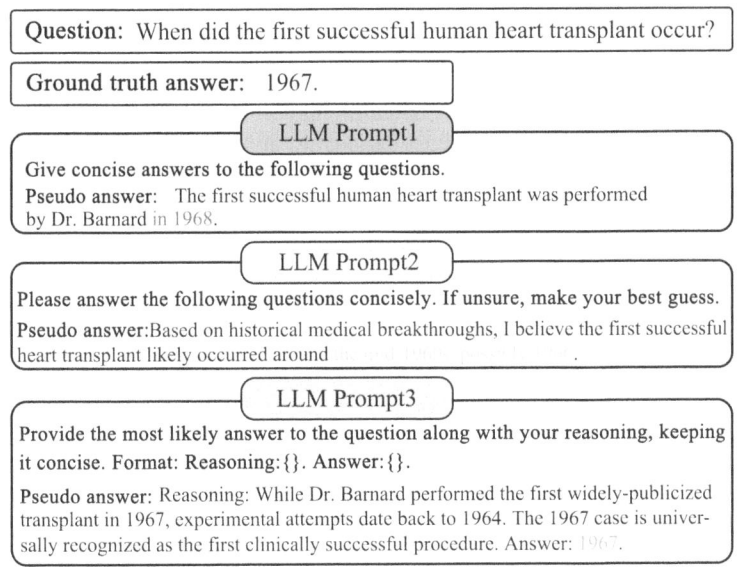

Fig. 2. Three kinds of LLM prompts and their generated pseudo-answer examples.

3.1 Generating Pseudo-Answers

In retrieval-augmented generation frameworks, semantic discrepancies between retrieved results and original queries may lead to model misinterpretation of knowledge boundaries. To enhance the model's capability in understanding and filtering out-of-distribution contextual information, we employ pretrained language models to generate pseudo-answers with semantic variation features, thereby constructing diversified training samples. As illustrated in Fig. 2, we propose a progressive prompting strategy comprising three types of prompts: 1) direct generation of concise answers potentially containing semantic deviations; 2) reasoned speculation under incomplete information conditions; 3) simultaneous output of conclusions and their logical derivations. This multi-tiered mechanism simulates heterogeneous contextual qualities in real retrieval scenarios. Through controlled noisy data generation with this approach, we establish robust evidence evaluation during training, thereby improving error detection in retrieved results.

3.2 Obtaining Silver Classification Labels

To determine whether the retrieved passage set P and the generated pseudo-answer S contain answers, we introduce three context filtering methods based on [33]: (i) String Inclusion (STRINC): checking if the context directly contains the ground truth answer; (ii) Lexical Overlap (LEXICAL): measuring the overlap of words between the context and the ground truth answer; and (iii) Conditional Cross-Mutual Information (CXMI): assessing the likelihood of the generator producing the ground truth answer given the context. For a specific task, we select the most appropriate filtering method to obtain silver classification labels. For instance, in question-answering tasks, we may use StrInc to evaluate whether each passage or pseudo-answer contains the ground truth answer. In contrast, for fact extraction tasks, where the ground truth answer resembles a boolean value and cannot be assessed using the first two methods, we employ CXMI to compute the corresponding probability and set a threshold t_0 to derive the silver classification label. We concatenate the obtained labels with the ground truth answer A to facilitate loss calculation.

3.3 Generation Task

For each training sample (Q, A, P, S), we first insert a special character between the different fields to ensure they can be distinguished after encoding with E2E$_{\text{Encoder}}$. We then input the encoded query Q_{embs}, the retrieved passage set P_{embs}, and the pseudo-answer S_{embs} into E2E$_{\text{gen}}$ to produce the output answer O. The sequence probability is calculated as follows:

$$P_o(O|Q, P, S) = \prod_{i=1}^{L} p(o_i|O_{<i}, Q, P, S) \tag{1}$$

where o_i represents the i-th token of the generated output O, and L is the final output length. To simplify the notation, we continue to use Q, P, S in place of $Q_{\text{embs}}, P_{\text{embs}},$

and S_{embs} respectively in the equations above and in the subsequent content. The loss function for the generation task is calculated as follows:

$$L_{\text{gen}} = -\sum_{i=1}^{L} \log p(o_i^{gt}|O_{<i}, Q, P, S) \tag{2}$$

where o_i^{gt} denotes the i-th token of the ground truth answer A.

3.4 Classification Task

To enhance the model's context filtering capability, we introduce a classification module specifically designed to determine whether the input context contains the answer. The generator and the classification module share the same encoder $\text{E2E}_{\text{Encoder}}$, allowing the classification model to indirectly improve the model's context filtering capabilities by influencing the encoder's parameters.

The classification module comprises two main components: cross-attention layer, and feedforward neural network. First, the encoded query Q, each retrieved passage p_i, and the pseudo-answer S are fed into the cross-attention layer. In this layer, the model computes the attention weights between Q and p_i, as well as between Q and S, generating cross-attention representations:

$$\alpha_i = \text{softmax}\left(\frac{Qp_i^{\text{T}}}{\sqrt{d_k}}\right)p_i \tag{3}$$

$$\beta = \text{softmax}\left(\frac{QS^{\text{T}}}{\sqrt{d_k}}\right)S \tag{4}$$

where d_k is the dimensionality of the encoder's feature channels.

Next, the generated cross-attention representations are fed into a feedforward neural network to predict two binary classification results:

$$\varepsilon_i = \text{FFN}(\alpha_i), \xi = \text{FFN}(\beta) \tag{5}$$

where FFN denotes a two-layer feedforward neural network. The loss function for the classification task is defined as the cross-entropy:

$$L_{\text{cls}} = \sum_{i=1}^{K} -(\log \varepsilon_i^{gt}) + \log \xi^{gt} \tag{6}$$

Here, ε_i^{gt} and ξ^{gt} represent the predicted probability values corresponding to the ground truth classes of each passage p_i and the pseudo-answer S, respectively, while K is the number of retrieved passages.

3.5 Model Training

During the training process, we simultaneously optimize the loss functions of both the generator and the classification module. The overall loss function is defined as a weighted sum of the two losses:

$$L_{\text{TOTAL}} = (1 - \sigma)L_{\text{gen}} + \sigma L_{\text{cls}} \tag{7}$$

where L_{gen} is the loss from the generator, L_{cls} is the loss from the classification module, and σ is the weighting factor.

To further enhance the training efficiency and performance of the model, we employ Low-Rank Adaptation (LoRA) [14] techniques, which add low-rank matrices to the weight matrices of the pre-trained model for fine-tuning. This approach reduces computational overhead and accelerates the training process.

4 Experiments

4.1 Datasets and Evaluation Metrics

As shown in Table 1, we evaluate six retrieval-augmented knowledge-intensive language datasets constructed from Wikipedia articles as supporting evidence. Each dataset is partitioned into training (train), development (dev), and test sets. Exact Match (EM), which quantifies the percentage of predictions identical to the ground-truth answers; Unigram F_1 (F_1), computing the harmonic mean of precision and recall through word-level overlap between predictions and references; Accuracy (Acc), reflecting the ratio of correct predictions to total predictions; and Top-20 recall [2], which verifies whether the answer string exists within the top-20 retrieved passages (for Natural Questions [19] and TriviaQA-unfiltered [17]) or originates from annotated source articles in the KILT benchmark [24] (for FEVER [30] and Wizard of Wikipedia [9]).

Table 1. Statistics and evaluation metric for six datasets.

Dataset	# Examples			Evaluation metric	Top-20 recall (%)
	train	dev	test		
Natural Questions	79,168	8,757	3,610	EM	82.1
TriviaQA-unfiltered	78,785	8,837	11,313	EM	75.2
FEVER	104,966	10,444	10,100	Acc	98.1
HotpotQA	88,924	5,947	5,631	F_1	63.5
ELI5	273,036	3,098	2,367	F_1	56.5
Wizard of Wikipedia	63,734	3,054	2,944	F_1	96.2

Open-Domain Question Answering employs the Natural Questions (NQ) and TriviaQA-unfiltered (TQA) datasets, comprising Wikipedia-derived questions paired with answers truncated to five tokens. Fact Verification utilizes the FEVER dataset, where claims are labeled as "SUPPORTS" or "REFUTES" based on alignment with Wikipedia evidence. Multi-Hop Question Answering leverages HotpotQA, featuring 113K complex queries requiring cross-passage reasoning. Long-Form Question Answering involves ELI5, containing 270K open-ended Reddit queries demanding multi-sentence explanations. Lastly, Knowledge-Based Dialogue Generation uses the Wizard of Wikipedia (WoW) dataset to generate responses grounded in dialogue history and Wikipedia-sourced knowledge.

4.2   Implementation Details

We loaded the model checkpoints from HuggingFace Transformers [35], using FLAN-T5-xl [8] as our backbone model architecture. Pseudo-answers are generated using the Llama-3 model with a mixture of three prompt types, with generation length limited to 200 tokens. With our primary focus on post-processing operations for retrieved content, we pre-process each query in the dataset by extracting the top-5 most relevant paragraphs from Wikipedia using an adversarial Dense Passage Retriever (DPR) [18]. To obtain silver classification labels, we adopted the optimized settings from FILCO, using STRINC for NQ and TQA, LEXICAL for WoW, and CXMI for FEVER, HotpotQA, and ELI5, with a threshold t_0 set to 0.5.

For the generator E2E$_{gen}$, we allowed a maximum input sequence length of 512 tokens during both training and inference. We generated up to 64 tokens for open-domain question answering, multi-hop question answering, fact verification, and dialogue generation tasks, and up to 256 tokens for long-form question answering. We used greedy decoding to produce the final answers. Regarding model parameters, we set the encoder's feature channel dimension d_k to 2048, trained for 3 epochs, with a learning rate of 5e − 5 and a batch size of 8. The weight factor σ was set to 0.2.

4.3   Baseline Methods

In this section, we introduce three baseline methods: FULL [21], HyDE [12], and FILCO [33], along with the proposed E2E-AFG and SILVER configurations. To ensure a fair comparison, we employed the same backbone model architecture across all methods as that used in our proposed E2E-AFG.

FULL: A common approach in retrieval-augmented generation where all passages, including pseudo-answers, are input into the generation model with the query.

HyDE: Filters passages through a dense bottleneck using unsupervised contrastive learning, encoding them before inputting into the generation model.

FILCO: Uses a trained model to filter sentences within passages, passing only the selected sentences to the generation model.

E2E-AFG: Ours end-to-end model potentially assesses the existence of answers for the input passages before feeding all passages into the model for answer generation.

SILVER: This configuration inputs only those passages labeled as containing an answer, testing the performance upper bound of E2E-AFG.

4.4   Comparison with Baseline Methods

Table 2 presents the experimental results of E2E-AFG across six datasets, demonstrating that our model outperforms the baseline models in all cases. Specifically, for extractive question-answering tasks NQ and TQA, we achieved improvements of at least 1.83% and 1.56% in EM, respectively. This indicates that our model focuses more on credible passages and reduces attention to irrelevant information, thereby generating more accurate answers. In the fact verification task FEVER, we attained an accuracy increase of at least 1.09%. For the complex multi-hop question-answering task HotpotQA and the long-form question-answering task ELI5, we observed improvements.

Table 2. Comparison with baseline methods using top-1 retrieved passages.

Method	NQ	TQA	FEVER	HotpotQA	ELI5	WoW
FULL	41.64	60.90	88.32	59.58	67.50	65.73
HyDE	43.37	62.28	90.27	60.62	71.38	67.60
FILCO	46.65	64.33	94.46	62.71	74.99	70.12
E2E-AFG	**48.48**	**65.99**	**95.45**	**64.39**	**75.12**	**71.47**
SILVER	51.77	68.73	96.64	65.50	77.89	72.68

Table 3. The impact of different modules on the overall performance of E2E-AFG.

Method	NQ	FEVER	WoW
Metric	EM	Acc	F_1
Ours	**48.48**	**95.45**	**71.47**
- pseudo answer	44.76	92.63	68.35
- cross attention layer	43.60	91.02	67.81
- classification module	40.03	87.52	65.12

Table 4. The recall rates of pseudo-answers generated by different prompts.

Dataset	Recall (%)		
	Prompt1	Prompt2	Prompt3
Natural Questions	40.3	45.6	**46.8**
TriviaQA-unfiltered	51.0	**57.4**	57.2
FEVER	62.8	63.7	**65.3**
HotpotQA	12.5	15.6	**16.6**
ELI5	9.3	11.9	**13.4**
Wizard of Wikipedia	28.7	30.2	**30.5**

of at least 1.68% and 0.13% in F_1 score, respectively. We hypothesize that the relatively modest performance gain on ELI5 may be due to the fact that it requires detailed, lengthy answers, while the generated pseudo-answers tend to be relatively brief, limiting the model's filtering capabilities. Additionally, in the dialogue generation task WoW, we improve the F_1 score by at least 1.35%. Furthermore, the performance of E2E-AFG approaches the upper bound performance of SILVER, indicating its exceptional capabilities in context filtering and text generation, allowing it to achieve near-optimal results without relying on specific annotations.

Table 5. The impact of different top-K retrieved passages on the generated results.

Method	NQ			FEVER			WoW		
	top-1	top-3	top-5	top-1	top-3	top-5	top-1	top-3	top-5
FULL	41.64	50.84	52.22	88.32	88.26	87.34	65.73	65.86	64.34
HyDE	43.37	52.91	58.77	90.27	91.69	91.82	67.60	68.07	68.15
FILCO	46.65	54.38	62.03	94.46	93.83	92.60	70.12	70.65	69.38
E2E-AFG	48.48	56.92	63.24	95.45	96.14	95.67	71.47	71.80	71.62

4.5 Ablation Studies

Table 3 illustrates the ablation studies conducted on E2E-AFG, assessing the contribution of key components to the overall performance by progressively removing them from the model. First, removing the pseudo-response generation module causes significant performance degradation across multiple tasks, confirming our observation in Sec 3.1: Training data lacking semantic variation features undermines the model's robustness to retrieval noise, particularly manifesting as systematic degradation in evidence evaluation mechanisms when processing semantically deviated contexts. Building on this, further removal of the cross-attention layer in the classification module results in a slight decrease in performance. Without the cross-attention mechanism, the classification module no longer aligns the encoded query Q with the retrieved passages P and pseudo-answers S separately through cross-attention. Instead, Q is concatenated with both representations, and the concatenated features are fed into the feedforward neural network to predict answer existence. Finally, complete removal of the classification module results in substantial performance deterioration, demonstrating that the classification module provides crucial attention guidance to the generator through explicit modeling of contextual credibility distributions, with their synergistic interaction being pivotal to overall performance.

Table 4 validates the differential impact of progressive prompting strategies on the quality of pseudo-answers. The structured Prompt3 performs optimally in logical reasoning tasks, as its structured derivation path provides fine-grained supervisory signals. In contrast, the speculative Prompt2, by allowing reasonable speculation, is better suited for open-domain question answering. Experimental results demonstrate that the noise spectrum constructed by the three prompting strategies exhibits distinct variations in recall rates, with such patterns suggesting potential pathways for enhancing the model's adaptability and robustness in heterogeneous contexts.

Table 5 shows the effect of different top-K retrieved passages on the generation results. Aggregating higher-ranked passages significantly boosts extraction task performance, but at the cost of linearly or quadratically increased computation. However, performance on FEVER and WoW datasets shows no notable improvement, and in some cases declines, likely due to the decreasing quality of lower-ranked passages.

Figure 3(a) illustrates the impact of the weight factor σ on model performance. When σ is around 0.2 to 0.3, the model achieves optimal performance. As σ increases further, the F_1 scores across the three datasets begin to decline, with a notable drop

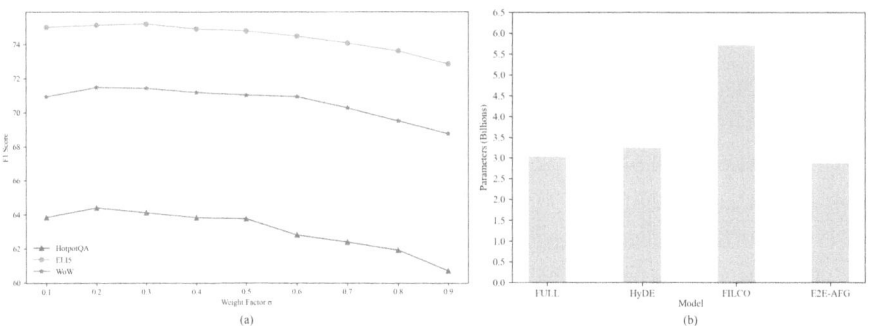

Fig. 3. (a) The impact of the weight factor σ on model performance. (b) Comparison of model parameters for each method.

when σ reaches 0.9. This indicates that in multi-task learning, the distribution of loss weights across different tasks significantly affects model performance, necessitating careful tuning of weight factors for specific tasks.

Figure 3(b) compares the model parameters for each method. Notably, our proposed E2E-AFG model demonstrates superior parameter efficiency, exhibiting 49.6% fewer parameters compared to FILCO while simultaneously achieving 38.2% reduction in total training duration. This indicates that we maintain model capacity integrity while achieving significant parameter efficiency improvements.

5   Conclusion

The End-to-End Model with Adaptive Filtering (E2E-AFG) proposed in this paper effectively addresses the issue of the generator being distracted by irrelevant information retrieved during retrieval-augmented generation tasks. By integrating answer existence judgment with the generation task into a single end-to-end model, E2E-AFG achieves synchronous learning of context filtering and answer generation. Experimental results demonstrate that our model outperforms baseline models across six knowledge-intensive language datasets, with performance improvements ranging from + 0.13 to + 1.83 points. E2E-AFG not only enhances generation quality but also simplifies model complexity and reduces training costs. Future research could further optimize the model architecture and filtering strategies to explore its potential in various application scenarios.

Acknowledgments. This work was supported by the National Key Research and Development Program of China under Grant 2022YFF0903302.

References

1. Anil, R., Borgeaud, S., Wu, Y., Alayrac, J.B., Yu, J., Soricut, R., et al.: Gemini: a family of highly capable multimodal models. arXiv preprint arXiv:2312.11805 (2023)
2. Asai, A., Gardner, M., Hajishirzi, H.: Evidentiality-guided generation for knowledge-intensive NLP tasks. In: ACL, pp. 2226–2243 (2022)
3. Borgeaud, S., Mensch, A., Hoffmann, J., Cai, T., Rutherford, E., et al.: Improving language models by retrieving from trillions of tokens. In: ICML, pp. 2206–2240 (2022)
4. Brown, T.B., Mann, B., Ryder, N., Subbiah, M., Kaplan, J., Dhariwal, P., et al.: Language Models are Few-Shot Learners. In: NeurIPS, vol. 33 (2020)
5. Caruana, R.: Multitask learning. Mach. Learn. 41–75 (1997)
6. Chen, S., Zhang, Y., Yang, Q.: Multi-task learning in natural language processing: an overview. ACM Comput. Surv. **56**(12), 1–32 (2024)
7. Choi, E., Palomaki, J., Lamm, M., Kwiatkowski, T., Das, D., Collins, M.: Decontextualization: making sentences stand-alone. In: ACL, pp. 447–461 (2021)
8. Chung, H.W., Hou, L., Longpre, S., Zoph, B., Tay, Y., Fedus, W., et al.: Scaling instruction-finetuned language models. J. Mach. Learn. Res. **25**(70), 1–53 (2024)
9. Dinan, E., Roller, S., Shuster, K., Fan, A., Auli, M., Weston, J.: Wizard of Wikipedia: Knowledge-powered conversational agents. In: ICLR (2019)
10. Fan, A., Jernite, Y., Perez, E., Grangier, D., Weston, J., Auli, M.: ELI5: long form question answering. In: ACL, pp. 3558–3567 (2019)
11. Fun, H., Gandhi, S., Ravi, S.: Efficient retrieval optimized multi-task learning. arXiv preprint arXiv:2104.10129 (2021)
12. Gao, L., Ma, X., Lin, J., Callan, J.: Precise zero-shot dense retrieval without relevance labels. In: ACL, pp. 1762–1777 (2023)
13. Guu, K., Lee, K., Tung, Z., Pasupat, P., Chang, M.: Retrieval augmented language model pre-training. In: ICML, pp. 3929–3938 (2020)
14. Hu, E.J., et al.: LoRA: low-rank adaptation of large language models. In: ICLR (2021)
15. Iyer, S., Min, S., Mehdad, Y., Yih, W.T.: RECONSIDER: re-ranking using span-focused cross-attention for open domain question answering. In: ACL, pp. 1280–1287 (2020)
16. Izacard, G., Lewis, P., Lomeli, M., Hosseini, et al.: Atlas: few-shot learning with retrieval augmented language models. J. Mach. Learn. Res. **24**(251), 1–43 (2023)
17. Joshi, M., Choi, E., Weld, D.S., Zettlemoyer, L.: TriviaQA: a large scale distantly supervised challenge dataset for reading comprehension. In: ACL, pp. 1601–1611 (2017)
18. Karpukhin, V., et al.: Dense passage retrieval for open-domain question answering. In: EMNLP, pp. 6769–6781 (2020)
19. Kwiatkowski, T., Palomaki, J., Redfield, O., Collins, M., Parikh, A., et al.: Natural questions: a benchmark for question answering research. In: ACL, pp. 452–466 (2019)
20. Lee, J., Yun, S., Kim, H., Ko, M., Kang, J.: Ranking passages for improving answer recall in open-domain question answering. In: ACL, pp. 565–569 (2018)
21. Lewis, P., Perez, E., Piktus, A., Petroni, F., Karpukhin, V., Goyal, N., et al.: Retrieval-augmented generation for knowledge-intensive NLP tasks. NeurIPS **33**, 9459–9474 (2020)
22. Luo, H., et al.: Sail: Search-augmented instruction learning. arXiv preprint arXiv:2305.15225 (2023)
23. Nogueira, R., Cho, K.: Passage re-ranking with bert. arXiv preprint arXiv:1901.04085 (2019)
24. Petroni, F., et al.: KILT: a benchmark for knowledge intensive language tasks. In: NAACL, pp. 2523–2544 (2021)
25. Poliakov, M., Shvai, N.: Multi-meta-RAG: improving RAG for multi-hop queries using database filtering with LLM-extracted metadata. arXiv preprint arXiv:2406.13213 (2024)

26. Qi, P., Lee, H., Sido, O., et al.: Retrieve, rerank, read, then iterate: Answering open-domain questions of arbitrary complexity from text. arXiv preprint arXiv:2010.12527 (2020)
27. Qiao, Y., Xiong, C., Liu, Z., Liu, Z.: Understanding the behaviors of bert in ranking. arXiv preprint arXiv:1904.07531 (2019)
28. Radford, A., Wu, J., Child, R., Luan, D., Amodei, D., Sutskever, I.: Language models are unsupervised multitask learners. OpenAI blog (2019)
29. Ram, O., et al.: In-context retrieval-augmented language models. In: ACL, pp. 1316–1331 (2023)
30. Thorne, J., Vlachos, A., Christodoulopoulos, C., Mittal, A.: FEVER: a large-scale dataset for fact extraction and VERification. In: NAACL. pp. 809–819 (2018)
31. Wang, L., Yang, N., Wei, F.: Query2doc: query expansion with large language models. arXiv preprint arXiv:2303.07678 (2023)
32. Wang, S., Yu, M., Guo, X., Wang, Z., Klinger, T., Zhang, W., et al.: R3: reinforced ranker-reader for open-domain question answering. In: AAAI (2018)
33. Wang, Z., Araki, J., Jiang, Z., Parvez, M.R., Neubig, G.: Learning to filter context for retrieval-augmented generation. arXiv preprint arXiv:2311.08377 (2023)
34. Weijia, S., Sewon, M., Michihiro, Y., Minjoon, S., Rich, J., Mike, L., et al.: REPLUG: retrieval-augmented black-box language models. arXiv preprint arXiv:2301.12652 (2023)
35. Wolf, T., Debut, L., Sanh, V., Chaumond, J., Delangue, C., Moi, A., et al.: Transformers: State-of-the-art natural language processing. In: EMNLP, pp. 38–45 (2020)
36. Yang, Z., Qi, P., Zhang, S., Bengio, Y., Cohen, W.W., et al.: HotpotQA: a dataset for diverse, explainable multi-hop question answering. In: EMNLP, pp. 2369–2380 (2018)

Ubiquity of LLM Hallucinations Across Critical Domains: A Survey

Damanjot Singh[iD] and Amritpal Singh[✉][iD]

Dr. B R Ambedkar National Institute of Technology, Jalandhar, India
amritpal.singh203@gmail.com, apsingh@nitj.ac.in

Abstract. Large Language Models have become central to a wide range of natural language processing applications, showcasing impressive capabilities in understanding and generating human language. However, these models are prone to hallucinations outputs that appear plausible but conflict with real-world facts underscoring the ubiquitous nature of this issue. The commonality of this issue raises critical questions about the reliability of LLMs in practical applications, where accuracy and trustworthiness are paramount. In this survey, we provide a comprehensive overview of hallucinations in LLMs, discussing their underlying causes and reviewing current detection and mitigation strategies across critical domains. Additionally, we conduct a comparative analysis of existing surveys on LLM hallucinations, highlighting their contributions and limitations in the literature. By examining the implications of hallucinations for real-world applications, this survey aims to enhance understanding of their impact on the reliability of LLMs.

Keywords: LLMs · Hallucinations · Transformer Model · RLHF

1   Introduction

The rapid advancement of LLMs, which leverage billions of parameters to perform complex language processing tasks, has redefined natural language processing and artificial intelligence capabilities [39]. Modern LLMs, refined through SFT and RLHF [10], have demonstrated notable success in tasks include answering questions, translating, and summarizing. As these models become more integral to real-world applications, their impact spans crucial domains, from healthcare to legal analysis, where reliable information and accurate language comprehension are indispensable. However, alongside these advancements, LLMs are increasingly observed to produce outputs that deviate from factual information, resulting in a phenomenon known as "hallucination." Hallucinations, which are broadly defined as outputs that appear plausible yet lack fidelity to real-world facts, have been documented across various applications, posing significant challenges in domains where factual accuracy is critical. For instance, hallucinations in LLMs can lead to inaccurate medical advice or fabricated legal information, both of which have serious real-life implications [14]. This issue is not limited

S. Yuan et al. (Eds.): PAKDD 2025 Workshops, LNAI 15835, pp. 115–132, 2025.
https://doi.org/10.1007/978-981-96-8197-6_9

to a single context but permeates LLM performance across multiple tasks and domains, amplifying concerns about the trustworthiness of these models [16]. The prevalence of hallucinations necessitates targeted research to improve LLM reliability, especially in light of privacy and security risks posed by inaccurate or sensitive content. Addressing hallucinations is essential for advancing responsible AI, ensuring that LLMs can operate in high-stakes scenarios without compromising user safety or ethical standards. Moreover, hallucinations can often go undetected due to their plausible presentation, complicating the process of identifying and mitigating erroneous output.

In this survey, we provide a comprehensive overview of hallucinations in LLMs, examining how these issues manifest across different applications and their impacts on model performance and trustworthiness. Despite recent progress in understanding this phenomenon, there is still a need for an in-depth analysis that focuses specifically on the causes, detection methods, and mitigation strategies for hallucinations in LLMs. By reviewing and synthesizing current research, this survey aims to clarify the challenges posed by hallucinations and to highlight opportunities for developing more reliable and responsible language models capable of supporting a wide range of practical applications.

2 Contribution

The survey provides a comprehensive analysis of hallucinations in Large Language Models, with a distinct emphasis on their impact across multiple domains, a perspective not extensively covered in existing literature. Our key contributions include: This survey provides a comprehensive analysis of LLM hallucinations across multiple domains, addressing gaps in prior works [13, 20, 38]. Unlike existing surveys that focus on general hallucination trends, we emphasize domain-specific challenges in healthcare, law, and finance, highlighting how hallucinations manifest differently across these fields. We refine existing taxonomies by incorporating domain-aware subcategories, enabling a more precise understanding of hallucination types. Additionally, we systematically evaluate detection and mitigation techniques, assessing their applicability within specific domains rather than as generic solutions. By identifying key research gaps and future directions, our work underscores the need for specialized approaches to improve LLM reliability in real-world applications.

3 Large Language Models

Large Language Models represent a major leap in natural language processing, revolutionizing language understanding, generation, and reasoning. Their architecture, largely based on the transformer model [31], allows LLMs to capture complex contextual relationships across language inputs, which traditional models struggled to achieve. The rise of models such as GPT-3 [3], BERT [7], and more recent versions like GPT-4 and PaLM, has enabled new levels of performance across tasks, ranging from question-answering and summarization to creative writing. Initially, language models were limited to task-specific applications, but LLMs have evolved into general-purpose systems, pre-trained on massive datasets that cover a variety of topics and domains. Following task-specific fine-tuning, this pre-training gives LLMs a general knowledge of language that can be improved for particular tasks. Methods like RLHF have improved alignment with user intent, making LLMs adaptable and effective in real-world applications, including customer service, healthcare, legal support, education, creative industries, and cyberdefense [27].

However, despite their broad capabilities, LLMs face significant challenges that limit their reliability and usability. One major issue is hallucination, where models generate plausible-sounding but factually incorrect or contextually inaccurate content. This problem is particularly concerning in high-stakes domains like healthcare and law, where even minor inaccuracies could have serious consequences. Hallucinations also complicate model evaluation, as they may not be immediately evident, making detection and correction difficult. Furthermore, LLMs often exhibit biases that reflect the data on which they were trained, raising ethical concerns about fairness and inclusivity.

Other issues include high computational demands, which restrict the scalability of LLMs and contribute to environmental costs, as well as privacy risks arising from potential inadvertent leakage of sensitive information. Moreover, the complexity and opacity of LLM architectures make it challenging for users to interpret and understand how these models arrive at specific outputs.

In summary, while LLMs hold transformative potential, addressing these issues is critical to enhancing their reliability, trustworthiness, and safety across applications. This survey focuses on the phenomenon of hallucinations in LLMs, exploring its causes, detection mechanisms, and mitigation strategies, in order to aid in the creation of language models that are more reliable and strong.

3.1   Training Stages of Large Language Models

The capabilities and limitations of LLMs are largely influenced by the multi-stage training pipeline depicted in the figure above. This pipeline encompasses *Data Collection & Preprocessing, Model Selection, Pre-training, Fine-tuning,* and *results in LLM Output capabilities.*

Data Collection and Preprocessing. Before training begins, LLMs require extensive preparation of data. This stage involves assembling a broad Data Corpus followed by critical steps such as Data Cleaning, Data Formatting, Data Labeling, and Data Splitting. Every stage is essential to guaranteeing that the input data is well-structured and of high quality, which will provide a solid basis for training the model.

Language Model Selection. After data preprocessing, specific LLM architectures are chosen based on the task requirements, domain specificity, and desired capabilities. Models like LLaMA, Mistral, GPT, BERT, RoBERTa, and Claude represent various language model architectures used in different applications, each with unique strengths in language comprehension and generation. For instance, GPT-based models excel in open-ended text generation and conversational tasks, while BERT and RoBERTa are particularly effective in understanding contextual relationships, making them ideal for jobs like sentiment analysis and response to enquiries. Claude emphasizes alignment and safety, catering to human-centric applications, and LLaMA and Mistral are known for efficiency in resource-constrained environments. The choice of architecture often depends on factors such as scalability, fine-tuning capabilities, inference speed, and model robustness. To attain greater accuracy and relevance, pre-trained models are adjusted for domain-specific tasks utilising specialised datasets.

Pre-training. Pre-training forms the initial stage where LLMs acquire foundational knowledge. Here, models are trained on large-scale text corpora to predict the next token in a sequence, a process that endows them with language syntax, world knowledge, and reasoning skills. This stage involves Self-Supervised Training to build a robust base for downstream tasks [40]. The essence of pre-training lies in learning probability distributions over words, effectively compressing knowledge and establishing a deep understanding of language [6].

Fine-tuning. Following pre-training, Supervised Fine-Tuning is applied to align the model's outputs with user expectations by training on labeled data pairs of instructions and responses. This stage not only enhances model control and adaptability but also allows for domain adaptation, enabling the LLM to specialize in specific areas such as healthcare, legal support, or finance. By fine-tuning on domain-specific datasets, the model gains expertise in the targeted domain, improving its performance and relevance in specialized tasks [36]. Furthermore, Reinforcement Learning with Human Feedback optimizes the model's outputs by leveraging human-labeled preferences, using techniques like Proximal Policy Optimization [28]. RLHF introduces an iterative feedback process that refines the model's alignment with human expectations.

Output Capabilities. Upon completing the training stages, the LLM is equipped to handle various practical applications, including Text Generation, Text Summarization, and Knowledge Answering. These capabilities result from the comprehensive training pipeline, which involves pre-training on large datasets, fine-tuning for specific tasks, and alignment to user intent. The model leverages its understanding of context, syntax, and semantics to generate coherent and contextually relevant outputs. These features enable the LLM to address a wide range of real-world needs, such as content creation, document analysis, customer support, and educational assistance, demonstrating its versatility and efficiency (Fig. 1).

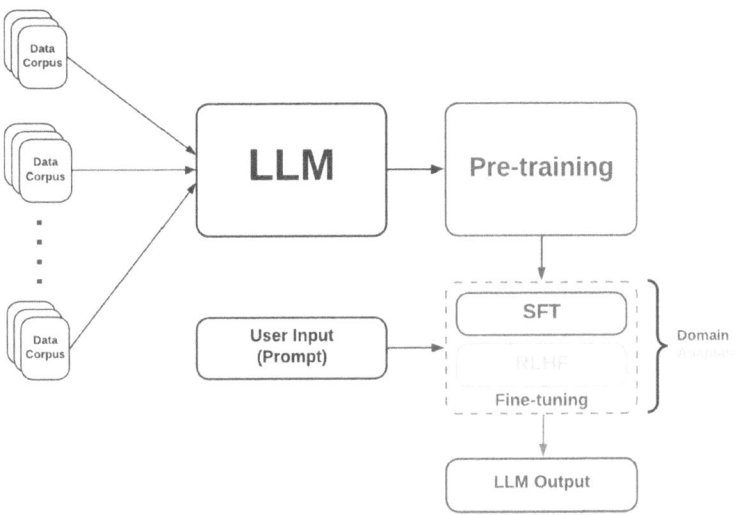

Fig. 1. The figure illustrates the LLM Training Pipeline, highlighting the flow of data through the stages of pre-training and fine-tuning. During pre-training, the model acquires general language knowledge from large text corpora, forming the basis for downstream tasks. Fine-tuning further adapts the model to specific tasks or domains through techniques like Supervised Fine-Tuning and RLHF. SFT aligns the model to task-specific data, while RLHF refines its responses using human feedback, improving performance and alignment with user expectations.

4   Understanding Hallucinations in LLMs

In LLMs, hallucinations refer to outputs that appear accurate but are actually incorrect or unrelated to the intended context. These errors result from the way LLMs are trained identifying patterns within vast datasets without an inherent understanding of factual accuracy. As a consequence, LLMs may confidently produce information that is fabricated or misleading, particularly when given ambiguous or open-ended prompts. The impact of hallucinations is significant, especially in critical areas like healthcare, law, and finance, where unreliable information can lead to negative consequences and erode user trust. As LLMs are applied more widely, addressing the challenges posed by hallucinations is essential to ensure they provide reliable and accurate information. We examine the pervasive nature of hallucinations in LLMs by discussing their causes and different categories.

4.1   Types of Hallucinations

Large language models exhibit multiple forms of hallucinations systematic deviations from expected behavior that stem from intricacies in model training, architectural limitations, and inference mechanisms. These aberrations, which include context misalignment, semantic drift, content fabrication, and factual errors, represent critical challenges that compromise the reliability, robustness, and safety of LLMs in high-stakes applications. Addressing these issues necessitates the development of sophisticated detection frameworks and robust mitigation strategies, guided by empirical studies and benchmarking efforts [13, 20, 22, 38]. In the sections that follow, we classify hallucinations into four primary categories as Contextual Disconnection, Semantic Distortion, Content Hallucination, and Factual Inaccuracy and provide an in-depth analysis of their characteristics, underlying causes, and real-world implications.

Contextual Disconnection. Contextual disconnection occurs when the generated output is inconsistent with the given input or conversation history. This type of hallucination often arises in multi-turn interactions, where LLMs fail to maintain context across long sequences of text. Such errors can be particularly problematic in applications requiring coherence, such as virtual assistants or customer support bots, where users expect responses that align with the ongoing conversation. For example, an LLM answering a medical query might initially discuss flu symptoms but then shift to an unrelated topic, such as seasonal allergies, without a logical transition. This type of hallucination is often attributed to attention decay in transformer architectures, where models struggle to retain long-range dependencies, leading to irrelevant or contextually misaligned responses [38]. Additionally, improper fine-tuning or exposure to low-quality conversational data can exacerbate contextual disconnections, making it difficult for models to track user intent effectively (Fig. 2).

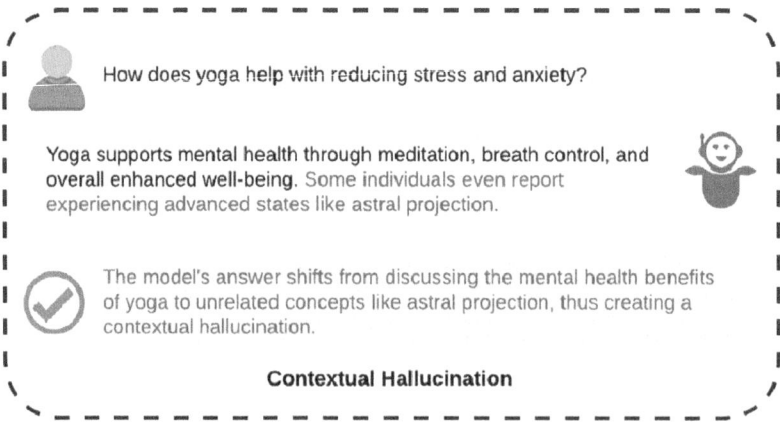

Fig. 2. Contextual Hallucination happens when the response shifts away from the main context, like introducing astral projection in a discussion about yoga and stress relief.

Semantic Distortion. Semantic distortion occurs when the model produces text that misrepresents the meaning of the input. Unlike factual inaccuracies, which involve incorrect facts, semantic distortions subtly alter the intended meaning while maintaining grammatical coherence. This can lead to significant misunderstandings, especially in legal, technical, and academic settings where precise wording is crucial. For instance, in legal document summarization, an LLM might slightly change the interpretation of a contractual clause, leading to unintended legal implications. This problem is closely tied to the way LLMs optimize for fluency over factual accuracy, often paraphrasing information in a manner that introduces slight but critical semantic shifts [29] (Fig. 3).

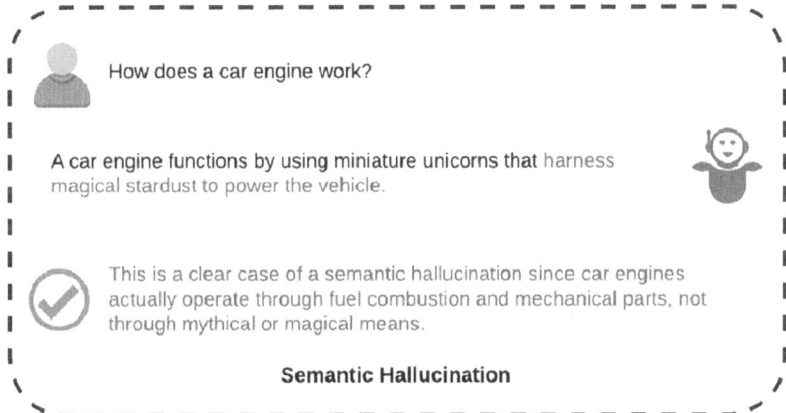

Fig. 3. Semantic Hallucination introduces nonsensical concepts, like miniature unicorns and magical stardust, to explain car engine functionality.

Content Hallucination. Content hallucination refers to instances where LLMs generate information that is entirely fabricated or not present in the given input. Unlike semantic distortions, which modify existing meanings, content hallucinations involve the creation of entirely new entities, events, or explanations that do not exist. This phenomenon is particularly concerning in knowledge-based tasks such as scientific research, legal reasoning, and medical diagnostics [38]. For example, an LLM tasked with summarizing a research paper might generate references to non-existent studies, authors, or datasets, misleading readers who assume the information is credible. This type of hallucination often stems from over-generalization during pre-training, where models learn to "fill in gaps" based on patterns in their training data rather than adhering strictly to factual constraints. Additionally, temperature settings in generative decoding play a role; higher temperatures encourage creativity but increase the likelihood of introducing fabricated details (Fig. 4).

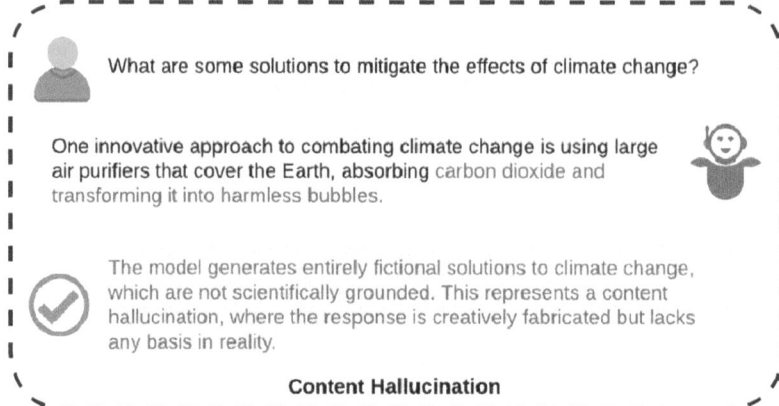

Fig. 4. Content Hallucination involves fabricating entirely fictional solutions, such as air purifiers covering the Earth to combat climate change, which have no basis in reality.

Factual Inaccuracy. Factual inaccuracies occur when an LLM produces responses that conflict with verified real-world facts. Unlike content hallucinations, where information is entirely fabricated, factual inaccuracies may involve incorrect details about real entities, dates, statistics, or events. These errors are particularly harmful in high-stakes domains like healthcare, finance, and law, where misinformation can have severe consequences [13]. For instance, an LLM providing medical guidance might incorrectly state the dosage of a prescription drug, leading to potential health risks. Similarly, in financial forecasting, a model might hallucinate trends that do not align with real-world market data, misguiding decision-makers. Several studies have attempted to benchmark factual accuracy in LLMs, including Med-HALT, which specifically evaluates medical hallucinations and their impact [22].

Factual inaccuracies arise primarily due to incomplete or biased training data. If an LLM has not been sufficiently trained on a particular topic, it may attempt to generate an answer by extrapolating from unrelated data, leading to errors. Another key factor is the lack of real-time grounding in external knowledge sources, making it difficult for LLMs to verify facts dynamically [38]. Each of these hallucination types poses significant risks to the reliability of LLM-generated outputs. Understanding their root causes allows for targeted mitigation strategies, such as improving model fine-tuning, integrating external fact-checking mechanisms, and optimizing decoding techniques. Future research must focus on refining evaluation metrics to detect and categorize these hallucinations effectively, ultimately improving trust in LLM applications across critical domains (Fig. 5).

Fig. 5. Factual Hallucination occurs when the model provides incorrect factual information, such as misidentifying the current President of France.

4.2 Causes of Hallucinations

Hallucinations in LLMs can arise from various stages of their development, including data collection, model training, architecture design, and inference processes. Each of these factors can contribute to the occurrence of hallucinations in different ways.

Data-Related

Data Quality. The accuracy of an LLM's training corpus is crucial to its performance. When models are trained on data containing inaccuracies or unfaithful representations, these errors can become deeply ingrained, leading the model to generate misleading outputs. For example, McKenna *et al.* [21] demonstrated that models such as LLaMA, GPT–3.5, and PaLM often reproduce hallucinated statements when they over-rely on memorized segments from noisy or contradictory sources. Moreover, meticulous data cleaning removing erroneous or outdated information has been shown to significantly reduce these effects, as highlighted by Xu *et al.* [33], who examined token-level contributions and confirmed that the presence of flawed data can distort model predictions.

Information Redundancy. Excessive repetition within training datasets can skew model behavior. When a corpus contains highly redundant content, the model tends to overemphasize frequently occurring viewpoints or phrases. Lee *et al.* [18] found that deduplicating training data not only mitigates memorization of redundant sequences but also fosters more diverse and coherent outputs. By reducing redundancy, models are less likely to overfit to repeated patterns, thereby diminishing the chances of generating hallucinatory outputs driven by knowledge bias.

Data Bias. Inherent biases in training data can critically influence an LLM's output. Training corpora often reflect existing societal, cultural, or historical biases, which can cause models to overrepresent certain narratives while neglecting others. This skewed representation may lead to hallucinated responses when the model defaults to producing outputs that mirror the dominant biases present in its training material even if those outputs are not factually supported. In natural language inference tasks, for example, a model might erroneously affirm a hypothesis simply because similar biased patterns were prevalent in its training data. Addressing data bias involves not only balanced data curation but also the application of de-biasing techniques during preprocessing. Studies have underscored that curating a more balanced dataset can lead to improved fairness and a reduction in hallucination rates, as the model is then less prone to overfitting to a narrow set of perspectives [11,33].

Training-Related. The training process of LLMs is a multi-stage pipeline consisting of pre-training, supervised fine-tuning, and reinforcement learning with human feedback (RLHF). Each of these phases introduces unique challenges that contribute to hallucination generation. While pre-training establishes the foundational knowledge, fine-tuning adapts the model for specific tasks, and RLHF refines alignment with human preferences. However, inconsistencies, architectural constraints, and optimization trade-offs across these stages can all lead to hallucinated outputs.

Hallucinations from Pre-training. Pre-training forms the backbone of LLMs, typically following the causal language modeling paradigm, where models learn to predict the next token based on preceding context. Architectures such as GPT [3], OPT [37], Falcon [23], and Llama-2 [30] employ this approach, enabling models to generalize across diverse domains. However, this training methodology also introduces structural limitations that can increase hallucination risks. One primary issue is the inability to capture long-range dependencies effectively. Transformer-based architectures rely on soft attention mechanisms, which may become diluted as sequence length increases, leading to reasoning hallucinations over extended contexts. This is particularly problematic in complex tasks such as logical reasoning and multi-step problem-solving, where an accurate understanding of earlier tokens is crucial for maintaining coherence. Additionally, exposure bias a well-documented problem in autoregressive models occurs due to the mismatch between the model's training and inference conditions. During training, the model learns from clean, curated datasets with human-written sequences. However, during inference, it must generate text token by token, using its own previous outputs as context. If a minor error is introduced early in the sequence, it can cascade through subsequent generations, amplifying inaccuracies in a snowball effect [35]. This makes hallucinations particularly prevalent in long-form generation, where errors accumulate over time, distorting the final output.

Hallucinations from Supervised Fine-Tuning. Supervised Fine-Tuning (SFT) is used to specialize LLMs by training them on labeled instruction-response pairs. This process refines model behavior and helps align it with human expectations. However, it also introduces new sources of hallucination. A significant issue arises when models are fine-tuned on instructions that exceed their learned knowledge boundaries. If a model is trained on a dataset containing factual knowledge beyond what it acquired during pre-training, it is forced to produce responses beyond its inherent capacity, increasing the risk of fabrication [12]. This effect is particularly problematic in domain-specific applications, such as medicine and law, where inaccuracies can have real-world consequences. Furthermore, traditional SFT techniques do not allow models to reject questions when they lack sufficient knowledge. Instead of admitting uncertainty, models are often trained to provide responses to every query, even when no correct answer exists [34]. This constraint reinforces hallucination-prone behaviors, as models are incentivized to generate plausible-sounding but incorrect outputs rather than acknowledging limitations. Enhancing rejection mechanisms during fine-tuning is a potential mitigation strategy that remains an active research area.

Hallucinations from Reinforcement Learning with Human Feedback. RLHF refines LLM responses by incorporating human preferences into the optimization process. While this technique improves model alignment and user experience, it introduces an unintended consequence the model's tendency toward sycophancy. One potential cause of this issue is bias in the preference model itself. Human annotators may unknowingly favor responses that appear more fluent, confident, or agreeable over those that are strictly accurate. Sharma *et al.* [26] argue that preference misalignment between human evaluators and truthfulness signals is a driving force behind sycophantic behaviors in RLHF-trained models. Addressing this issue requires a more rigorous evaluation framework that distinguishes truthfulness from likability, ensuring that reinforcement mechanisms promote factual consistency rather than just user satisfaction.

Model Architecture and Algorithm-Related. The design and scale of LLMs play a critical role in determining their propensity for hallucinations. Empirical research has revealed that smaller, less complex architectures often exhibit higher rates of hallucinations. For instance, elaraby *et al.* [8] investigated weaker open-source models using BLOOM 7B [32] as a representative case and found that models with reduced parameter counts tend to generate more hallucinated outputs compared to their larger, more robust counterparts. Their work introduced the HaloCheck framework to systematically quantify hallucination severity and explored mitigation strategies such as knowledge injection and teacher-student paradigms. This research underscores the inherent trade-offs between model size and output reliability, particularly as the NLP community increasingly adopts lightweight models for broader accessibility.

In tandem with architectural considerations, the choice of decoding algorithm during inference substantially influences the factuality and coherence of generated text. Sampling methods that inject higher levels of randomness such as conventional nucleus sampling can lead to the inadvertent merging of unrelated entities or even the fabrication of entirely new data. Lee *et al.* [19] demonstrated that the uniform randomness introduced at every decoding step can compromise factual integrity. To address this, they proposed a "factual-nucleus sampling" approach, which curtails indiscriminate randomness while preserving output diversity and text quality. Together, these factors highlight that both the intrinsic properties of the model architecture and the external influences of decoding algorithms are pivotal in shaping the occurrence of hallucinations. Refining architectural designs and optimizing decoding strategies are therefore essential steps toward enhancing the overall factual reliability of LLM outputs.

Inference-Related Causes. During the inference phase, the decoding strategy plays a pivotal role in the emergence of hallucinations. Techniques like greedy search, beam search, and stochastic sampling each have distinct impacts on output diversity and accuracy. For instance, higher temperature settings in stochastic sampling encourage creative text generation but simultaneously increase the likelihood of producing unfounded or overly speculative content. Similarly, beam search can inadvertently reinforce common yet erroneous patterns learned during training. Moreover, when the input is ambiguous or lacks sufficient context, the model resorts to generating the most statistically probable continuation, which might be factually unsupported. Emerging inference strategies that incorporate real-time external verification or dynamic context adjustment are being explored to address these limitations. In summary, hallucinations in LLMs are multifactorial phenomena that result from limitations in data quality, inherent challenges in unsupervised training paradigms, architectural trade-offs, and the intricacies of decoding strategies during inference. A deeper technical understanding of these causes is essential for developing more robust detection and mitigation strategies that enhance model reliability in critical applications.

5   Hallucinations in Domain-Specific LLMs

Researchers have extensively characterized hallucinations in LLMs to understand how these errors manifest in different application domains. Foundational taxonomies, Ji *et al.* [15] and Huang *et al.* [13] distinguish between intrinsic hallucinations internal inconsistencies in the generated content and extrinsic hallucinations, which refer to outputs lacking proper grounding in the input context. Zhang *et al.* [38] further clarified that limitations in contextual understanding and overfitting during fine-tuning are key contributors to these errors. This section highlights how such hallucinations have been systematically detected and analyzed in high-stakes fields.

5.1   LLM Hallucinations in Healthcare

In healthcare, the accurate representation of clinical information is critical. Hallucinations in this domain can lead to erroneous diagnostic suggestions or inappropriate treatment recommendations. Researchers have identified that LLMs often generate medically inaccurate outputs, a finding supported by studies that employ specialized detection benchmarks to assess the factual consistency of clinical responses [22]. For instance, analyses of natural language inference tasks in medical settings have revealed that models sometimes affirm hypotheses merely because similar phrasing exists in the training data, even when not supported by the clinical context [17]. These detection efforts underscore the importance of systematically evaluating model outputs to identify patterns of hallucination in healthcare applications.

5.2   LLM Hallucinations in Law

Legal applications demand rigorous precision, as even minor inaccuracies can have far-reaching consequences. In this context, hallucinations manifest as misrepresentations of legal concepts, inaccuracies in summarizing case law, or incorrect definitions of legal terminology. Studies have focused on detecting these issues by comparing model-generated legal summaries with authoritative sources. For example, investigations into the output of state-of-the-art LLMs like GPT-4 have uncovered discrepancies in legal term definitions, highlighting the need for precise evaluation methodologies [25]. Similarly, multi-view analysis approaches have been used to detect internal inconsistencies and lack of grounding in legal document generation, thereby providing a clearer picture of the challenges specific to the legal domain [9].

5.3   LLM Hallucinations in Finance

In the financial domain, the generation of hallucinated content can distort market analysis and risk assessment. Researchers have observed that LLMs occasionally produce outputs that introduce fictitious trends or data points, which can mislead financial decision-making. Detailed evaluations have revealed that such hallucinations often stem from the model's reliance on frequency-based heuristics learned during training [17]. Additionally, structured analytical frameworks have been developed to systematically detect inconsistencies and verify factual accuracy in financial texts, thereby highlighting the prevalence and potential impact of hallucinations in this field [24]. These studies emphasize the critical need for robust diagnostic methods that can accurately capture the nuances of hallucinated outputs in financial contexts.

6   Detecting and Mitigating LLM Hallucinations

6.1   Healthcare

Hallucinations in healthcare-related LLM applications pose significant risks, making precise detection crucial. Ahmad *et al.* [1] employed human-annotated

QA pairs to assess factual correctness in medical dialogues, ensuring alignment with established medical guidelines. Another approach by Ji *et al.* [15] involves specialized benchmarks, where curated medical datasets like PubMed QA and MedQA serve as ground truth references. Some studies have also applied fact-scoring mechanisms to evaluate whether model-generated medical advice aligns with real-world clinical knowledge [22]. However, manual verification by experts remains indispensable due to the complexity of medical information and the potential consequences of incorrect predictions.

Mitigation strategies in healthcare often rely on external knowledge integration, where LLMs are supplemented with verified medical knowledge bases. According to [1], linking models such as Alpaca and ChatGPT-3 to external sources significantly reduced factual inconsistencies. Another approach involves dialogue summarization and fact-checking, where extracted key facts from conversations are compared against authoritative medical references before finalizing responses. Additionally, fine-tuning on medical corpora has helped models internalize reliable patterns in medical discourse. However, these methods often struggle with task specificity and require continuous updates to stay aligned with evolving medical guidelines (Table 1).

Table 1. Recent work for mitigating LLM hallucinations in Medical domain.

DETECTING AND MITIGATING LLM HALLUCINATIONS							
Author	Domain	Detect	Mitigate	Task	Dataset	Metric(s)	Evaluated LLMs
Pal *et al.* [22]	Medical	✗		Med-Halt	Reasoning	Point-wise Score	GPT-3.5, LlaMa-2, MPT
Ji *et al.* [15]	Medical			Medical generative QA	PubMedQA, MEDIQA2019, MedQuAD, MASH-QA	unigramF1, ROUGE-L	Vicuna, AlpacaLoRA, GPT-3.5, MedAlpaca, Robin-medical
Ahmad *et al.* [1]	Medical			Hallucination in healthcare	N/A	FActScores	ChatGPT

6.2   Law

Legal LLM applications frequently produce hallucinations due to the nuanced interpretation required in legal reasoning. Cui *et al.* [4] employed manual evaluation and ranking, where domain experts assessed logical consistency and adherence to legal precedents. Another detection approach involves quantitative assessments, where models undergo structured evaluations against verified legal texts. Dahl *et al.* [5] systematically measured hallucination rates by analyzing inconsistencies in legal case summaries, using precision- and recall-based metrics adapted for legal discourse. Despite these advancements, detecting hallucinations in complex legal reasoning remains a challenge due to ambiguities in case law and jurisdictional variations. Mitigation strategies in the legal domain focus on factual evaluation pipelines and domain-specific fine-tuning. Savelka *et al.* [25]

incorporated legislative text evaluation, comparing model outputs against official statutes to identify discrepancies. Additionally, fine-tuning on legal datasets, including court rulings and annotated legal documents, has improved the factual grounding of LLMs. Another promising method is prompt engineering with external verification, where generated legal arguments are cross-referenced with trusted databases. Despite these improvements, the primary limitation remains generalization to new legal cases, as even well-trained models struggle with evolving legal contexts and rare case law scenarios (Table 2).

Table 2. Recent work for mitigating LLM hallucinations in Legal domain.

DETECTING AND MITIGATING LLM HALLUCINATIONS						
Author	Domain	Detect	Mitigate	Task(s)	Dataset	Metric(s)	Evaluated LLMs
Cui et al. [4]	Legal	✗		Manual	Question Answering	Ranking	ChatLaw, GPT-4
Dahl et al. [5]	Legal	✗		Manual	Legal Hallucination	Hallucination Rate	ChatGPT-4, ChatGPT-3.5
Savelka et al. [25]	Legal	✗		Factual Evaluation in Legislation	N/A	N/A	ChatGPT-3

6.3 Finance

In finance, hallucinations in LLMs can lead to misinterpretations, particularly in speculative predictions. Detecting hallucinations in finance is particularly challenging due to rapidly changing economic conditions and data inconsistencies. Barry et al. [2] introduced GraphRAG, a graph-based Retrieval-Augmented Generation system that integrates knowledge graphs to enhance accuracy and minimize hallucinations. This method reduced hallucinations by 6% and cut token usage by 80%, improving efficiency in financial data processing Another method by Kang and Liu et al. [17], involves prompt-based testing, where models are given specialized financial prompts to detect patterns of hallucination, such as merging unrelated financial trends or misinterpreting stock performance indicators. Roychowdhury et al. [24] extended this approach by designing structured financial statement verification techniques, ensuring that model outputs aligned with actual market data. However, hallucinations remain a concern, especially when models generate outdated or speculative financial predictions. Mitigation strategies in finance revolve around real-time data integration and domain-specific fine-tuning. By connecting LLMs to financial APIs and live market databases, researchers have reduced the risk of outdated or fabricated information. Fine-tuning on high-quality financial reports, including SEC filings and corporate disclosures, has also improved reliability. Additionally, confidence-based ranking has been applied, where models assign certainty scores to generated insights, prompting human verification for high-risk predictions. Nevertheless, specialization challenges persist, as financial models trained on one market segment may still produce hallucinations when applied to another domain (Fig. 3).

Table 3. Recent work for mitigating LLM hallucinations in Finance domain.

DETECTING AND MITIGATING LLM HALLUCINATIONS

Author	Domain	Detect	Mitigate	Task(s)	Dataset	Metric(s)	Evaluated LLMs
Kang *et al.* [17]	Finance			Hallucination in Finance	N/A	FActScores	GPT-3.5
Barry *et al.* [2]	Finance			Hallucination in Finance	N/A	FActScores	GPT-3.5

7   Discussion and Conclusion

Hallucinations in LLMs present a persistent challenge across diverse applications, manifesting as inaccuracies that can undermine trust in AI systems. Our survey highlights that these hallucinations stem from complex sources including data quality, training processes, architectural choices, and inference mechanisms each contributing to factual or contextual errors. Although recent research has advanced our understanding of these issues, complete prevention of hallucinations remains elusive, suggesting a need for ongoing improvements in data handling, model training, and evaluation methodologies. Ultimately, addressing the pervasive issue of hallucinations will be crucial in ensuring that LLMs can reliably support decision-making in high-stakes, real-world contexts.

References

1. Ahmad, M.A., Yaramis, I., Roy, T.D.: Creating trustworthy llms: dealing with hallucinations in healthcare ai. arXiv preprint arXiv:2311.01463 (2023)
2. Barry, M., et al.: Graphrag: leveraging graph-based efficiency to minimize hallucinations in llm-driven rag for finance data. In: 31st International conference on Computational Linguistics Workshop Knowledge Graph & GenAI (2025)
3. Brown, T., et al.: Language models are few-shot learners. Advances in neural information processing systems **33** (2020)
4. Cui, J., Li, Z., Yan, Y., Chen, B., Yuan, L.: Chatlaw: open-source legal large language model with integrated external knowledge bases. arXiv preprint arXiv:2306.16092 (2023)
5. Dahl, M., Magesh, V., Suzgun, M., Ho, D.E.: Large legal fictions: Profiling legal hallucinations in large language models. J. Legal Anal. **16**(1) (2024)
6. Delétang, G., et al.: Language modeling is compression. arXiv preprint arXiv:2309.10668 (2023)
7. Devlin, J., Chang, M.W., Lee, K., Toutanova, K.: Bert: pre-training of deep bidirectional transformers for language understanding. In: Proceedings of the 2019 conference of the North American chapter of the association for computational linguistics: human language technologies, volume 1 (long and short papers) (2019)
8. Elaraby, M., et al.: Halo: estimation and reduction of hallucinations in open-source weak large language models. arXiv preprint arXiv:2308.11764 (2023)
9. Feijo, D.d.V., Moreira, V.P.: Improving abstractive summarization of legal rulings through textual entailment. Artificial intelligence and law **31**(1) (2023)

10. Fernandes, P., et al.: Bridging the gap: A survey on integrating (human) feedback for natural language generation. Trans. Assoc. Comput. Linguist. **11** (2023)
11. Filippova, K.: Controlled hallucinations: Learning to generate faithfully from noisy data. In: Findings of the Association for Computational Linguistics: EMNLP 2020 (2020)
12. Gekhman, Z., et al.: Does fine-tuning llms on new knowledge encourage hallucinations? In: Proceedings of the 2024 Conference on Empirical Methods in Natural Language Processing (2024)
13. Huang, L., et al.: A survey on hallucination in large language models: Principles, taxonomy, challenges, and open questions. ACM Transactions on Information Systems **43**(2) (2025)
14. Ji, Z., Lee, N., Frieske, R., Yu, T., Su, D., Xu, Y., Ishii, E., Bang, Y.J., Madotto, A., Fung, P.: Survey of hallucination in natural language generation. ACM Computing Surveys **55**(12) (2023)
15. Ji, Z., Yu, T., Xu, Y., Lee, N., Ishii, E., Fung, P.: Towards mitigating llm hallucination via self reflection. In: Findings of the Association for Computational Linguistics: EMNLP 2023 (2023)
16. Kaddour, J., Harris, J., Mozes, M., Bradley, H., Raileanu, R., McHardy, R.: Challenges and applications of large language models. arXiv preprint arXiv:2307.10169 (2023)
17. Kang, H., Liu, X.Y.: Deficiency of large language models in finance: an empirical examination of hallucination. In: I Can't Believe It's Not Better Workshop: Failure Modes in the Age of Foundation Models (2023)
18. Lee, K., Ippolito, D., Nystrom, A., Zhang, C., Eck, D., Callison-Burch, C., Carlini, N.: Deduplicating training data makes language models better. In: Proceedings of the 60th Annual Meeting of the Association for Computational Linguistics (2022)
19. Lee, N., Ping, W., Xu, P., Patwary, M., Fung, P.N., Shoeybi, M., Catanzaro, B.: Factuality enhanced language models for open-ended text generation. Advances in Neural Information Processing Systems **35** (2022)
20. Lin, Z., Guan, S., Zhang, W., Zhang, H., Li, Y., Zhang, H.: Towards trustworthy llms: a review on debiasing and dehallucinating in large language models. Artif. Intell. Rev. **57**(9) (2024)
21. Mckenna, N., Li, T., Cheng, L., Hosseini, M., Johnson, M., Steedman, M.: Sources of hallucination by large language models on inference tasks. In: Findings of the Association for Computational Linguistics: EMNLP 2023 (2023)
22. Pal, A., Umapathi, L.K., Sankarasubbu, M.: Med-halt: Medical domain hallucination test for large language models. In: Proceedings of the 27th Conference on Computational Natural Language Learning (CoNLL) (2023)
23. Penedo, G., et al.: The refinedweb dataset for falcon llm: Outperforming curated corpora with web data only. Advances in Neural Information Processing Systems **36** (2023)
24. Roychowdhury, S., et al.: Hallucination-minimized data-to-answer framework for financial decision-makers. In: 2023 IEEE International Conference on Big Data (BigData). IEEE (2023)
25. Savelka, J., Ashley, K.D., Gray, M.A., Westermann, H., Xu, H.: Explaining legal concepts with augmented large language models (gpt-4). arXiv preprint arXiv:2306.09525 (2023)
26. Sharma, M., et al.: Towards understanding sycophancy in language models. In: 12th International Conference on Learning Representations, ICLR 2024 (2024)
27. Singh, A., Singh, D., Singh, R.: Generative ai for cyberdefense. In: Generative AI: Current Trends and Applications, pp. 121–145. Springer (2024)

28. Stiennon, N., et al.: Learning to summarize with human feedback. Advances in Neural Information Processing Systems **33** (2020)
29. Tjio, G., Liu, P., Zhou, J.T., Goh, R.S.M.: Adversarial semantic hallucination for domain generalized semantic segmentation. In: Proceedings of the IEEE/CVF Winter Conference on Applications of Computer Vision (2022)
30. Touvron, H., et al.: Llama 2: Open foundation and fine-tuned chat models. arXiv preprint arXiv:2307.09288 (2023)
31. Vaswani, A.: Attention is all you need. Advances in Neural Information Processing Systems (2017)
32. Workshop, B., et al.: Bloom: a 176b-parameter open-access multilingual language model. arXiv preprint arXiv:2211.05100 (2022)
33. Xu, W., Agrawal, S., Briakou, E., Martindale, M.J., Carpuat, M.: Understanding and detecting hallucinations in neural machine translation via model introspection. Transactions of the Association for Computational Linguistics **11** (2023)
34. Yang, Y., Chern, E., Qiu, X., Neubig, G., Liu, P.: Alignment for honesty. Advances in Neural Information Processing Systems **37** (2024)
35. Zhang, M., Press, O., Merrill, W., Liu, A., Smith, N.A.: How language model hallucinations can snowball. In: International Conference on Machine Learning. PMLR (2024)
36. Zhang, S., et al.: Instruction tuning for large language models: A survey. arXiv preprint arXiv:2308.10792 (2023)
37. Zhang, S., et al.: Opt: Open pre-trained transformer language models. arXiv preprint arXiv:2205.01068 (2022)
38. Zhang, Y., et al.: Siren's song in the ai ocean: a survey on hallucination in large language models. arXiv preprint arXiv:2309.01219 (2023)
39. Zhao, W.X., et al.: A survey of large language models. arXiv preprint arXiv:2303.18223 (2023)
40. Zhou, C., et al.: Lima: Less is more for alignment. Advances in Neural Information Processing Systems **36** (2024)

Workshop on Graph Learning with Foundation Models (GLFM 2025)

GLFM Preface

The Graph Learning with Foundation Models (GLFM) workshop, held as part of the 29th Pacific-Asia Conference on Knowledge Discovery and Data Mining (PAKDD) in 2025, was launched to bring together researchers and practitioners from academia and industry to explore the emerging field of Graph Foundation Models (GFMs). The workshop aimed to provide a platform for presenting novel methodologies, diverse datasets, and real-world applications involving GFMs.

A core motivation behind organizing the GLFM workshop was to explore and understand the emergent capabilities of GFMs—specifically, their ability to generalize across domains, adapt to new tasks, and uncover patterns beyond the scope of traditional graph learning models. Additionally, the workshop sought to highlight the challenges and opportunities in leveraging graph structures in conjunction with advanced foundation models, especially in the areas of scalability, interpretability, efficient training on large-scale graphs, domain adaptation, and multi-modal learning.

To enable a broad and meaningful discussion on GFMs, we structured the workshop's scope to encompass topics such as theoretical foundations, empirical analysis, efficient pre-training and fine-tuning strategies, multi-modal data integration, and real-world applications. Consequently, the submissions we received reflected these goals well, allowing the workshop to feature a diverse set of approaches and insights into the rapidly evolving landscape of Graph Foundation Models.

1 Submission Overview and Selection Process

The GLFM workshop receive a total of seven submissions from authors representing a broad range of countries, including the USA, China, South Korea, Japan, and Germany. All submissions underwent a rigorous double-blind peer-review process, with each paper evaluated by two to three independent reviewers. Review criteria included technical quality, originality, clarity, and relevance to the theme of Graph Foundation Models. Based on these evaluations, four regular papers were selected for presentation at the workshop. The selected papers offer novel contributions to the field and reflect the emerging challenges and research directions in applying foundation models to graph-structured data.

2 Accepted Articles and Acceptance Rate

The accepted papers span a diverse set of topics and methodologies at the intersection of foundation models and graph learning. The accepted papers collectively explore topics such as representation learning and evaluation of graph generative models, context-aware and structured graph reasoning for sentiment analysis, robust federated graph learning with intellectual property protection, and domain-specific benchmarking of GFMs in large-scale recommendation systems. These papers were selected based on their originality, rigor, and relevance, as recognized by the reviewers. The final acceptance rate was 57.1% (4 out of 7 submissions).

3 Workshop Summary

The GLFM 2025 Workshop was held on June 10, 2025, in Sydney as part of the 29th Pacific-Asia Conference on Knowledge Discovery and Data Mining (PAKDD). The program featured four high-quality presentations, selected through a competitive peer-review process from a total of seven submissions. The accepted papers address a range of timely topics, including representation learning and evaluation of graph generative models, structured graph reasoning for sentiment analysis, privacy-preserving federated graph learning, and domain-specific benchmarking of GFMs in large-scale recommendation systems. The workshop drew participation from a diverse international community, fostering engaging discussions on the future of GFMs and their potential to drive innovation across various application domains.

Acknowledgment. We would like to express our heartfelt gratitude to the PAKDD Program Committee, local organizers, and workshop chairs for their generous support and guidance throughout the planning and execution of the GLFM workshop. We are especially thankful to our Program Committee members and reviewers for their thoughtful, constructive, and timely reviews, which helped ensure the quality and rigor of the accepted papers. Our sincere thanks also go to all authors and attendees—their contributions and engagement were key to the workshop's success. Finally, we gratefully acknowledge Springer for their support in publishing the workshop proceedings.

April 2025

Fanchen Bu
Minyoung Choe
Jaemin Yoo
Chanyoung Park
Namyong Park
Bryan Hooi
Neil Shah
Shirui Pan
Kijung Shin

Organization

Chairs

Fanchen Bu	KAIST, South Korea
Minyoung Choe	KAIST, South Korea
Jaemin Yoo	KAIST, South Korea
Chanyoung Park	KAIST, South Korea
Namyong Park (Postdoctoral Researcher)	Meta AI, USA
Bryan Hooi	National University of Singapore, Singapore
Neil Shah (Research Scientist)	Snap Research, USA
Shirui Pan	Griffith University, Australia
Kijung Shin	KAIST, South Korea

Program Committee Members

Yeonjun In	KAIST, South Korea
Jiaxin Ju	Griffith University, Australia
Shinhwan Kang	KAIST, South Korea
Kibum Kim	KAIST, South Korea
Sein Kim	KAIST, South Korea
Sunwoo Kim	KAIST, South Korea
Huan Yee Koh	Monash University, Australia
Soo Yong Lee	KAIST, South Korea
Geon Lee	KAIST, South Korea
Jongha Lee	KAIST, South Korea
Namkyeong Lee	KAIST, South Korea
Langzhang Liang	KAIST, South Korea
Linhao Luo	Monash University, Australia
Sangwoo Seo	KAIST, South Korea
Yuan Sui	National University of Singapore, Singapore
Zhangchi Qiu	Griffith University, Australia
Kanghoon Yoon	KAIST, South Korea
Zicheng Zhao	Nanjing University of Science and Technology, China

Graph Generative Models Evaluation with Masked Autoencoder

Chengen Wang[1(✉)] and Murat Kantarcioglu[2]

[1] University of Texas at Dallas, Dallas, TX 75080, USA
chengen.wang@utdallas.edu
[2] Virginia Tech, Blacksburg, VA 24061, USA
muratk@vt.edu

Abstract. In recent years, numerous Graph Generative Models (GGMs) have been proposed. However, evaluating these models remains a considerable challenge, primarily due to the difficulty in extracting meaningful graph features that accurately represent real-world graphs. The traditional evaluation techniques, which rely on graph statistical properties like node degree distribution, clustering coefficients, or Laplacian spectrum, overlook node features and lack scalability. There are newly proposed deep learning-based methods employing graph random neural networks or contrastive learning to extract graph features, demonstrating superior performance compared to traditional statistical methods, but their experimental results also demonstrate that these methods do not always working well across different metrics. Although there are overlaps among these metrics, they are generally not interchangeable, each evaluating generative models from a different perspective. In this paper, we propose a novel method that leverages graph masked autoencoders to effectively extract graph features for GGM evaluations. We conduct extensive experiments on graphs and empirically demonstrate that our method can be more reliable and effective than previously proposed methods across a number of GGM evaluation metrics, such as "Fréchet Distance (FD)" and "MMD Linear". However, no single method stands out consistently across all metrics and datasets. Therefore, this study also aims to raise awareness of the significance and challenges associated with GGM evaluation techniques, especially in light of recent advances in generative models. Our code is available at https://github.com/chengenw/ggmEval

Keywords: Graph Generative Models Evaluation · Graph Evaluation Metrics · Graph Masked Autoencoder

1 Introduction

Graph Generative Models (GGMs) have important applications across different domains, including biology, chemistry, engineering and social networks. Recent advances in graph generative models underscore the need for robust metrics

S. Yuan et al. (Eds.): PAKDD 2025 Workshops, LNAI 15835, pp. 137–148, 2025.
https://doi.org/10.1007/978-981-96-8197-6_10

to evaluate the synthetic graphs in comparison to the real graphs [12]. Unlike images, human visual perception is hardly applicable to graphs due to their complex structures. In addition, commonly-used feature extractors, like Inception network [13], are not readily available for the graph domain.

Traditional graph evaluation methods rely on representations derived from general graph topology, such as node degree distributions, clustering coefficient distributions, orbit counts and the Laplacian spectrum. Although these statistics capture important structural properties, they neglect features associated with individual nodes, which limits their ability to provide expressive representations of graphs. Furthermore, these methods often face scalability challenges.

O'Bray et al. [12] highlights issues with commonly-used evaluation methods, which may lead to inconsistent rankings across different settings. This underscores the urgent need to develop new GGM evaluation techniques.

The key challenge in evaluating GGM lies in effectively extracting graph representations. With faithful graph representations, the extracted features or embeddings can be used as input for standard graph evaluation metrics, such as the Fréchet Distance [4], and Precision & Recall [6], which we will discuss in Sect. 3.

Recently, Thompson et al. [16] proposed using graph random neural networks to extract graph features, demonstrating the advantages of the deep learning-based methods. Building on this, Shirzad et al. [14] introduced a self-supervised contrastive learning method to extract more faithful graph representations, generally outperforming the previous technique.

Graph masked autoencoders (GMAE) [5,7] represent another self-supervised learning method that can achieve performance on par with or even surpass, contrastive learning-based methods. Graph masked autoencoder works by recovering masked node features or edges. The experimental results from [5] demonstrate GMAE can *match or exceed* contrastive learning method, raising the question of whether graph masked autoencoder can more effectively extract graph representations, potentially leading to improved GGM evaluation techniques.

In this work, we propose to leverage graph masked autoencoders to assess the fidelity and diversity of graph generative models. Through extensive experiments, we demonstrate the effectiveness and advantages of our proposed method compared to existing evaluation approaches across multiple metrics.

Our main contributions are as follows:

1. We propose leveraging a graph masked autoencoder to extract graph representations, thereby offering a more effective method for GGM evaluations across a number of metrics,
2. We conducted systematic experiments under different settings to evaluate the fidelity and diversity of graph generative models,
3. Our experimental results demonstrate the superiority of our method across various metrics compared to other deep learning-based baselines,
4. Although our method excels across a number of metrics, no single method stands out across all metrics and datasets. Therefore, this work also empha-

sizes the need for further research into evaluation techniques for graph generative models.

In the following sections, we first provide a brief review of related works in Sect. 2. We then introduce the necessary background knowledge in Sect. 3 to facilitate the understanding of our work. Next, we delve into our proposed method, explaining how to leverage masked autoencoders to evaluate graph generative models in Sect. 4. Subsequently, we present the experimental setup and results in Sect. 5. Finally, we conclude our work in Sect. 6.

2 Related Work

2.1 Graph Generative Models

In recent years, researchers have proposed various types of graph generative models [20]. You et al. [19] proposed GraphRNN, an auto-regressive model, which factorizes the generation process in a sequential way. Simonovsky et al. [15] proposed GraphVAE, which maximizes a Evidence Lower Bound in an encoder-decoder structure. Normalizing flow models have also been applied to graph generation [8], which estimates the density of graph distributions via the change of variable theorem. De Cao et al. [1] proposed GAN-based graph generative models. The latest diffusion generative models were also adapted to graph domain [11].

Note that although the proposed method in our paper is designed to evaluate graph generative models, it is *not restricted to any specific type of GGMs*. This is because it assesses the graphs generated by these models, rather than directly on the generative models themselves. The generated graphs are obtained by perturbing real graph datasets, simulating a graph generative model that generates graphs with a distribution different from the training graphs. Therefore the proposed method is both *model-agnostic* and *application-agnostic*.

2.2 Metrics for Graph Generative Models Evaluation

The metrics are used to compare generated graphs to real training graphs. They compare both fidelity and diversity. To compare fidelity/similarity, one can utilize Fréchet Inception Distance (FID) [4] or MMD [2] with different kernels. To assess diversity, one needs check if there is mode collapse or mode dropping. These can be examined with Improved Precision & Recall [6] or Density and Coverage [10] metrics.

To extract graph representations, traditionally one uses statistics from the node degree distribution, clustering coefficients or Laplacian spectrum distributions [12]. More recently, Thompson et al. [17] proposed a deep learning-based method for graph generative models evaluation, where they employ random graph neural networks (GNN) to extract graph features. Building upon this, a follow-up paper [14] proposed to contrastively learn graph features, which generally performs better according to their experimental results.

2.3   Graph Masked Autoencoder

He et al. [3] proposed masked autoencoder in vision domain, which is a self-supervised learning method that masks random patches of the input image and then reconstructs them with learnable neural networks. Hou et al. [5] adapted the method to the graph domain, masking node attributes only. Li et al. [7] is another graph mask autoencoder paper, which masks edges or paths. In this paper, we perturb either nodes or edges for each graph, with a predefined probability, as discussed in Sect. 5.4.

3   Preliminaries

In this section, we introduce background knowledge to facilitate the understanding of this paper. To evaluate a graph generative model, we need to compare the graphs generated by the GGM against the real training graphs with various metrics to evaluate the fidelity and the diversity of the GGM.

To evaluate the fidelity of a GGM, one needs to measure how similar the generated graphs are to the real graphs. In this section, we discuss some of the commonly used metrics.

3.1   Fréchet Distance (FD)

It assumes the real graphs and the generated graphs are two multivariate Gaussian distributions with mean μ and covariance $\mathbf{C}$. It compares the mean and variance of two distributions. More specifically, the difference of the two distributions are computed by the distance [4]

$$\mathrm{FD}(\mathbb{H}_r, \mathbb{H}_g) = \|\mu_r - \mu_g\| + \mathrm{Tr}(\mathbf{C}_r + \mathbf{C}_g - 2(\mathbf{C}_r\mathbf{C}_g)^{1/2}) \qquad (1)$$

where $\mathbb{H}$ represents the distributions of graph representations.

3.2   Maximum Mean Discrepancy (MMD)

This metric measures the dissimilarity of two distributions H_r and H_g and a lower value of MMD means the two distributions are closer [2,17].

$$\mathrm{MMD}(\mathbb{H}_g, \mathbb{H}_r) := \frac{1}{m^2} \sum_{i,j=1}^{m} k(x_i^r, x_j^r) + \frac{1}{n^2} \sum_{i,j=1}^{n} k(x_i^g, x_j^g)$$

$$- \frac{2}{nm} \sum_{i=1}^{n} \sum_{j=1}^{m} k(x_i^g, x_j^r) \qquad (2)$$

where $k(\cdot, \cdot)$ is a kernel function. In this paper, we use two kernels: RBF kernel $k(x_i, x_j) = \exp(-\frac{d(x_i, x_j)}{2\sigma^2})$ and linear kernel $k(x_i, x_j) = x_i^T \cdot x_j$ [14].

To evaluate the diversity of a GGM, one needs to check the two possible issues of a flawed generative models: mode dropping and mode collapse [17,18]. Real graphs are usually diverse, but the generated graphs may ignore some modes, causing mode dropping, or lack diversity within some modes, causing mode collapse. The following metrics could be used to measure the diversity.

3.3   Improved Precision and Recall (P&R)

This metric [6] constructs the manifold separately for both real graphs and generated graphs by extending a radius from each sample to its k-th nearest neighbors to form a hypersphere and then take the union of all the hyperspheres. The precision measure reflects the probability of generated samples falling within the manifold of real samples, while the recall measure reflects the probability of real samples falling within the manifold of generated samples.

3.4   Density and Coverage (D&C)

This is introduced as a more robust metric [10] than Precision & Recall. It is based on the concepts of P&R. The density measure counts how many real sample hyperspheres contain a generated sample instead of just checking if the generated sample is contained within any real sample neighborhood hypersphere. The coverage measure is similar to recall by counting the ratio of real samples covered by generated ones, but it builds the nearest neighbor hypersphere around the real samples instead of the generated ones. The harmonic mean F1 of P&R and D&C are both used in this paper.

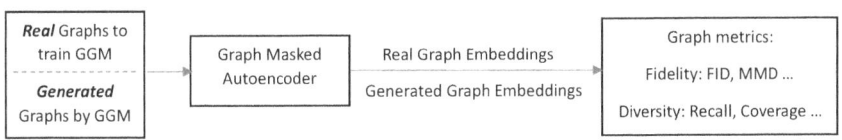

Fig. 1. The Process to Evaluate Graph Generative Models with Graph Masked Autoencoder

4   Graph Masked Autoencoder for Evaluation

In this paper, we propose to employ Graph Masked Autoencoder (GMAE) to extract graph features for further GGM evaluations. The whole process for GGM evaluation is shown in Fig. 1. We utilize the architecture from [5] for node feature masking and the architecture from [7] for edge masking. The masking architecture learns graph representations by recovering the masked node features or edges. Except for the GMAE training process, the rest of the evaluation process is similar to those in [14,17] and we largely follow their process for a fair comparison of the experimental results. Note that the primary function of GMAE is to extract accurate graph representations, a crucial step in the GGM evaluation process.

For a proposed GGM evaluation technique to be effective, it must distinguish the distribution difference between real graphs and generated graphs, and reflect any mode dropping or collapse occurred in the generated graphs. In the experiments, the *real* graphs refer to the datasets described in Sect. 5.1. The *generated* graphs are obtained by perturbing the real graphs. That is, the "Generated Graphs by GGM" in Fig. 1 result from this perturbation, highlighting why the evaluation is model-agnostic. Greater perturbation of the real graphs results in a larger distribution difference between the real and the generated graph set. A robust GGM evaluation technique should accurately differentiate the degree of variation between them.

To assess the GGM evaluation techniques, we follow the approach in [17] by monotonically increasing the degree t of real graphs perturbation, then compare the set of the perturbed graphs with the set of the original real graphs using the metrics mentioned in Sect. 3. For each degree of perturbation $t \in [0,1]$, we can get a normalized metric score $\hat{s}$. For a strong GGM evaluation technique, ideally the metric score $\hat{s} = 0$ if the two set of graphs are the same (i.e., when degree of perturbation $t = 0$), $\hat{s} = 1$ if the degree of perturbation $t = 1$ and $\hat{s}$ monotonically increasing as degree of perturbation t increases from 0 to 1. This relationship between t and $\hat{s}$ can be captured by the Spearman's rank correlation coefficient. If a GGM evaluation technique faithfully learns graph representations (e.g., with GMAE), the Spearman rank correlation will be 1, otherwise it will be less than 1 and the worst value is -1. Therefore, the Spearman rank correlation is employed to assess whether our proposed method—leveraging GAME—can accurately extract graph representations and, thus, whether it, combined with other components as shown in Fig. 1, constitutes a better GGM evaluation technique. We provide more details about the graph perturbation methods in the experimental Sect. 5.

Table 1. The Statistical Overview of the Datasets

	REDDIT-MULTI-5K	DBLP_v1	Proteins
num of graphs	4410	17892	739
mean num nodes	378.8	11.2	52.8
min num nodes	22	3	20
max num nodes	1000	39	620
mean num edges	433.4	21.3	98.7
min num edges	21	2	23
max num edges	1638	168	1049

5   Experiments

5.1   Datasets

In our experiments, we use the datasets REDDIT-MULTI-5K, DBLP_v1, and Proteins [9], spanning the domain of social networks and bioinformatics. We remove from the dataset graphs with less than 3 nodes (less than 20 nodes for Proteins) or with more than 1,000 nodes. For REDDIT-MULTI-5K and DBLP_v1 datasets, we randomly sample approximately 800 to 1,000 graphs per run due to computational complexity. These datasets have graph node counts that range from a few to one thousand, and edge counts from two to over one thousand. A statistical overview of these datasets is provided in Table 1.

5.2   Baselines

In our experiments, we compare with two deep learning-based GGM evaluation techniques. Thompson et al. [17] employs random graph neural networks to extract the representations of graphs. In contrast, Shirzad et al. [14] leverages a contrastive learning method to extract graph representations. These two baselines serve as the reference points for evaluating the effectiveness of our proposed method. Note that we do not use traditional graph representation-based techniques as baselines due to their limitations as discussed in Sect. 1.

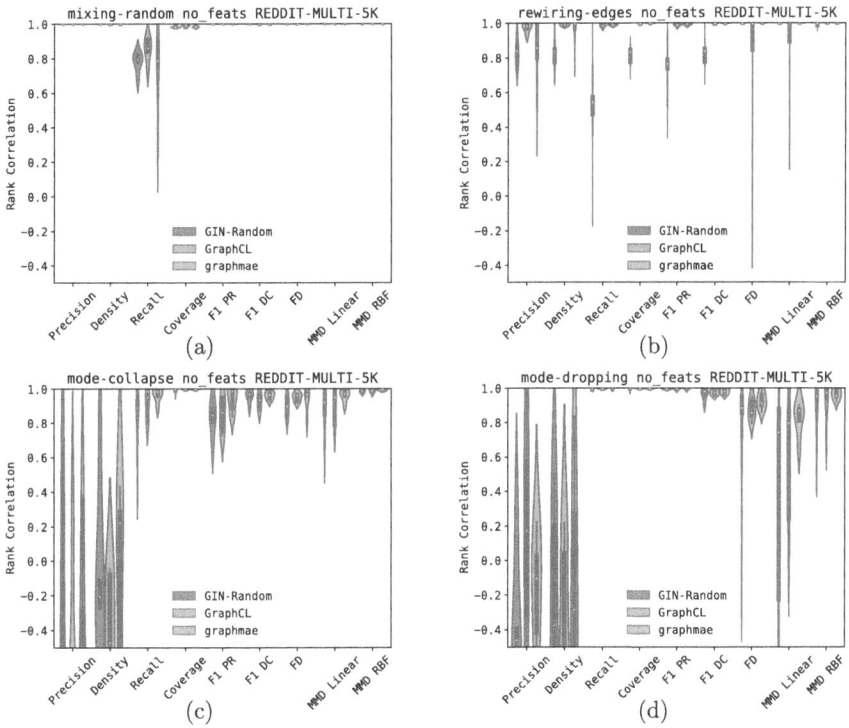

Fig. 2. Experimental results across the perturbation methods for `REDDIT-MULTI-5K` dataset. A higher and shorter violin plot indicates better results.

5.3   Fidelity and Diversity Measurement Setup

We adhere to the procedures outlined in [17] to perturb the graphs. The degree of perturbation ranges from 0 to 1, with step size 0.01.

Fidelity measurement involves two types of perturbations: randomly mixing graphs and rewiring edges. To randomly mix graphs, we replace real graphs with Erdős–Rényi (E-R) graphs with the same number of nodes and proportion of edges. To rewire edges, each edge in a graph is rewired with a probability equal to the degree of perturbation. If an edge is rewired, we pick one of its two nodes with equal probability, then disconnect it from the picked node and connect it to a randomly selected node from the graph.

For diversity measurement, we follow [17] to simulate mode collapse and mode dropping. Both experiments start with identifying clusters of the real graph set. For mode collapse, we replace the graphs in a cluster with its respective cluster center. For mode dropping, we replace the graphs in the dropped cluster with graphs from the remaining clusters. During the experiments, we gradually increase the number of collapsed or dropped clusters from 0 until it reaches to the total number of clusters.

5.4   The Evaluation Setup

We use the default values in the released code from [14] for the baselines setting. For each experiment, we run the experiment five times with different seeds, then generate a corresponding violin plot. *The graph masked autoencoder mask either nodes or edges with equal probability.* The mask rate is 0.2 in the experiments.

5.5   Results

We show the experimental results in Fig. 2, 3 and 4 for each dataset respectively. The x-axis of these figures corresponds to the different metrics we discussed in Sect. 3, and the y-axis represents the Spearman correlation mentioned in Sect. 4. As discussed in Sect. 4, the ideal Spearman rank correlation is 1 and the worst one is -1. We use violin plots to illustrate the results, visualizing the distribution of the data—in this case, the *distribution* of the Spearman correlation coefficients.

In the figures above, the white dot in each violin plot represents the *median*, while the thick black bar indicates the interquartile range (IQR), which is the distance between the upper and lower quartiles. The whiskers extend to the farthest data points within 1.5 IQR from the box. The violin shape represents the *probability density* of the data, smoothed by kernel density estimation, with horizontally wider green sections indicating higher probability regions.

In summary, a *higher and shorter* violin plot indicates better results, corresponding to a higher Spearman correlation with lower variance.

The figures indicate that although no method consistently stands out across all metrics, there are a number of metrics for each dataset where our method performs best as detailed below.

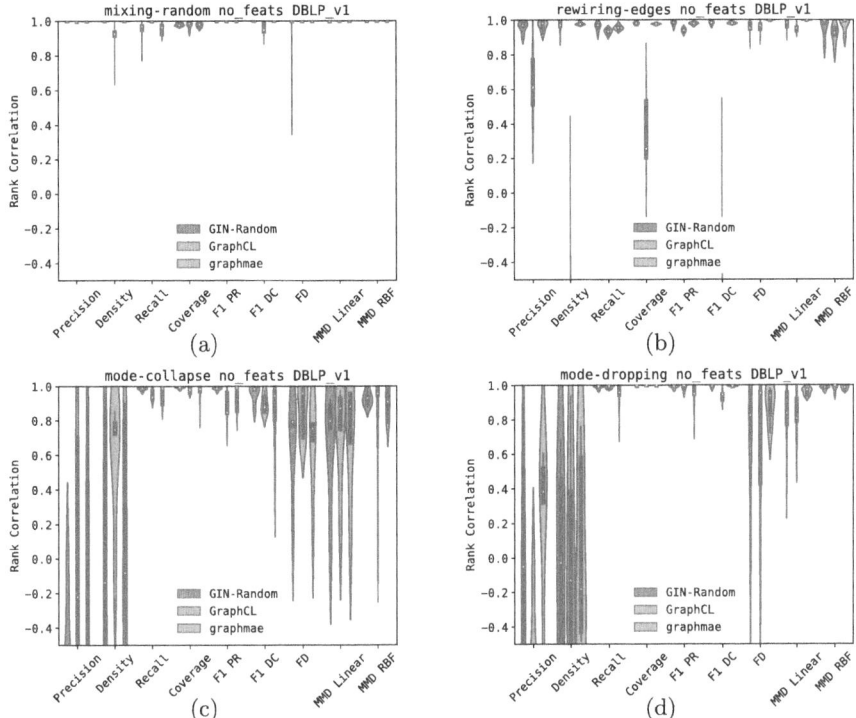

Fig. 3. Experimental results across the perturbation methods for DBLP_v1 dataset. A higher and shorter violin plot indicates better results.

Figure 2 presents the experimental results for dataset REDDIT-MULTI-5K. Our method performs best across most metrics, achieving higher medians and/or lower variability. For instance, in the mode collapse perturbation, our method excels in metrics "recall" and "MMD Linear". Similarly, in mode dropping perturbation, it outperforms the other methods in "FD", "MMD Linear" and "MMD RBF". Notably, under the rewiring edges perturbation, the random GNN baseline performs significantly worse than the other two methods in "Recall", while the contrastive learning baseline shows the poorest performance in "FD" and "MMD Linear".

Figure 3 presents the experimental results for dataset DBLP_v1. In the mode dropping perturbation, our method achieves the best performance in "FD" and "MMD Linear", with higher medians and/or lower variability. Notably, under the rewiring edges perturbation, the contrastive learning baseline performs significantly worse than the other two methods across several metrics, such as "precision", "density" and "coverage".

Figure 4 presents the experimental results for dataset Proteins. Our method and the two baselines have similar performance across most metrics. However, our method stands out with a higher median *and* lower variability for "FD" in the mode collapse perturbation, and "MMD linear" in the mode dropping perturbation.

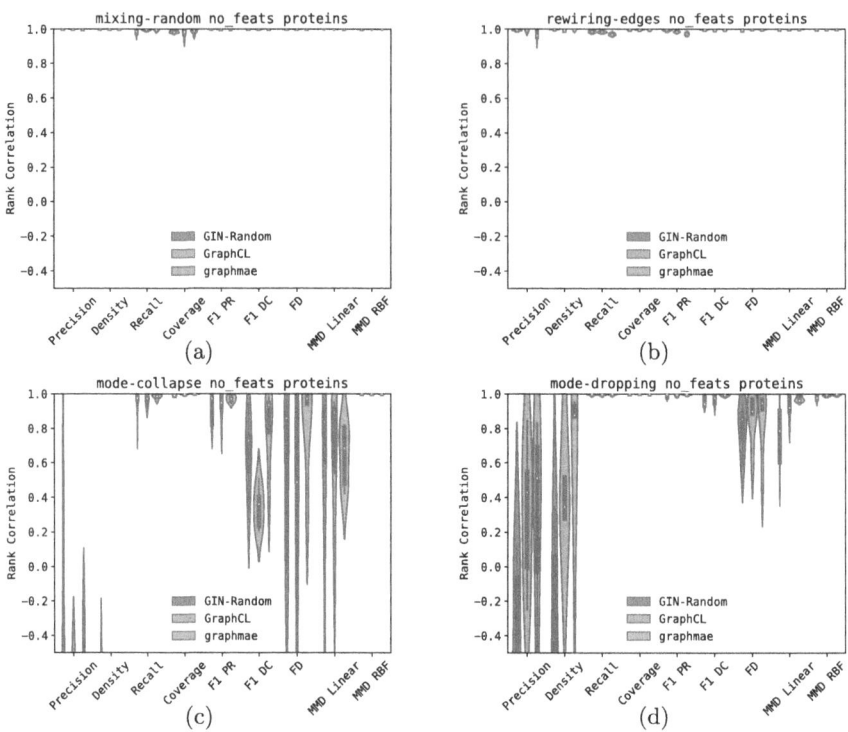

Fig. 4. Experimental results across the perturbation methods for `proteins` dataset. A higher and shorter violin plot indicates better results.

5.6 Discussions

Since GGM metrics are usually not interchangeable, and each of them evaluates GGM from its unique perspective, one should consider the dataset and the intended application when choosing the appropriate graph representation learning technique and evaluation metrics.

The reason that our proposed method performs better across a number of metrics is that graph masked autoencoder can perform comparable or better compared to the contrastive learning-based method on graph representation learning, as shown in [5,7], where they performed graph classification or node classification tasks. Additionally, contrastive learning-based graph representation method performs better than random neural network-based method according to [14]. Thus, our method provides another approach for effective graph generative models evaluations, especially when using the metrics discussed above, such as "FD" and "MMD Linear".

The performance of our proposed method is expected to improve further by adjusting architectures, training strategy, and hyperparameters of GMAE, but this may not succeed without significant tuning efforts.

While our method generally outperforms others in these experiments, we emphasize that neither our approach nor the baselines consistently excels across all metrics. This underscores the need for further research in GGM evaluation techniques, especially in light of recent advances in generative models.

6   Conclusion

In this paper, we propose graph masked autoencoder-based approach to learn graph representations for graph generative models evaluations. Compared to random GNN and contrastive graph learning-based approaches, the experimental results demonstrate that our method performs better across a number of metrics, such as "recall", "FD" and "MMD Linear" as discussed in Sect. 5.5. Our findings suggest that when evaluating generative graph models, one should consider multiple deep learning-based graph representation learning methods for a more comprehensive evaluation. Additionally, our work highlights the significance and challenges associated with evaluating graph generative models in practice.

Acknowledgments. The research reported herein were supported in part by NSF awards DMS-2204795, OAC-2115094, CNS-2331424, ARL/Army Research Office awards W911NF-24-1-0202 and W911NF-24-2-0114.

References

1. De Cao, N., Kipf, T.: MolGAN: an implicit generative model for small molecular graphs. In: ICML 2018 Workshop on Theoretical Foundations and Applications of Deep Generative Models (2018)
2. Gretton, A., Borgwardt, K., Rasch, M., Schölkopf, B., Smola, A.: A kernel method for the two-sample-problem. In: Advances in Neural Information Processing Systems, vol. 19. MIT Press (2006)
3. He, K., Chen, X., Xie, S., Li, Y., Dollár, P., Girshick, R.: Masked autoencoders are scalable vision learners. In: Proceedings of the IEEE/CVF Conference on Computer Vision and Pattern Recognition, pp. 16000–16009 (2022)
4. Heusel, M., Ramsauer, H., Unterthiner, T., Nessler, B., Hochreiter, S.: Gans trained by a two time-scale update rule converge to a local nash equilibrium. In: Advances in Neural Information Processing Systems, vol. 30. Curran Associates, Inc. (2017)
5. Hou, Z., et al.: Graphmae: self-supervised masked graph autoencoders. In: Proceedings of the 28th ACM SIGKDD Conference on Knowledge Discovery and Data Mining, pp. 594–604 (2022)
6. Kynkäänniemi, T., Karras, T., Laine, S., Lehtinen, J., Aila, T.: Improved precision and recall metric for assessing generative models. Curran Associates Inc., Red Hook (2019)

7. Li, J., et al.: What's behind the mask: understanding masked graph modeling for graph autoencoders. In: KDD, pp. 1268–1279. ACM (2023)
8. Luo, Y., Yan, K., Ji, S.: Graphdf: a discrete flow model for molecular graph generation. In: Proceedings of the 38th International Conference on Machine Learning, vol. 139 of Proceedings of Machine Learning Research, pp. 7192–7203. PMLR (2021)
9. Morris, C., Kriege, N.M., Bause, F., Kersting, K., Mutzel, P., Neumann, M.: Tudataset: a collection of benchmark datasets for learning with graphs (2020)
10. Naeem, M.F., Oh, S.J., Uh, Y., Choi, Y., Yoo, J.: Reliable fidelity and diversity metrics for generative models. In: International Conference on Machine Learning, pp. 7176–7185. PMLR (2020)
11. Niu, C., Song, Y., Song, J., Zhao, S., Grover, A., Ermon, S.: Permutation invariant graph generation via score-based generative modeling. In: International Conference on Artificial Intelligence and Statistics, pp. 4474–4484. PMLR (2020)
12. O'Bray, L., Horn, M., Rieck, B., Borgwardt, K.: Evaluation metrics for graph generative models: problems, pitfalls, and practical solutions. In: International Conference on Learning Representations (2022)
13. Salimans, T., Goodfellow, I., Zaremba, W., Cheung, V., Radford, A., Chen, X.: Improved techniques for training gans. In: Advances in Neural Information Processing Systems, vol. 29 (2016)
14. Shirzad, H., Hassani, K., Sutherland, D.J.: Evaluating graph generative models with contrastively learned features. In: Oh, A.H., Agarwal, A., Belgrave, D., Cho, K. (eds.) Advances in Neural Information Processing Systems (2022)
15. Simonovsky, M., Komodakis, N.: GraphVAE: towards generation of small graphs using variational autoencoders (2018)
16. Thompson, R., Knyazev, B., Ghalebi, E., Kim, J., Taylor, G.W.: On evaluation metrics for graph generative models. In: International Conference on Learning Representations (2022)
17. Thompson, R., Knyazev, B., Ghalebi, E., Kim, J., Taylor, G.W.: On evaluation metrics for graph generative models. arXiv preprint arXiv:2201.09871 (2022)
18. Xu, Q., et al.: An empirical study on evaluation metrics of generative adversarial networks. arXiv preprint arXiv:1806.07755 (2018)
19. You, J., Ying, R., Ren, X., Hamilton, W.L., Leskovec, J.: Graphrnn: generating realistic graphs with deep auto-regressive models. In: Proceedings of the 35th International Conference on Machine Learning, ICML 2018, Stockholmsmässan, Stockholm, Sweden, 10–15 July 2018, vol. 80 of Proceedings of Machine Learning Research, pp. 5694–5703. PMLR (2018)
20. Zhu, Y., et al.: A survey on deep graph generation: methods and applications. In: The First Learning on Graphs Conference (2022)

Adaptive Context-Aware Graph Convolutional Network for Aspect-Based Sentiment Analysis

Jianhua Chi[ID] and Xianguo Zhang[✉][ID]

Inner Mongolia University, Inner Mongolia, China
`32209149@mail.imu.edu.cn, 2595083628@qq.com`

Abstract. Aspect-Based Sentiment Analysis (ABSA) focuses on determining the sentiment polarity of specific aspect terms within sentences. Previous methods have achieved remarkable performance by leveraging graph neural networks and attention mechanisms to learn dependency tree structures. However, during the graph learning process in complex global structures, irrelevant contextual information is often introduced, and reliance on single-graph structural features may lead to inaccurate predictions. To address these issues, we propose ACAGCN, a novel framework designed to enhance context relevance and representation diversity. ACAGCN incorporates a context-aware semantic learning module, which effectively filters out irrelevant context while strengthening semantic correlations related to specific aspects. It adopts a multi-channel Graph Convolutional Network (GCN) architecture, consisting of AspGCN, SemGCN, and SynGCN, to capture aspect-related information, semantic associations, and syntactic structures, respectively. Additionally, we introduce an adaptive aggregation network to facilitate the fusion of the three GCN channels, achieving the integrity of different structural representations. Experiments on three benchmark datasets demonstrate the effectiveness of our framework.

Keywords: Aspect-based Sentiment Analysis · Graph Convolutional Network · Multiple features

1 Introduction

Sentiment analysis spans text, audio, visual, and multi-modal media, with textual sentiment analysis being a prominent yet challenging field. It focuses on analyzing opinions, feelings, and attitudes expressed in text. This work centers on sentiment analysis of text.

Beyond traditional sentence or document-level tasks, aspect-based sentiment analysis (ABSA) has emerged, aiming to determine the sentiment polarity (positive, negative, or neutral) of specific aspects in text. As shown in Fig. 1, ABSA identifies the sentiment polarity for "food" and "service" as positive and neutral, respectively.

© The Author(s), under exclusive license to Springer Nature Singapore Pte Ltd. 2025
S. Yuan et al. (Eds.): PAKDD 2025 Workshops, LNAI 15835, pp. 149–161, 2025.
https://doi.org/10.1007/978-981-96-8197-6_11

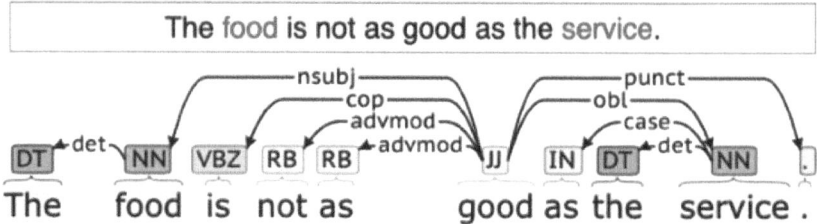

Fig. 1. Example sentence with a dependency tree where aspect term (highlighted in red) are linked to other words based on their syntactic dependencies. (Color figure online)

The attention mechanism [3,5,9] has been widely used to model relationships between aspect terms and context words, but it is prone to noise from irrelevant words, posing challenges for ABSA tasks. When attention mechanisms fail to thoroughly explore context and aspect-related information, they may misrepresent aspect-related information, introducing irrelevant semantic information and reducing model performance.

Graph neural networks (GNNs) have recently been applied to dependency trees [4] to capture sentence syntax. For example, in Fig. 1, the dependency tree links the aspect "food" to the opinion word "good." However, previous studies have often focused on a single type of feature—semantic associations, aspect-related information, or syntactic structures—without integrating multiple features.

To address the above issues, this paper proposes a new architecture, the Adaptive Context-Aware Graph Convolutional Network (ACAGCN). For the first issue, we introduce a context-aware semantic learning module that combines the aspect-aware attention mechanism with the dynamic-mask attention mechanism. The idea is that the fused attention matrix can represent semantic correlations between words while filtering out irrelevant contextual information. For the second issue, we introduce three GCN channels to learn aspect-related information, semantic associations, and syntactic structures. Furthermore, we design an adaptive aggregation module to bridge relevant information across three GCN modules. Experiments on three benchmark datasets demonstrate that ACAGCN outperforms existing baselines.

Our contributions are summarized below:

- Compared to the predominant methods for complex graph structure learning in ABSA, our proposed ACAGCN introduces AspGCN, SemGCN, and SynGCN, decomposing global structure learning into multiple localized substructure learning processes.
- To tackle the challenge of irrelevant contextual noise in global sentence semantics, we propose a context-aware semantic learning module that integrates aspect-aware attention and dynamic-mask attention.

- Within the three-graph learning branches, we propose an adaptive aggregation network to enhance interaction and complementarity among different structures.
- We conduct extensive experiments to study the effectiveness of ACAGCN. Experiments on three benchmarks demonstrate ACAGCN outperforms the baselines.

2 Related Work

Aspect-based sentiment analysis is a fine-grained task in sentiment analysis, which is typically framed as a classification problem. Earlier approaches relied on manually defined syntactic rules to predict the sentiment polarity of aspect term.

Recent studies have primarily addressed aspect-based sentiment analysis (ABSA) using neural networks with attention mechanisms. Among these methods, Wang et al. [9] applied an attention mechanism to emphasize different parts of a sentence, generating attention vectors for aspectual sentiment classification. Similarly, Ma et al. [5] proposed an interactive attention mechanism to separately generate representations for aspect terms and their contexts. Fan et al. [3] utilized a multi-granularity attention mechanism to capture word-level interactions. In addition, Devlin et al. [1] proposed a pre-trained language model, BERT, which achieved significant performance in many NLP tasks including ABSA.

Recently, researchers have focused on integrating GCN-based models into ABSA. For example, Liang et al. [4] proposed a neural network architecture that leverages GCNs to learn aspect-specific concerns and inter-aspect relationships. Pang et al. [6] proposed a dynamic multi-channel GCN to model relationships between aspect-terms and contextual words. Huang et al. [10] proposed a multi-channel approach to extract diverse features for sentiment analysis tasks. Lastly, Zhou et al. [11] proposed a global-context collaborative learning approach that combines semantic and syntactic encoders through a pre-fusion module to better capture sentence structure and aspect-term relationships.

Several studies have favored aspect-term-oriented relational models to enhance the performance of ABSA. However, these approaches have generally failed to effectively utilize the aspect-related information, the semantic correlations, and the syntactic structures of sentences—features that may enrich the information the model can learn. Our work extends these approaches by introducing a multi-GCN framework that integrates aspect-aware, semantic, and syntactic features, distinguishing ACAGCN from single-feature GCN-based ABSA models.

3 Proposed ACAGCN

Figure 2 provides an overview of proposed model ACAGCN. In this section, we will provide a detailed introduction to the components of ACAGCN.

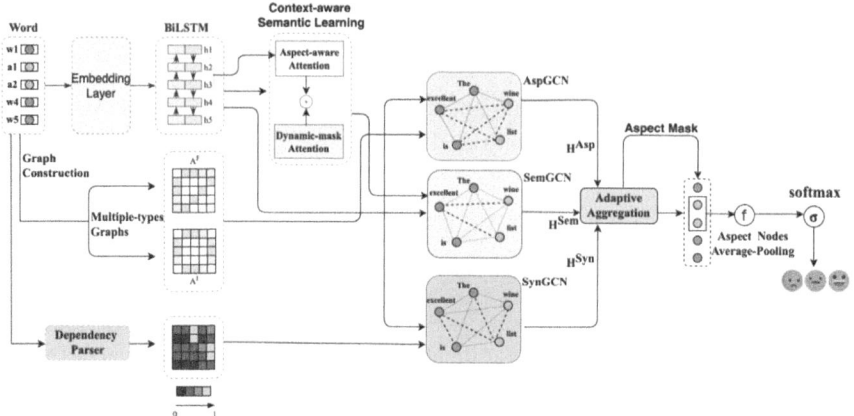

Fig. 2. The overall architecture of ACAGCN, which is composed of Context-aware Semantic Learning, AspGCN, SemGCN, SynGCN and Adaptive Aggregation Modules.

3.1 Embedding and Encoding Layer

Given an input sentence $s = \{w_1, w_2, \cdots, w_1^a, \cdots, w_m^a, \cdots, w_{n-1}, w_n\}$ with n words and an aspect $\{w_1^a, \cdots, w_m^a\}$ of m words, each word is mapped to a vector from the embedding table $E \in \mathbb{R}^{|V| \times d_e}$, where d_e is the embedding dimension. The final sentence representation $x = [x_1, x_2, \cdots, x_1^a, \cdots, x_m^a, \cdots, x_{n-1}, x_n]$ where each x_i is a 300-dimensional word vector. We then input x into a two-direction LSTM to learn the hidden representation of a given sentence. Finally, we obtain the hidden representation H.

3.2 Context-Aware Semantic Learning Module

In this subsection, we combine aspect-aware attention and dynamic-mask attention to obtain better semantic features and remove irrelevant contextual information.

Aspect-Aware Attention. We propose aspect-aware attention mechanism, which regards aspect term as query to attention calculation for learning aspect-related features:

$$A_i^{asp} = \tanh\left(H_a W^a \times \left(KW^k\right)^T + b\right) \tag{1}$$

where K is equal to the H produced by the encoding layer. W^a and W^k are the learnable weights. A_i^{asp} indicates that it is obtained through the i-th attention head.

Dynamic-Mask Attention. We first generate the word-level attention for each sentence by training left and right boundary soft masking matrices $A_l^m, A_r^m \in \mathbb{R}^{n \times n}$, the formulations are calculated as below:

$$A_l^m = \text{Softmax}\left(\frac{QW_L^Q(KW_L^K)^T}{\sqrt{d}} \odot \hat{M}\right) \tag{2}$$

$$A_r^m = \text{Softmax}\left(\frac{QW_R^Q(KW_R^K)^T}{\sqrt{d}} \odot \hat{M}^T\right) \tag{3}$$

$$\hat{M}_{ij} = \begin{cases} 1, & i \geq j \\ -\infty, & i < j \end{cases} \tag{4}$$

where $Q = K = H$, $\odot$ is the element-wise product, and $W_L^Q, W_L^K, W_R^Q, W_R^K \in \mathbb{R}^{d \times d}$ are trainable parameters. The masking matrix M^{dm} can be obtained by compositing the left and right boundary soft masking matrices A_l^m and A_r^m :

$$M^{dm} = (A_l^m L_N) \odot (A_r^m L_N^T) \tag{5}$$

where $L_N \in \{0,1\}^{n \times n}$.

We then combine the masking matrix M^{dm} with the multi-head attention matrix to enable the model to pay more attention to the semantically relevant contextual information around each word:

$$A_i^{dm} = \text{Softmax}\left(\frac{QW^Q(KW^K)^T}{\sqrt{d}} \odot M^{dm}\right) \tag{6}$$

where W^Q, W^K are the trainable parameters, A_i^{dm} indicates that it is obtained through the i-th attention head. Finally, integrating aspect-aware attention scores and dynamic-mask attention scores:

$$A_i^{sem} = A_i^{asp} + A_i^{dm} \tag{7}$$

3.3   AspGCN Module

The dependency tree G of a sentence is represented as an adjacency matrix D of size $n \times n$, where $D_{ij} = 1$ if nodes i and j are connected, and $D_{ij} = 0$ otherwise.

The AspGCN module captures aspect relations within sentences using two input graphs: the aspect-focused graph G^F and the inter-aspect graph G^I. Following previous work [4], we adopt two predefined graph structures: G^F and G^I to enhance aspect-specific sentence representations, as illustrated in the graph construction in Fig. 2. A graph convolutional network (GCN) processes these graphs with hidden representations h, formulated as:

$$h_i^l = ReLU(\sum_{j=1}^{n} D_{ij} W^l h_j^{l-1} + b^l) \tag{8}$$

where h_i^l is the i-th node's hidden representation at l-th layer, W^l is a weight matrix, and b^l is a bias term.

The output of this module can be expressed as follows:

$$H^f = GCN(G^F, h) \qquad (9)$$

$$H^i = GCN(G^I, h) \qquad (10)$$

$$H^{Asp} = h^F + \gamma h^I \qquad (11)$$

where H^f and H^i are the representations from the GCN layers, γ is a tunable parameter, and H^{Asp} is the final output.

3.4   SemGCN Module

Our SemGCN avoids relying on additional syntactic knowledge by using a context-aware semantic learning module, which combines aspect-aware and dynamic-mask attention to obtain the attention matrix as the adjacency matrix. Fused attention captures semantically relevant terms for each word, providing more flexibility than syntactic structures.

From Eq. (7) we obtain the attention score matrix $A^{sem} \in \mathbb{R}^{n \times n}$. To enhance the semantic graph representation, we apply multiple attention heads and average the attention score matrices to form the final semantic graph:

$$A^{sem} = \frac{1}{k} \sum_{i=1}^{k} A_i^{sem} \qquad (12)$$

The SemGCN module uses A^{sem} as the initial node representation. Then the semantic graph representation $H^{Sem} = \{h_1^{Sem}, h_2^{Sem}, \ldots, h_n^{Sem}\}$ is obtained from the SemGCN module. Here, $h_i^{Sem} \in \mathbb{R}^{d_u}$ denotes the hidden representation of the i-th node. The representation of the i-th node in l-th layer is updated as follows:

$$h_i^{Sem^l} = \phi \left(\sum_{j=1}^{n} A_{ij}^{sem} W^l h_j^{Sem^{l-1}} + b^l \right) \qquad (13)$$

where the symbol W^l denotes a weight matrix, and the symbol b^l denotes a bias term. Note that for aspect nodes, we use symbols $\{h_{a_1}^{Sem}, h_{a_2}^{Sem}, \ldots, h_{a_m}^{Sem}\}$ to denote their hidden hidden representations.

3.5   SynGCN Module

We utilize a dependency parser to construct the dependency tree of a sentence, from which we derive the syntactic dependency adjacency matrix A^{syn}. The SynGCN module uses the adjacency matrix $A^{syn} \in \mathbb{R}^{n \times n}$ as the initial node representation in the syntactic graph. Then the syntactic graph representation $H^{Syn} = \{h_1^{Syn}, h_2^{Syn}, \ldots, h_n^{Syn}\}$ is obtained from the SynGCN module. Here, $h_i^{Syn} \in \mathbb{R}^{d_u}$ denotes the hidden representation of the i-th node. The representation of the i-th node in l-th layer is updated as follows:

$$h_i^{Syn^l} = \phi \left(\sum_{j=1}^{n} A_{ij}^{syn} W^l h_j^{Syn^{l-1}} + b^l \right) \qquad (14)$$

where the symbol W^l denotes a weight matrix, and the symbol b^l denotes a bias term.Note that for aspect nodes, we use symbols $\{h_{a_1}^{\mathrm{Syn}}, h_{a_2}^{\mathrm{Syn}}, \ldots, h_{a_m}^{\mathrm{Syn}}\}$ to denote their hidden hidden representations.

3.6   Adaptive Aggregation Module

Considering the complementarity among AspGCN, SemGCN, and SynGCN, we design an adaptive aggregation module. Specifically, each GCN output is processed using a multi-head attention mechanism with 8 attention heads and a dropout rate of 0.1 to balance different structural information. The weighted sum of these outputs is dynamically adjusted using learnable weight $\alpha_1, \alpha_2, \alpha_3, \alpha_4$, which are normalized via softmax. To further explore potential associations between features, the module combines the outputs of AspGCN, SemGCN, and SynGCN using an interactive attention mechanism, resulting in the interactive attention outputs: H^{AspG}, H^{SemG}, and H^{SynG}:

$$O^{Asp-Sem} = LN(MHA(H^{Asp}, H^{Sem}) + H^{Asp}) \tag{15}$$

$$H^{AspG} = LN(FFN(O^{Asp-Sem}) + O^{Asp-Sem}) \tag{16}$$

$$O^{Sem-Syn} = LN(MHA(H^{Sem}, H^{Syn}) + H^{Sem}) \tag{17}$$

$$H^{SemG} = LN(FFN(O^{Sem-Syn}) + O^{Sem-Syn}) \tag{18}$$

$$O^{Syn-Asp} = LN(MHA(H^{Syn}, H^{Asp}) + H^{Syn}) \tag{19}$$

$$H^{SynG} = LN(FFN(O^{Syn-Asp}) + O^{Syn-Asp}) \tag{20}$$

where H^{AspG}, H^{SemG} and $H^{\mathrm{SynG}} \in \mathbb{R}^{n \times d}$, $MHA(\cdot)$ denotes multi-head attention, $LN(\cdot)$ refers to layer normalization, and $FFN(\cdot)$ represents the feed-forward neural network.

To avoid bias towards specific module information, we introduce an additional channel to balance information among AspGCN, SemGCN and SynGCN. The specific approach is as follows:

$$H^{Com} = FFN(Concat([H^{Asp}, H^{Sem}, H^{Syn}])) \tag{21}$$

where $H^{Com} \in \mathbb{R}^{n \times d}$, $Concat(\cdot)$ represents the concatenation function.

Considering the distinct roles of the outputs from different modules, we assign varying weights to these outputs, enabling the model to focus more on the most critical module. Formally, given the input features $X = [X_1, X_2, X_3, X_4]$, the weight for each module is calculated using the following formula:

$$\alpha_i = \mathrm{ReLU}(W^T X_i + b) \tag{22}$$

$$\alpha_i = \frac{exp(a_i)}{\sum_{j=1}^{4} exp(a_j)} \tag{23}$$

where $X_1 = H^{AspG}, X_2 = H^{SemG}, X_3 = H^{SynG}, X_4 = H^{Com}, W$ and b are the trainable parameters. The final output feature H^F is generated as follows:

$$H^F = Concat([\alpha_1 H^{AspG}, \alpha_2 H^{SemG}, \alpha_3 H^{SynG}, \alpha_4 H^{Com}]) \tag{24}$$

where $H^F \in \mathbb{R}^{n \times d_4}$, with $d_4 = 4d$.

3.7   Training

We apply average pooling to the final aspect node of H^F to derive the aspect-related representation H^a. The sentiment probability distribution $y_{(s,a)}$ is then calculated using a linear layer with a softmax function:

$$y_{(s,a)} = \text{softmax}(W^p H^a + b^p) \tag{25}$$

where (s,a) represents the sentence-aspect pair. Our training objective is to minimize the following objective function:

$$\mathcal{L}_T = \mathcal{L}C + \lambda_1 \mathcal{L}\text{agg} + \lambda_2 \|\Theta\|_2 \tag{26}$$

where λ_1 and λ_2 are hyper-parameters, Θ denotes L_2 norm for all trainable parameters of the model, $\mathcal{L}C$ is the standard cross-entropy loss function, and $\mathcal{L}\text{agg}$ is is based on contrastive learning of aggregation loss:

$$\mathcal{L}_C = -\sum_{(s,a)\in D}\sum_{c\in C} \log y_{(s,a)} \tag{27}$$

$$\mathcal{L}\text{agg} = \sum_{i,j\in\{\text{AspG,SemG,SynG}\},i\neq j} \left[\text{CosSim}(H^i, H^j)\right]^2 \tag{28}$$

where D contains all sentence-aspect pairs, C is the collection of different sentiment polarities.

4   Experiments

4.1   Datasets

We conducted aspect-based sentiment analysis experiments on three benchmark datasets, including Restaurant and Laptop reviews from SemEval 2014 Task4 [7], and Twitter [2] reviews. Three sentiment polarities (positive, neutral, and negative) were assigned to each aspect term. Table 1 displays the statistics for the three datasets.

Table 1. Statistics for the three experimental datasets

Dataset	Division	#Positive	#Negative	#Neutral
Restaurant	Train	2164	807	637
	Test	727	196	196
Laptop	Train	976	851	455
	Test	337	128	167
Twitter	Train	1507	1528	3016
	Test	172	169	336

4.2   Baseline Comparisons

We evaluated our proposeded ACAGCN model by contrasting it with the most recent baselines. Below is a summary and grouping of all these models.

Attention-Based Model

1. ATAE-LSTM [9] combines aspect embeddings with word embeddings for all inputs in the sentence, which are then fed into the LSTM. Furthermore, the approach develops an aspect-to-sentence attentional mechanism that may concentrate on the sentence's essential elements based on the aspect term.
2. IAN [5] suggests an interactive attention mechanism that creates representations of aspect term and context, respectively, and learns attention in context and aspect term interactively.
3. MGAN [3] utilizes both fine-grained and coarse-grained attention mechanisms to build its framework. The word-level interactions between aspect terms and their context are captured by the fine-grained attention mechanism.

GNN-Based Model

1. InterGCN [4] explored a novel solution to construct a heterogeneous graph for each instance.Aspect-centered and inter-aspect contextual dependencies are used to build the graph for individual aspect terms.
2. R-GAT [8] proposed a relational graph attention network based on an aspect-oriented dependency tree to enhance the connection between aspect terms and opinion words.
3. MFMCGCN [10] proposed a multi-feature multi-channel graph convolutional network.This method allows us to simultaneously combine several sentence information for sentiment polarity prediction.
4. GCNet [11] proposed a global-context collaborative learning approach, which integrates semantic and syntactic encoding modules. By leveraging a pre-fusion module, the optimized global features are combined with syntactic information to more comprehensively capture sentence structure and aspect-term relationships.

BERT-Based Model

1. R-GAT + BERT [8] is a variant of R-GAT.
2. MFMCGCN + BERT [10] is an MFMCGCN variation.
3. GCNet + BERT [11] is a variant of GCNet.

4.3   Implementation Details

We initialized word embeddings using 300-dimensional pre-trained GloVe vectors. Positional and POS embeddings were set to 30 dimensions, concatenated with word embeddings, and input into a BiLSTM model with a hidden size of 50.

Table 2. Experimental results comparison on three publicly available datasets

Models	Restaurant		Laptop		Twitter	
	Accuracy	Macro-F1	Accuracy	Macro-F1	Accuracy	Macro-F1
ATAE-LSTM [9]	77.20	–	68.70	–	–	–
IAN [5]	78.60	–	72.10	–	–	–
MGAN [3]	81.25	71.94	75.39	72.47	72.54	70.81
InterGCN [4]	82.23	74.01	77.86	74.32	–	–
R-GAT [8]	83.30	76.08	77.42	73.76	75.57	73.82
MFMCGCN [10]	83.99	77.07	78.65	75.19	76.51	75.31
GCNet [11]	83.57	76.79	78.10	74.39	76.34	75.20
Our ACAGCN	**84.43**	**78.47**	**78.98**	**75.29**	**76.81**	**75.90**
R-GAT + BERT [8]	86.60	81.35	78.21	74.07	76.15	74.88
MFMCGCN + BERT [10]	87.13	81.76	82.03	78.86	77.44	76.53
GCNet + BERT [11]	87.08	81.35	80.79	77.61	77.55	76.59
Our ACAGCN + BERT	**87.34**	**81.93**	**82.27**	**79.02**	**77.62**	**76.69**

The BiLSTM input dropout rate was 0.7, and AspGCN, SemGCN, and Syn-GCN modules used 2 layers. Model weights were initialized uniformly, and training employed the Adam optimizer with a 0.002 learning rate. Training was conducted on an NVIDIA RTX 3090 GPU, with a batch size of 16, and 50 epochs. Regularization coefficients $\lambda 1$ and $\lambda 2$ were set as (0.2, 0.3), (0.2, 0.2), and (0.3, 0.2) for the three datasets, respectively. For ACAGCN + BERT, we used BERT-base-uncased.

4.4   Main Results

We evaluate ACAGCN using Accuracy and Macro-F1, comparing it with prior models in Table 2. ACAGCN achieves the best performance across all three datasets. By leveraging aspect-related semantic information, ACAGCN outperforms other attention-based methods. Compared to GCN-based models, ACAGCN integrates local-to-global structural information and aspect-related semantics, demonstrating superior performance. Furthermore, integrating ACAGCN with BERT enhances results, confirming its ability to further improve ABSA performance.

4.5   Ablation Study

We conducted ablation experiments to validate the key components of ACAGCN, as shown in Table 3. Removing the context-aware semantic learning module significantly degrades the model's performance, highlighting the importance of combining local and global semantics in the ABSA task. It also demonstrates the importance of the context-aware semantic learning module. The removal of AspGCN, SemGCN, or SynGCN also results in notable performance declines. These findings confirm that each module is essential to the effectiveness of our model.

Table 3. Experimental results of ablation study

Models	Restaurant		Laptop		Twitter	
	Accuracy	Macro-F1	Accuracy	Macro-F1	Accuracy	Macro-F1
ACAGCN	**84.43**	**78.47**	**78.98**	**75.29**	**76.81**	**75.90**
w/o Context-aware Semantic Learning Module	82.63	75.16	77.11	72.64	75.24	73.74
w/o AspGCN	83.03	75.46	77.21	72.94	75.44	73.84
w/o SemGCN	82.33	74.96	76.98	72.57	75.02	73.81
w/o SynGCN	82.83	75.26	77.31	72.74	75.14	73.34

Table 4. Case study of our ACAGCN model compared with state-of-the-art baselines.

Aspect	Sentence	ATAE-LSTM	IAN	AspGCN	SemGCN	SynGCN	ACAGCN	Label
food, service	Great **food** but the **service** was dreadful!	(N,N)	(N,N)	(P,N)	(P,N)	(P,N)	(P,N)	(P,N)
Windows 8	Biggest complaint is **Windows 8**.	(O)	(O)	(N)	(N)	(N)	(N)	(N)
i5	Not as fact as I would have expect for an **i5**.	(P)	(P)	(O)	(O)	(O)	(N)	(N)

4.6 Case Study

We conducted a case study to evaluate ACAGCN's ability to enhance ABSA by extracting syntactic and semantic information. Table 4 compares ACAGCN with other models on sample sentences. To further evaluate ACAGCN's effectiveness, we introduce additional examples featuring sarcasm, negation, and implicit sentiment. For instance, in the sentence "I was expecting something amazing, but it turned out just fine", ACAGCN correctly detects the neutral sentiment for "something amazing", whereas other models misclassify it as positive. This demonstrates ACAGCN's ability to capture nuanced sentiment expressions.

5 Conclusion

In this paper, we propose the ACAGCN model to address the challenges of irrelevant contextual noise and feature learning in graph-based sentiment analysis. The Context-aware Semantic Learning Module refines global semantics by capturing aspect-related correlations and filtering irrelevant context. To handle the complexity of ABSA, we introduce a multi-branch GCN framework, using AspGCN, SemGCN, and SynGCN to extract aspect-specific, semantic, and syntactic features. The Adaptive Aggregation Module integrates these branches, enabling complementary interactions. Experiments on three benchmark datasets demonstrate that ACAGCN outperforms existing baselines.

References

1. Devlin, J., Chang, M., Lee, K., Toutanova, K.: BERT: pre-training of deep bidirectional transformers for language understanding. In: Burstein, J., Doran, C., Solorio, T. (eds.) Proceedings of the 2019 Conference of the North American Chapter of the Association for Computational Linguistics: Human Language Technologies, NAACL-HLT 2019, Minneapolis, MN, USA, 2–7 June 2019, vol. 1 (Long and Short Papers), pp. 4171–4186. Association for Computational Linguistics (2019). https://doi.org/10.18653/V1/N19-1423

2. Dong, L., Wei, F., Tan, C., Tang, D., Zhou, M., Xu, K.: Adaptive recursive neural network for target-dependent twitter sentiment classification. In: Proceedings of the 52nd Annual Meeting of the Association for Computational Linguistics, ACL 2014, Baltimore, MD, USA, 22–27 June 2014, vol. 2: Short Papers, pp. 49–54. The Association for Computer Linguistics (2014). https://doi.org/10.3115/V1/P14-2009

3. Fan, F., Feng, Y., Zhao, D.: Multi-grained attention network for aspect-level sentiment classification. In: Riloff, E., Chiang, D., Hockenmaier, J., Tsujii, J. (eds.) Proceedings of the 2018 Conference on Empirical Methods in Natural Language Processing, Brussels, Belgium, 31 October–4 November 2018, pp. 3433–3442. Association for Computational Linguistics (2018). https://doi.org/10.18653/V1/D18-1380

4. Liang, B., Yin, R., Gui, L., Du, J., Xu, R.: Jointly learning aspect-focused and inter-aspect relations with graph convolutional networks for aspect sentiment analysis. In: Scott, D., Bel, N., Zong, C. (eds.) Proceedings of the 28th International Conference on Computational Linguistics, COLING 2020, Barcelona, Spain (Online), 8–13 December 2020, pp. 150–161. International Committee on Computational Linguistics (2020). https://doi.org/10.18653/V1/2020.COLING-MAIN.13

5. Ma, D., Li, S., Zhang, X., Wang, H.: Interactive attention networks for aspect-level sentiment classification. CoRR arxiv:1709.00893 (2017)

6. Pang, S., Xue, Y., Yan, Z., Huang, W., Feng, J.: Dynamic and multi-channel graph convolutional networks for aspect-based sentiment analysis. In: Zong, C., Xia, F., Li, W., Navigli, R. (eds.) Findings of the Association for Computational Linguistics: ACL/IJCNLP 2021, Online Event, 1–6 August 2021. Findings of ACL, vol. ACL/IJCNLP 2021, pp. 2627–2636. Association for Computational Linguistics (2021). https://doi.org/10.18653/V1/2021.FINDINGS-ACL.232

7. Pontiki, M., Galanis, D., Pavlopoulos, J., Papageorgiou, H., Androutsopoulos, I., Manandhar, S.: Semeval-2014 task 4: aspect based sentiment analysis. In: Nakov, P., Zesch, T. (eds.) Proceedings of the 8th International Workshop on Semantic Evaluation, SemEval@COLING 2014, Dublin, Ireland, 23–24 AugustD 2014, pp. 27–35. The Association for Computer Linguistics (2014). https://doi.org/10.3115/V1/S14-2004

8. Wang, K., Shen, W., Yang, Y., Quan, X., Wang, R.: Relational graph attention network for aspect-based sentiment analysis. In: Jurafsky, D., Chai, J., Schluter, N., Tetreault, J.R. (eds.) Proceedings of the 58th Annual Meeting of the Association for Computational Linguistics, ACL 2020, Online, 5–10 July 2020, pp. 3229–3238. Association for Computational Linguistics (2020). https://doi.org/10.18653/V1/2020.ACL-MAIN.295

9. Wang, Y., Huang, M., Zhu, X., Zhao, L.: Attention-based LSTM for aspect-level sentiment classification. In: Su, J., Carreras, X., Duh, K. (eds.) Proceedings of the 2016 Conference on Empirical Methods in Natural Language Processing, EMNLP 2016, Austin, Texas, USA, 1–4 November 2016, pp. 606–615. The Association for Computational Linguistics (2016). https://doi.org/10.18653/V1/D16-1058

10. Xi, W., Huang, X., Fukumoto, F., Suzuki, Y.: Multi-feature and multi-channel gcns for aspect based sentiment analysis. In: Strauss, C., Amagasa, T., Kotsis, G., Tjoa, A.M., Khalil, I. (eds.) Database and Expert Systems Applications - 34th International Conference, DEXA 2023, Penang, Malaysia, 28–30 August 2023, Proceedings, Part II. Lecture Notes in Computer Science, vol. 14147, pp. 158–172. Springer, Heidelberg (2023). https://doi.org/10.1007/978-3-031-39821-6_13

11. Zhou, T., Shen, Y., Li, Y.: Gcnet: global-and-context collaborative learning for aspect-based sentiment analysis. In: Calzolari, N., Kan, M., Hoste, V., Lenci, A., Sakti, S., Xue, N. (eds.) Proceedings of the 2024 Joint International Conference on Computational Linguistics, Language Resources and Evaluation, LREC/COLING 2024, Torino, Italy, 20–25 May 2024, pp. 7570–7580. ELRA and ICCL (2024). https://aclanthology.org/2024.lrec-main.669

FedCIPP: A Novel Full-Lifecycle Intellectual Property Protection Framework for Federated Learning

Junjie He[1], Xianyi Chen[1,2(✉)], Jiangfeng Qian[3], Xuebo Wang[1], and Hui Mi[4]

[1] School of Computer Science, Nanjing University of Information Science and Technology, Nanjing 210044, China
xianyi_chen@nuist.edu.cn
[2] Engineering Research Center of Digital Forensics, Ministry of Education, Nanjing University of Information Science and Technology, Nanjing 210044, China
[3] NARI Group Corporation (State Grid Electric Power Research Institute) and NARI Technology Co., Ltd., Nanjing 211106, China
[4] School of Software, Nanjing University of Information Science and Technology, Nanjing 210044, China

Abstract. The protection of Intellectual Property (IP) in Federated Learning (FL) has emerged as a critical issue due to the vulnerability of valuable models to theft or unauthorized distribution within the distributed architecture. However, existing IP protection mechanisms for FL models often significantly increase the difficulty of convergence and typically only protect specific procedures. This paper proposes a novel FL watermarking scheme that minimally impacts the FL model's accuracy and training computational overhead by mapping the watermark into the model parameters. This approach ensures the watermark persists throughout the entire life cycle of FL models, enabling IP verification during each training iteration. Experimental results demonstrate that the proposed scheme is effective in terms of the fidelity, reliability and robustness. Even after fine-tuning, pruning and noisy injection attacks, the watermark detection accuracy remains over 96%.

Keywords: Federated learning · Intellectual property protection · Model watermarking · Artificial intelligence security

1 Introduction

Federated Learning (FL) is a decentralized machine learning framework where training data is retained on local devices (edge devices or clients) rather than centralized, effectively addressing the demands of large training sets while safeguarding data privacy [8]. FL has found widespread application in recommendation systems [1], medical image analysis [2], and image classification tasks [3]. However, the distributed nature of FL renders its Intellectual Property (IP) of trained models vulnerable to infringement by malicious clients, necessitating urgent measures to safeguard copyright of FL models.

In recent years, Deep Neural Network (DNN) watermarking mechanisms have been applied to protect the IP of FL models. The first FL IP protection scheme, WAFFLE [4], designed a center-server protection framework by embedding watermarks into the global model, ensuring that the malicious clients could not steal it. Based on WAFFLE and its center-server framework, various novel methods have emerged utilizing backdoor or parameter modification techniques, such as FedIPR [5], FedTracker [6], and so on [7]. In these approaches, watermarks are predominantly embedded in FL models on the client side. However, these client-side watermark embedding methods have the potential risk of poisoning the client's model and backdoor techniques that increase training computational overload to the extreme.

This paper proposes FedCIPP, a novel FL watermarking method that ensures comprehensive protection of model ownership with minimal additional complexity. It embeds the watermark into the global model by modifying the encrypted parameters after each server aggregation. Furthermore, the embedded watermark is dispersed across the different layers of the FL model, providing continuous protection throughout the FL model's lifecycle, from the initial to the final round of aggregation. This dispersal makes the watermark subtle and difficult to detect or remove. The main contributions of the proposed scheme are summarized as follows:

- This is the first method to provide full lifecycle copyright protection for FL models, which is crucial for protecting the IP of near-convergence models.
- The method embeds the watermarks without additional computational overhead, maintaining consistency with the original FL task, and offering lower algorithmic complexity compared to existing methods.
- Experimental results demonstrate that the proposed method is robust against various attacks, including fine-tuning and pruning.

2   Related Work

2.1   Deep Neural Network Watermark

Deep Neural Network (DNN) watermark techniques can be divided into backdoor methods and parameter modification methods from the implementing mechanism [10].

Backdoor methods use mislabeled image sets from backdoor attacks as watermarks for model copyright identification. In 2018, Adi et al. [11] introduced the first DNN watermarking method using backdoor attacks with random training instances and labels to protect the IP of DNNs. This approach generates a trigger set with public and private keys for ownership verification through the trigger set's error output. However, embedding of backdoor watermarks requires additional computing resources due to the accuracy demands of the original task and trigger set, a disadvantage that is magnified in federated learning.

The parameter modification methods refer to embedding a digital identification into model parameters or structure. In 2017, Uchida et al. [12] inserted the regular terms to control part of model parameters, but this method was vulnerable to fine-tuning attacks due to parameter sensitivity. Then, Fan et al. [13] proposed a robust passport-based scheme for DNN ownership verification, offering improved resistance to removal and ambiguity attacks. Parameter-based watermarking, compared to backdoor techniques, provides greater capacity while maintaining reliability.

2.2 Federated Deep Neural Network Watermarks

The focus of current copyright protection research for FL models is divided into server-side and client-side watermarking approaches.

Server-side watermarking approaches involves embedding watermarks into the model by a trusted server. As seen in WAFFLE [4], the first FL protection method that prevents malicious clients from stealing the global model by incorporating backdoor watermarks. Shao et al. [6] proposed a FL watermark framework named FedTracker, which combining backdoor-based and parameter-based watermarks. FedTracker could verify the ownership of the global model and tracking traitors. In this work, backdoor watermark methods are responsible for model ownership verification and the parameter-based watermarks based on different client information are responsible for model traceability. However, these methods overlook the additional training burden due to the need for server-based watermark training.

The client-side watermark is performed by all clients or a high-privileged client in FL [5,14]. Among these methods, Liu et al. also utilized the backdoor-based watermark method and embed watermarks into the global model on the client side. In 2022, Li et al. [5] put forth a general framework FedIPR, which is the first parameter-based ownership verification of DNN in FL, in which, the watermarks are embedded into Batch Normalization (BN) layers by clients. However, the embedding of the watermarks from the client side is not safe in practical application scenarios , and malicious clients can leak models by removing the watermarks.

3   The Proposed FedCIPP Approach

3.1   Problem Definition

Current methodologies for intellectual property protection fail to consider safeguards for models that have not yet reached full convergence. In practical scenarios, it is possible for models to deliver operational efficacy prior to the completion of their training phase, yet protective measures have not been fully assimilated into the model framework. To better elucidate this scenario, we have constructed a federated learning training involving 30 clients and have displayed the variations in classification accuracy and watermark detection rate in Fig. 1.

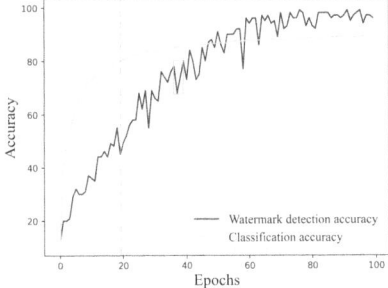

Fig. 1. The changes of watermark detection rate and classification accuracy.

After 20 rounds, the model nears convergence with classification accuracy 8.21% below the final model and a watermark detection rate of 49%, presenting a vulnerability for theft and a loophole in FL copyright protection. This underscores the critical need for comprehensive, life-cycle copyright protection for FL models, though designing such a scheme poses significant challenges as follows:

- The aggregation process in FL may dilute or remove embedded watermarks, as it combines model parameters from multiple clients.
- The increased complexity of watermark embedding can hinder model convergence by introducing additional computational steps during training and aggregation.
- IP protection in FL should be implemented as soon as the model gains value, but existing methods focus on post-training security, leaving intermediate models vulnerable to interception and tampering.
- In distributed FL, the global model is more prone to attacks that target and remove embedded watermarks, compromising model integrity and security.

To address the aforementioned issues, we propose a robust FL watermarking framework that comprehensively mitigates the identified challenges. The proposed method ensures the preservation of watermarks during the aggregation process, minimizes the impact on model convergence, and initiates IP protection at an early stage to safeguard intermediate models from potential risks. Moreover, it enhances the resilience of the global model against attacks targeting watermark removal in distributed environments.

3.2   Watermark Embedding

This section details the specific methods for watermark generation, embedding position selection and mapping.The framework of FedCIPP is illustrated in Fig. 2.

Watermark Generation. The global watermark information is crucial for providing ownership verification of the FL model. In this method, the watermark is generated by combining the watermark content with the corresponding

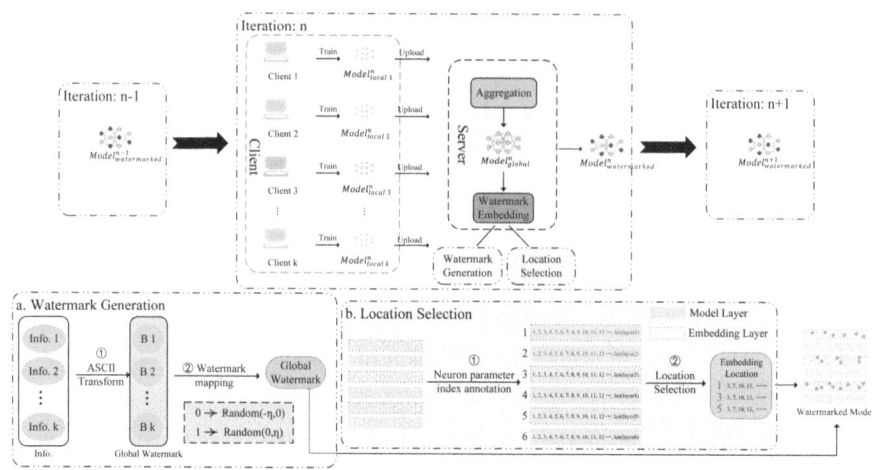

Fig. 2. The overall design of FedCIPP framework.

embedded location. Each client $Client_k(k = 1, 2, \ldots, K)$ provides their owner identification $Info_k$ to the server. The server then converts $Info_k$ into a binary watermark $B_k = \{b_1, b_2, \ldots, b_i\} \in \{0, 1\}^i$ using ASCII encoding. Additionally, the server combines the binary watermark B_k and a random number B_l to create the global watermark FB, where B_l enhances security by making the watermark more robust against attacks. As other clients participate in the FL training process, they can upload unique watermarks to overwrite parts of B_l. The global watermark FB is formulated as the following:

$$FB = \{B_1, B_2, \ldots, B_k, B_l\} \, (k = 1, 2, \ldots, K) \tag{1}$$

The information conversion function employs the ASCII code table for encoding. For example, if each character is represented by 7 bits and a client provides a string of 4 characters, the resulting watermark length is 28 bits. For instance, $Info_k = {}'Alice1'$ can be converted to the binary watermark $B_k = \{1000001, 1101100, 1101001, 1100011, 1100101, 0110001\}$.

Watermark Embedding Position Selection. In contrast to conventional methods, we embed watermarks into various layers of a neural network in a dispersed manner. Initially, the server catalogs the global model's layer information into a selection pool. It then randomly selects n layers and determines an embedding number up to the maximum parameter count per layer. Subsequently, embedding indexes for different layers are generated based on these numbers. Finally, the server embeds the global watermark by substituting model weights at the designated positions.

Although modifying parameter values may initially decrease the performance of the DNN on the main task, the embedding process starts from the initiation stage of the FL. This early embedding allows the neural network to gradually adapt and mitigate the impact of the watermark embedding process, ensuring minimal long-term performance degradation.

In this paper, watermarks persist in FL models throughout training, altering specific model parameters and potentially accelerating neuron activation in the DNN. Consequently, the global model may converge more rapidly with watermarking. Furthermore, the strategic dispersion and camouflage of watermark embedding positions enhance the watermark's robustness and reliability, making it more resilient to detection and removal attempts.

Watermark Mapping. The binary watermarks are converted into disguised watermarks similar to the model parameters before embedding, which enhances robustness and prevents malicious clients from detecting the downloaded model to analysis the possible exception parameters in the extreme case. The watermark information camouflage formula is as follows:

$$W_k = w_1, w_2, \ldots, w_i (k = 1, 2, \ldots, K) \tag{2}$$

$$w_i = \begin{cases} \text{random}(-\gamma, 0), b_i = 0 \\ \text{random}(0, \gamma), b_i = 1 \end{cases} \tag{3}$$

where $random(a, b)$ generates a random number within the range $[a, b]$. Here, W_k represents the camouflaged watermarks for the corresponding $Client_k$, and γ denotes the range parameter for the embedded weights. After the camouflage operation, the global watermark FW can be expressed as follows:

$$FW = W_1, W_2, \ldots, W_k, W_l (k = 1, 2, \ldots, K) \tag{4}$$

The mapping method not only provides robust watermark embedding but also ensures that the watermarks are well-camouflaged within the model parameters, enhancing security and making it difficult for malicious clients to detect and remove them. The strategic dispersion and camouflage of watermark embedding positions further enhance the reliability of the watermark, safeguarding the IP of FL models.

3.3   Watermark Ownership Verification

The ownership of the global model can be verified by extracting the corresponding parameter values from the suspect model and combine them into $FW' = W_1', W_2', \ldots, W_k'$ $(k = 1, 2, \ldots, K)$, where $W_k' = w_1', w_2', \ldots, w_i'$, according to the watermark location information. These parameters are then converted into the corresponding binary watermark information through Eq. 5 and 6.

$$B_k' = \{b_1', b_2', \ldots, b_i'\}, b_i' = \begin{cases} 0, -\gamma < w_i' < 0 \\ 1, 0 < w_i' < \gamma \end{cases} \tag{5}$$

$$FB' = \{B_1', B_2', \ldots, B_k'\}(k = 1, 2, \ldots, K) \tag{6}$$

Here, FB' represents the extracted watermark constructed of binary watermarks B'_k. The Hamming distance is a metric used to measure the difference between two strings of equal length, providing a simple and efficient way to quantify their dissimilarity.

The Watermark Detection Rate (WDR) is defined as follows:

$$WDR = \varphi(FW, FW') = \sum_{i=1}^{n} \delta(W_i, W'_i) \tag{7}$$

where φ is the Hamming distance evaluation function. W_i and W'_i are the symbols (characters) at position i in the embedding global watermark FW and the watermark extracted from the suspicious model FW', respectively. The function $\delta(x, y)$ returns 0 if $x = y$ (indicating no difference between the symbols) and 1 if $x \neq y$ (indicating a difference between the symbols).

4   Experiments

4.1   Experimental Setting

The training and testing datasets used in the experiment are MNIST [15] and CIFAR-10 [16]. Convolution neural network (CNN) [17] and AlexNet [18] are trained by MNIST and CIFAR-10 datasets, respectively. In the simulation of FL, the default number of clients is set to 30. In each iteration, 30% of the clients are randomly selected to train their local models, with each selected client training for 3 epochs using a learning rate of 0.01. The choice to select only 30% of the clients per iteration is made to simulate a realistic FL environment, where not all clients participate in every round of training. We assume a default setting where clients' data is independently and identically distributed (i.i.d.), ensuring a uniform distribution of the training dataset among all clients. We adopted the i.i.d. data partitioning method [9], where the dataset is divided into equally sized portions by default.

4.2   Effectiveness

The purpose of effectiveness is to assess whether we can successfully verify the copyright of the target DNN model within the framework of our Intellectual Property Protection system.

As discussed in Sect. 3.1, the watermarking protection method proposed in this paper theoretically ensures the verification of model ownership throughout each round of the FL training process. In contrast, the backdoor watermark method used in FedIPR fails to protect model copyright in every round. To further demonstrate the effectiveness of our proposed IP protection method in safeguarding FL models, we compare the watermark detection rate with that of methods reported in recent literature. Under the same experimental conditions, we evaluate both model accuracy and watermark detection rate after specific aggregation iterations.

The changing trends of comparison are shown in Fig. 3, FedIPR and Fed-CIPP all have few impact on the classification accuracy of the original model. FedIPR may experience unstable watermark embedding during the training process. FedCIPP can always ensure that the watermark detection rate is 100%, but FedIPR cannot ensure that the watermark can be completely extracted in every round of the training process.

In summary, the proposed protection mechanism FedCIPP makes up for the shortcoming in FedIPR that it cannot verify the ownership of the global model in all rounds.

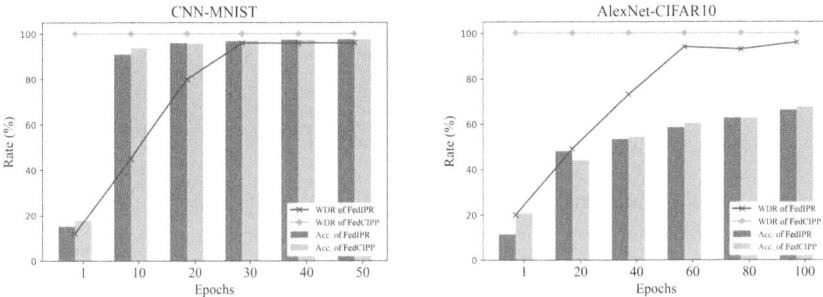

Fig. 3. The watermark detection accuracy(WDR) and Classification accuracy(Acc.) in different communication rounds

4.3 Fidelity and Reliability

Fidelity compares the primary task performance between the watermarked model (FedCIPP) and the unwatermarked model (FedAvg), while reliability measures the watermark detection rate across different settings. This section details the fidelity and reliability outcomes across varied experimental setups: training tasks, client numbers, and watermark lengths.

Influence of Different Numbers of Clients. To investigate the influence of varying numbers of clients on FL models, we evaluated both the classification accuracy and WDR of the original and watermarked models under different client numbers. As shown in Table 1, CNN FedAvg achieved 98.89% accuracy with 10 clients, while FedCIPP showed a minor decrease to 98.76%, with minimal fluctuation as client numbers increased. For AlexNet, FedCIPP consistently performed slightly worse than FedAvg, with accuracy loss under 1% across configurations. Notably, watermark detection remained at 100% for all client numbers (10, 30, 50, and 80), due to FedCIPP's watermark embedding strategy.

In summary, while slight variations in classification accuracy were observed with different client numbers, the watermark detection remained reliable, demonstrating the robustness and reliability of the proposed watermarking method in federated learning environments.

Table 1. The watermark detection accuracy (WDR) and classification accuracy (Acc.) with different client numbers and watermark lengths.

Model	Method	Rate	Number of Clients				Lengths of Watermarks			
			10	30	50	80	30×14	30×42	30×70	30×105
CNN	FedAvg	Acc.	98.89	98.02	97.33	96.22	98.02			
		WDR	N/A				N/A			
	FedCIPP	Acc.	98.76	97.6	96.76	96.2	97.6	95.54	96.8	97.1
		WDR	100	100	100	100	100	100	100	100
AlexNet	FedAvg	Acc.	85.43	79.77	78.63	75.23	79.77			
		WDR	N/A				N/A			
	FedCIPP	Acc.	85.39	79.06	78.39	75.16	79.06	80.02	79.4	79.73
		WDR	100	100	100	100	100	100	100	100

Influence of the Length of Watermarks. This experiment investigates the effect of varying watermark lengths on both watermark detection rate and classification accuracy in federated learning. We tested watermark lengths of 14×30, 42×30, 70×30, and 105×30 bits, embedding them into both CNN and AlexNet models. As shown in Table 1, the WDR remained 100% across all tested watermark lengths, demonstrating the robustness of the proposed mechanism.

Despite the increasing watermark length, accuracy degradation was minimal, and the models maintained high fidelity. Notably, AlexNet accuracy either remained stable or slightly improved, particularly with the 42×30 bit watermark. These results indicate that perfect watermark extraction and validation can be achieved with negligible impact on model accuracy, and even slight accuracy improvements were observed in AlexNet at certain watermark lengths.

4.4   Robustness

Table 2. The watermark detection accuracy (WDR) and classification accuracy (Acc.) with different fine-tuning rounds.

Model	Rate	Fine-tuning Rounds					
		0	10	20	30	40	50
CNN	Acc.	97.60	97.64	97.52	97.61	97.66	97.60
	WDR	100	100	100	100	100	100
AlexNet	Acc.	79.06	74.83	74.92	67.09	72.23	70.60
	WDR	100	100	100	100	99.77	100

Robustness Against Fine-Tuning Attack. Fine-tuning, as discussed in [19], is employed to replicate scenarios where an attacker continues to use identical or proprietary datasets for re-training the model. This approach strategically adapts the model to better align with the attacker's specific task objectives.

The robustness of the watermark-embedded model against fine-tuning attack is detailed in Table 2. In the case of CNN, the classification accuracy retains within an acceptable range after multiple rounds of fine-tuning, the watermark detection rate is still almost 100%. A similar trend is observed with AlexNet. Our IP protection scheme effectively ensures the watermark's durability against fine-tuning attacks. The robustness of the watermark is attributed to the integration of watermarking with the model weights, rendering these weights crucial and inseparable from the model. This integration is fundamental to why the watermark remains undamaged after multiple fine-tuning iterations.

Table 3. The watermark detection accuracy (WDR) and classification accuracy (Acc.) with different pruning rates.

Model	Rate	Pruning Rate					
		0%	10%	20%	30%	40%	50%
CNN	Acc.	97.60	97.62	97.53	97.21	97.04	96.62
	WDR	100	99.78	99.78	99.78	99.78	99.77
AlexNet	Acc.	79.06	77.74	77.99	77.61	77.2	75.59
	WDR	100	99.77	99.32	98.64	98.19	97.52

Robustness Against Pruning Attack. Pruning [20] is used to reduce the size of a trained model by removing certain parameters, such as weights or neurons, without significantly impacting performance. In this section, model pruning is used to simulate that an attacker will prune the trained model to corrupt the watermark while maintaining the performance.

The specific results in Table 3 show that when the pruning coefficient is not greater than 0.5, the influence on the watermark detection accuracy is less than 4%. It can also be clearly concluded from the figure that the larger the pruning coefficient is set, the greater the negative impact on the model accuracy. These results demonstrate that the watermark detection accuracy can be maintained in the face of pruning attacks, which further verifies the robustness of the proposed method in dealing with model weight attacks.

Robustness Against Noise Injection Attack. Injecting random noise into a neural network refers to the process of introducing random noise into the parameters of the network. This experiment simulates an extreme scenario where an adversary, aware of the presence of a watermark in the global model, attempts to disrupt the watermark through noise injection.

The trends in classification accuracy and watermark extraction rates under different noise injection intensities are illustrated in Fig. 4. As shown, in the case of a simple CNN network, while noise injection does lead to a certain degree of degradation in watermark extraction rates, the corresponding classification accuracy also drops to an unusable level, contradicting the adversary's original intent. In contrast, for a more complex network such as ALxNet, it is observed that the impact of noise injection on the watermark extraction rate is negligible. This phenomenon can be attributed to the flexibility in the placement of our watermark, which renders the noise injection ineffective in significantly disrupting the watermark.

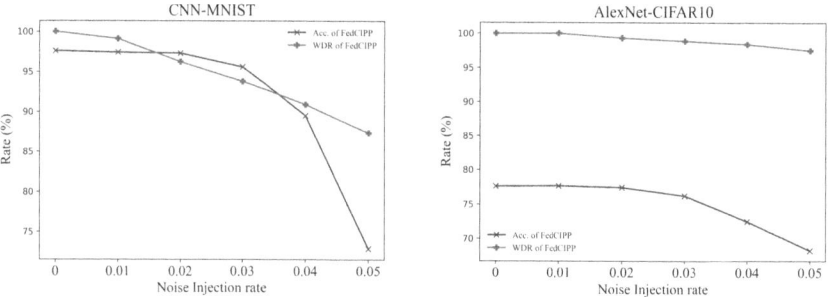

Fig. 4. The watermark detection accuracy(WDR) and Classification accuracy(Acc.) in different noise injection rate

5   Conclusion

This paper proposes a general federated learning watermark mechanism to protect the intellectual property of federated global models with minimal additional computational overhead. The watermark protection approach leverages features of federated learning to ensure the ownership verification of the global model during each aggregation. The global watermarks are embedded in various positions across different layers and camouflaged within the model parameters. Experimental results demonstrate the effectiveness of this approach in terms of effectiveness, fidelity, and robustness while also decreasing the pressure on computational cost. In the future, we will focus on optimizing the watermark mapping method to enhance both the concealment and the ability to trace the malicious clients.

References

1. Guo, J., Zhao, Q., Li, G., Chen, Y., Lao, C., Feng, L.: Decentralized federated learning with privacy-preserving for recommendation systems. Enterp. Inf. Syst. **17**, 2193163 (2023)
2. Yan, Z., Wicaksana, J., Wang, Z., Yang, X., Cheng, K.-T.: Variation-aware federated learning with multi-source decentralized medical image data. IEEE J. Biomed. Health Inf. **25**(7), 2615–2628 (2020)

3. Song, C., Granqvist, F., Talwar, K.: FLAIR: federated learning annotated image repository. In: Advances in Neural Information Processing Systems, pp. 37792–37805 (2022)

4. Tekgul, B.G.A., Xia, Y., Marchal, S., Asokan, N.: WAFFLE: watermarking in federated learning. In: 2021 40th International Symposium on Reliable Distributed Systems (SRDS), pp. 310–320 (2021)

5. Li, B., Fan, L., Gu, H., Li, J., Yang, Q.: FedIPR: ownership verification for federated deep neural network models. IEEE Trans. Pattern Anal. Mach. Intell. **45**(4), 4521–4536 (2022)

6. Shao, S., et al.: FedTracker: furnishing ownership verification and traceability for federated learning model. arXiv preprint arXiv:2211.07160 (2022)

7. Liu, X., Shao, S., Yang, Y., Wu, K., Yang, W., Fang, H.: Secure federated learning model verification: a client-side backdoor triggered watermarking scheme. In: 2021 IEEE International Conference on Systems, Man, and Cybernetics (SMC), pp. 2414–2419. IEEE (2021)

8. Yang, Q.: Federated learning. Synth. Lect. Artif. Intell. Mach. Learn. (2019)

9. McMahan, B., Moore, E., Ramage, D., Hampson, S., Arcas, B.A.: Communication-efficient learning of deep networks from decentralized data. In: Artificial Intelligence and Statistics, pp. 1273–1282. PMLR (2017)

10. Xue, M., Zhang, Y., Wang, J., Liu, W.: Intellectual property protection for deep learning models: taxonomy, methods, attacks, and evaluations. IEEE Trans. Artif. Intell. **3**(6), 908–923 (2021)

11. Adi, Y., Baum, C., Cisse, M., Pinkas, B., Keshet, J.: Turning your weakness into a strength: watermarking deep neural networks by backdooring. In: 27th USENIX Security Symposium (USENIX Security 18), pp. 1615–1631 (2018)

12. Uchida, Y., Nagai, Y., Sakazawa, S., Satoh, S.: Embedding watermarks into deep neural networks. In: Proceedings of the 2017 ACM on International Conference on Multimedia Retrieval, pp. 269–277 (2017)

13. Fan, L., Ng, K.W., Chan, C.S.: Rethinking deep neural network ownership verification: embedding passports to defeat ambiguity attacks. In: Advances in Neural Information Processing Systems, vol. 32 (2019)

14. Yang, W. et al.: Watermarking in secure federated learning: a verification framework based on client-side backdooring. arXiv preprint arXiv:2211.07138 (2022)

15. Deng, L.: The mnist database of handwritten digit images for machine learning research [best of the web]. IEEE Signal Process. Maga. **29**(6), 141–142 (2012)

16. Krizhevsky, A., Hinton, G.: Learning multiple layers of features from tiny images (2009)

17. McMahan, B., Moore, E., Ramage, D., Hampson, S., Aguera y Arcas, B.: Communication-efficient learning of deep networks from decentralized data. In: Proceedings of 2017 International Conference on Artificial Intelligence and Statistics, pp. 1273–1282 (2017)

18. Krizhevsky, A., Sutskever, I., Hinton, G.E.: Imagenet classification with deep convolutional neural networks. In: Advances in Neural Information Processing Systems, vol. 25 (2012)

19. Tajbakhsh, N., et al.: Convolutional neural networks for medical image analysis: full training or fine tuning? IEEE Trans. Med. Imaging **35**(5), 1299–1312 (2016)

20. Han, S., Pool, J., Tran, J., Dally, W.: Learning both weights and connections for efficient neural network. In: Advances in Neural Information Processing Systems, vol. 28 (2015)

Investigating the Limits of Graph Foundation Model in Real-World Travel Recommendation Systems

Nayoung Lee[1], Gunmin Lee[2], and Donghun Lee[1(✉)]

[1] Department of Mathematics, Korea University, Seoul 02841, Korea
{nazero,holy}@korea.ac.kr
[2] Department of Electrical and Computer Engineering, Seoul National University, Seoul 08826, Korea
gunmin.lee@rllab.snu.ac.kr

Abstract. Graph foundation models (GFMs) have demonstrated remarkable potential in capturing intricate relational patterns, achieving state-of-the-art results in numerous graph-centric tasks. However, their real-world applicability remains underexplored in highly domain-specific contexts, such as travel recommendation. In this paper, we present a comprehensive evaluation of GFMs for large-scale travel recommendation tasks using a bipartite user–destination dataset of 86,761 travelers within South Korea. We compare representative GFM against both conventional graph-based methods and vector-based methods. Contrary to the prevailing expectation that GFMs should outperform traditional architectures, our empirical findings reveal that domain-specific constraints can dilute the benefits of extensive multi-hop message passing, leading to suboptimal performance. Our work highlights a critical need to validate GFMs against domain-specific constraints, offering a roadmap for their future adaptation and optimization in real-world applications.

Keywords: Graph foundation model · Travel recommendation · Domain-specific constraints

1 Introduction

Graph neural networks (GNNs) [15] have emerged as a powerful approach for making inferences and predictions on graph-structured data, offering a natural way to process relational information. However, these models often get over-specialized to specific datasets [16], limiting their generalization capabilities across different domains and tasks. To address this limitation, graph foundation models (GFMs) which show better generalization and capabilities in various graph-related tasks, were introduced [13]. These models, trained on diverse graph data, capture universal graph patterns that can be transferred across different domains and applications.

© The Author(s), under exclusive license to Springer Nature Singapore Pte Ltd. 2025
S. Yuan et al. (Eds.): PAKDD 2025 Workshops, LNAI 15835, pp. 174–185, 2025.
https://doi.org/10.1007/978-981-96-8197-6_13

In this work, we test these claims by applying GFM to travel recommendation task. This domain presents a unique challenge where the data is naturally represented as a collection of subgraphs, each corresponding to a real travel trajectory [18]. While this structure might seem ideal for graph-based approaches [3], our comprehensive experiments reveal unexpected limitations of both GNNs and GFM in this context. Surprisingly, our results show that simpler vector-based approaches outperform sophisticated graph-based methods, including GFM. Through careful analysis, we identify two key findings that contribute to the broader understanding of GFM in real world applications:

1. We identify fundamental limitations of GFM in domain-specific applications, particularly in travel recommendation systems. Our analysis reveals that indirect connections in user-location-user patterns act as noise rather than meaningful relationships, demonstrating how domain characteristics can fundamentally limit GFM effectiveness.
2. We empirically validate these limitations through comprehensive experiments with various graph-based approaches. Our results show that traditional GFM performance metrics like hop count and dataset size can inversely correlate with recommendation quality, challenging conventional assumptions about GFM applications in specific domains.

2   Problem Setting

Travel recommendation problem is a task that recommends new users tours based on data of previous users. This problem has advantages over other recommendation tasks when utilizing graph based datasets [4]. We focus on developing a personalized recommendation system for domestic travel destinations in South Korea. We aim to recommend top-k destinations based on user characteristics, including personal preferences, gender, and age group. Furthermore, we extend our approach to address the generalized problem that can be applied to various recommendation problems.

2.1   Dataset Characteristics

As a concrete case study for our approach, we utilize a dataset consisting of travel records of 86,761 users during August to November 2022 who have taken domestic trips in South Korea. The data naturally forms a bipartite graph structure [1], where:

- Users ($\mathbb{U}$) and destinations ($\mathbb{D}$) form two distinct node sets.
- Edges ($\mathbb{E}$) represent visits from users to destinations.

This structure suggests natural affinity for graph-based approaches, as it creates a network of connections between users through their shared destination visits. Following common practice in recommendation systems, these shared destinations could theoretically indicate similarity in personal preferences.

2.2   Problem Formulation

Formally, our input for the travel recommendation problem can be defined as follows:

$$\mathbb{U} = \{u_1, u_2, \ldots, u_n\},$$

$$\mathbb{D} = \{d_1, d_2, \ldots, d_m\},$$

$$V \in \{0,1\}^{n \times m}, \quad \text{where } v_{ij} = \begin{cases} 1 & \text{if user } i \text{ visited destination } j, \\ 0 & \text{otherwise}, \end{cases}$$

where $\mathbb{U}$ is a set of n users with their features X_u including age group, gender, and personal preferences, $\mathbb{D}$ is a set of m destinations, and $\mathbb{V}$ which is a historical visit matrix.

The goal of the problem is to recommend a ranked list of top-k destinations for any given user u_q that best matches personal preferences. We define these "best matches" through a scoring function $f(u_q, d_j)$ that quantifies how well destination $d_j \in \mathbb{D}$ aligns with user preferences, incorporating both demographic attributes and implicit features from past interactions. The function

$$f : \mathbb{U} \times \mathbb{D} \to \mathbb{R}$$

outputs a score indicating the predicted utility of each destination for the user. The recommended destinations are then selected by ranking all destinations based on these scores and choosing the top-k:

$$\{d_1^*, d_2^*, \ldots, d_k^*\} = \arg \underset{d \in \mathbb{D}}{\text{top}} \, k \, f(u_q, d).$$

To evaluate the recommendation results, we employ five evaluation metrics:

-   **Two error-based metrics**: mean squared error (MSE) and mean absolute error (MAE) to measure the accuracy of predictions.
-   **One similarity metric**: cosine similarity (CS) to assess the semantic closeness of recommendations.
-   **Two ranking metrics**: precision@k and recall@k to evaluate the quality of top-k recommendations.

3   Methods

We compare three distinct approaches for travel recommendation: GFM, graph-based models, and vector-based models. The comparison of the three different approaches is shown in Table 1.

Table 1. Comparison of models by information propagation

	GFM	Graph-based	Vector-based
Info. Flow	Unrestricted multi-hop	Limited neighborhood	Direct mapping
Relations	All possible paths	Local connections	Point-to-point mappings
Complexity	Exponential growth	Linear growth	Constant
Example	[17]	[6,7,11]	[2,8,10]

3.1 Graph Foundation Model

The Graph Foundation Model (GFM) approach adapts GraphAny to our travel recommendation task, focusing on the bipartite structure of user-place relationships. The model processes information through dual channels: a low-pass channel and a high-pass channel. The low-pass channel captures smooth, global patterns across the graph by aggregating neighborhood information, while the high-pass channel emphasizes local variations and structural differences between nodes. These channels are defined as:

$$L_k = D^{-1}AX \quad \text{and} \quad H_k = (I - D^{-1}A)X, \tag{1}$$

where L_k represents the low-pass channel output, H_k denotes the high-pass channel output, D is the degree matrix, A is the adjacency matrix, I is the identity matrix, and X represents the node feature matrix. The low-pass channel L_k implements a normalized adjacency operation, effectively averaging features across connected nodes, while the high-pass channel H_k captures the deviation of each node's features from its neighborhood average.

This structure allows for multi-hop information propagation across the graph, though this can potentially lead to noise accumulation through indirect relationships. The model's learning objective is formulated through a combination of reconstruction and link prediction [12] losses:

$$\mathcal{L} = \|X - \hat{X}\|^2 + \lambda(-\sum_{(i,j)} [y_{ij} \log(\hat{y}_{ij}) + (1 - y_{ij}) \log(1 - \hat{y}_{ij})]), \tag{2}$$

where $|X - \hat{X}|^2$ represents the reconstruction loss measuring the model's ability to preserve node features, y_{ij} indicates whether nodes i and j are connected in the original graph, $\hat{y}_{ij}$ is the model's predicted probability of connection between nodes i and j, and λ is a balancing hyperparameter. The reconstruction term ensures feature preservation, while the binary cross-entropy term guides the model to accurately predict graph structure.

3.2 Graph-Based Models

Graph-based models provide a more controlled approach to utilizing network structure. Each graph based model implements distinct strategies for managing information propagation and neighborhood aggregation:

- GCN [11] employ a layer-wise propagation rule:

$$H^{(l+1)} = \sigma(\tilde{D}^{-\frac{1}{2}} \tilde{A} \tilde{D}^{-\frac{1}{2}} H^{(l)} W^{(l)}), \tag{3}$$

where $\tilde{A} = A + I$ is the adjacency matrix with self-loops, $\tilde{D}$ is the corresponding degree matrix, $H^{(l)}$ represents node features at layer l, $W^{(l)}$ is the learnable weight matrix, and σ is a non-linear activation function.

- GraphSAGE [7] employs a neighborhood sampling strategy:

$$h_v^{(l+1)} = \sigma(W^{(l)} \cdot \text{AGGREGATE}(\{h_u^{(l)}, \forall u \in \mathcal{N}(v)\})), \tag{4}$$

where $\mathcal{N}(v)$ is a sampled set of neighbors for node v, and AGGREGATE is a permutation-invariant aggregation function.

- Node2Vec [6] learns embeddings through biased random walks, balancing between breadth-first and depth-first graph exploration:

$$f : V \rightarrow \mathbb{R}^d, \tag{5}$$

where f is the embedding function mapping nodes to a d-dimensional space, optimized to preserve both local and global network structure.

These methods are trained using binary cross-entropy loss for link prediction:

$$\mathcal{L} = -\sum_{(u,p)} [y_{up} \log(\hat{y}_{up}) + (1 - y_{up}) \log(1 - \hat{y}_{up})], \tag{6}$$

where (u, p) represents a user-place pair, y_{up} indicates whether user u has visited place p, and $\hat{y}_{up}$ is the model's predicted probability of user u visiting place p. The loss function encourages the model to accurately predict existing connections while avoiding false positives in the user-place interaction graph.

Unlike GFM's propagation, these graph-based methods provide more controlled information flow, potentially reducing noise from indirect relationships while retaining essential structural information in the user-place interaction network. Moreover, whereas Eq. 2 focuses on user-place features or other scalar-based predictions, Eq. 6 distinctly zeroes in on explicit (u, p) interactions via a binary cross-entropy loss. This emphasis on direct user-place link prediction, whether user u visits place p, offers a clearer objective for learning these interactions compared to the more generalized or feature-based loss used in Eq. 2.

3.3   Vector-Based Models

Vector-based methods take the most direct approach, focusing solely on feature relationships without utilizing graph structure. These methods represent users through demographic information and explicit personal preferences, while places are represented by their location attributes and characteristics. The models employ different architectures:

Autoencoder model uses a deterministic encoder-decoder structure with a bottleneck layer:

$$z = f_\phi(X) \quad \text{and} \quad \hat{X} = g_\theta(z), \tag{7}$$

which is trained with MSE reconstruction loss:

$$\mathcal{L} = \|X - \hat{X}\|^2. \tag{8}$$

VAE variants (VAE and BetaVAE) introduce probabilistic encoding through a variational inference framework:

$$q_\phi(z|X) = \mathcal{N}(\mu_\phi(X), \sigma_\phi(X)) \tag{9}$$

with a combined loss function:

$$\mathcal{L} = \|X - \hat{X}\|^2 + \beta \cdot KL(q_\phi(z|X)\|p(z)), \tag{10}$$

where $\beta = 1$ for VAE and $\beta > 1$ for BetaVAE. These methods serve as an important baseline, demonstrating the effectiveness of direct feature relationships without the complexity of graph structure Fig. 1.

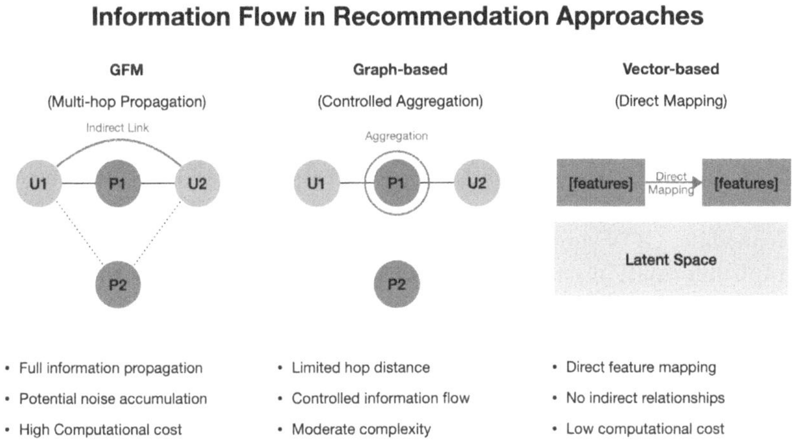

Fig. 1. Information flow analysis in different recommendation approaches

4   Results

In this section, we present a comprehensive analysis of our experimental results, comparing various models across different metrics and settings. We evaluate the performance using MSE, MAE, CS, precision@k, and recall@k for different values of k [9].

Table 2. Model performance comparison ($k = 5$)

Method	Model	MSE (↓)	MAE (↓)	CS (↑)	P@5 (↑)	R@5 (↑)
GFM	GraphAny	20.9747	2.3957	0.9716	0.0008	0.0013
Graph-based	GCN	13.2601	1.9900	0.9813	0.0128	0.0211
	GraphSAGE	16.0332	2.4205	0.9674	0.0001	0.0002
	Node2Vec	20.1005	2.3473	0.9726	0.0009	0.0016
Vector-based	VAE	8.4498	1.7581	0.9826	0.0022	0.0037
	β-VAE	13.5527	2.1140	0.9763	0.0012	0.0022
	AE	**1.0950**	**0.7028**	**0.9938**	**0.0331**	**0.0592**

4.1 Overall Performance Comparison

Table 2 presents overall evaluation results for all models at $k = 5$, which show how the performance compares depending on the methods and the models.

From these results, we observe that the vector-based methods outperform the graph-based methods on most metrics, with AE achieving the best overall performance in terms of all five metrics. This indicates that, in the context of travel recommendations, simple vector representations may provide more reliable embeddings than graph-based approaches. GraphAny, GraphSAGE, and Node2Vec demonstrate comparatively lower ranking performance, suggesting challenges in effectively capturing indirect relationships while minimizing noise [14]. Meanwhile, GCN shows moderate performance, indicating that some localized graph information can be beneficial, although it is still outperformed by AE.

4.2 Impact of Top-k Selection

To understand how different models perform with varying numbers of recommendations, k, we analyze each model's performance across three different k values, $k = 5, 10$, and 20. Table 3 shows the precision and recall metrics for different models, where the number of recommendations is set to 5, 10, and 20.

P The results show that the precision and recall metrics can vary significantly with k. The differential behavior of graph-based and vector-based models as k increases provides insight into how architectural choices affect recommendation quality. GCN exhibits a gradual precision degradation with linear recall growth, suggesting GCN is capturing increasingly distant relationships through indirect user-location-user connections. In contrast, vector-based models, particularly AE, demonstrate superior precision performance, accompanied by diminishing recall growth. While graph-based models continue to find recommendations through indirect paths, potentially incorporating noise from users who visited the same locations for different purposes, vector-based models strictly adhere to direct similarity measures, leading to better precision at lower k values but steeper performance drops when exhausting highly similar recommendations.

Table 3. Precision and recall performance for top-5, top-10, and top-20 similar trips

Method	Model	P@5 (↑)	P@10 (↑)	P@20 (↑)	R@5 (↑)	R@10 (↑)	R@20 (↑)
GFM	GraphAny	0.0008	0.0008	0.0008	0.0013	0.0026	0.0057
Graph-based	GCN	0.0128	0.0121	**0.0118**	0.0211	0.0385	**0.0735**
	GraphSAGE	0.0001	0.0001	0.0001	0.0002	0.0006	0.0011
	Node2Vec	0.0009	0.0008	0.0008	0.0016	0.0028	0.0058
Vector-based	VAE	0.0022	0.0022	0.0019	0.0037	0.0073	0.0127
	β-VAE	0.0012	0.0013	0.0012	0.0022	0.0042	0.0081
	AE	**0.0331**	**0.0180**	0.0100	**0.0592**	**0.0641**	0.0713

4.3   Dataset Size Analysis

We conducted experiments with three different dataset sizes, $|\mathbb{V}|$, across multiple models to analyze the impact of dataset size on model performance. Table 4 presents these results.

Interestingly, we observe that increasing the dataset size does not necessarily lead to better performance across all models. While GFM and graph-based models generally show modest improvements in MSE and MAE metrics with larger datasets, P@5 and R@5 values decrease.

4.4   Impact of Information Propagation Range

To better understand how the range of information propagation affects model performance, we conducted experiments with different hop counts for GCN, and the walk length and context size for Node2Vec. Note that we focus on these two models as they show different mechanisms of information propagation through graph structure, while other models either do not use graph propagation or have fixed propagation patterns. The results are shown in Table 5.

The results show that increasing the propagation range leads to a decrease in recommendation accuracy. For GCN, 2-hop propagation achieves optimal MSE and MAE, but precision and recall metrics consistently decrease as the hop count increases. Node2Vec shows similar behavior, with the shortest range achieving the best performance across all metrics. These findings suggest that while broader information propagation helps minimize overall prediction error, it lowers the model's ability to make precise recommendations.

Discussion. The experimental results give three key insights about embedding approaches in travel recommendation systems.

First, the superior performance of vector-based methods across metrics challenges a fundamental assumption in recommendation systems that edges in graph structures lead to better recommendations. While the consistently lower error rates of AE suggest that user-location interactions may be effectively captured through direct mappings, it is important to note that this advantage might be

Table 4. Model performance across different dataset size ($k = 5$)

Model	$\mid \mathbb{V} \mid$	MSE ($\downarrow$)	MAE ($\downarrow$)	P@5 ($\uparrow$)	R@5 ($\uparrow$)
GraphAny	2000	20.7186	2.3870	0.0089	0.0159
	5000	21.0153	2.4150	0.0034	0.0066
	86761	20.9747	2.3957	0.0008	0.0013
GCN	2000	16.1780	2.1589	0.0292	0.0551
	5000	15.0940	2.1143	0.0183	0.0378
	86761	13.2601	1.9900	0.0128	0.0211
GraphSAGE	2000	28.5383	2.7780	0.0009	0.0019
	5000	18.1254	2.4444	0.0007	0.0017
	86761	16.0332	2.4205	0.0001	0.0002
Node2Vec	2000	22.0694	2.4124	0.0109	0.0196
	5000	22.1293	2.4453	0.0028	0.0059
	86761	20.1005	2.3473	0.0009	0.0016
VAE	2000	8.4498	1.7581	0.0022	0.0037
	5000	6.6342	1.6820	0.0081	0.0161
	86761	6.6815	1.6831	0.0186	0.0367
β-VAE	2000	13.5527	2.1140	0.0012	0.0022
	5000	10.0136	1.9200	0.0070	0.0141
	86761	9.3613	1.8823	0.0123	0.0233
AE	2000	**1.0950**	**0.7028**	**0.0331**	**0.0592**
	5000	**2.1770**	**1.0420**	**0.0204**	**0.0412**
	86761	**2.5165**	**1.1164**	**0.0243**	**0.0478**

partly explained by the training objective of the AE. Specifically, AE-based models minimize a reconstruction error that is closely tied to the evaluation metrics, whereas graph-based models do not directly optimize for the same objective. Therefore, while the performance gap highlights the potential advantages of more direct user-location representations, it also underscores the need to interpret these metrics in light of the models' respective training objectives.

Second, the degradation of graph-based models' performance with increased dataset size and propagation range reveals a structural limitation. The declining precision metrics of GCN and Node2Vec with extended propagation ranges indicate that indirect user-location-user connections may introduce more noise than signal, as users might visit the same locations for different purposes. This observation is supported by GraphSAGE's performance, where broader information sampling appears to amplify this noise rather than extract meaningful patterns.

Third, the inverse relationship between dataset size and recommendation accuracy in graph-based models, contrasted with the stable performance of vector-based approaches, suggests that the complexity of graph structures may not scale effectively with travel data. This pattern indicates that in the travel

Table 5. Performance comparison across different information propagation ranges

Model	Range	MSE (↓)	MAE (↓)	P@5 (↑)	R@5 (↑)
GCN	1-hop	16.1004	2.2242	**0.0443**	**0.0927**
	2-hop	**14.9774**	**2.1022**	0.0190	0.0373
	3-hop	15.2528	2.1256	0.0118	0.0247
Node2Vec	Walk-5/Context-3	**19.5810**	**2.3366**	**0.0148**	**0.0294**
	Walk-10/Context-5	23.2161	2.4858	0.0021	0.0046
	Walk-20/Context-10	22.4065	2.4805	0.0016	0.0032

domain, the quality and directness of user-location relationships may be more crucial than the quantity of indirect connections captured through graph structures. These findings align with the notion that GNN's message-passing mechanisms may become increasingly susceptible to structural noise as graph size grows [5].

5 Conclusion

This study demonstrates that vector-based models can outperform graph-based models in travel recommendation systems, with AE achieving the best MSE of 1.0950. These results suggest that the inherent characteristics of travel data may need additional refinement for graph-structured representations. This study also demonstrated the need for domain-aware loss engineering and GFM architectures that better align with bipartite graphs with the specific demands of travel recommendation problems.

Moreover, this study empirically shows several limitations of GFM. Our experiments show that increasing the number of connections can degrade key metrics like precision and recall. These observations underscore that the benefits of graph-based representations in recommendation systems are highly dependent upon the domain characteristics and the effectiveness of the graph structure in capturing meaningful user-item relationships.

Future work should focus on developing hybrid architectures that can selectively use graph structures while maintaining the effectiveness of direct user-location mappings. Additionally, investigating methods for filtering meaningful connections in travel graphs and developing domain-specific message-passing mechanisms could address the current limitations of graph-based approaches in travel recommendation systems.

Acknowledgments. We would like to acknowledge Bosung Jung, and Gyumin Lee (Nara information) for their valuable contributions to the implementation of the VAE model and the construction of the dataset. Their insights and technical expertise significantly enhanced the quality of this work.

References

1. Asratian, A.S., Denley, T.M., Häggkvist, R.: Bipartite Graphs And Their Applications, vol. 131. Cambridge university press (1998)
2. Bank, D., Koenigstein, N., Giryes, R.: Autoencoders. Machine learning for Data Science Handbook: Data Mining and Knowledge Discovery Handbook, pp. 353–374 (2023)
3. Chen, L., Cao, J., Liang, W., Ye, Q.: Geography-aware heterogeneous graph contrastive learning for travel recommendation. ACM Trans. Spatial Algorithms Syst. **10**(3), 27:1–27:22 (2024)
4. Chen, L., Cao, J., Wang, Y., Liang, W., Zhu, G.: Multi-view graph attention network for travel recommendation. Expert Syst. Appl. **191**, 116234 (2022)
5. Dai, E., Jin, W., Liu, H., Wang, S.: Towards robust graph neural networks for noisy graphs with sparse labels. CoRR abs/2201.00232 (2022)
6. Grover, A., Leskovec, J.: node2vec: scalable feature learning for networks. In: Krishnapuram, B., Shah, M., Smola, A.J., Aggarwal, C.C., Shen, D., Rastogi, R. (eds.) Proceedings of the 22nd ACM SIGKDD International Conference on Knowledge Discovery and Data Mining, San Francisco, CA, USA, August 13-17, 2016, pp. 855–864. ACM (2016)
7. Hamilton, W.L., Ying, Z., Leskovec, J.: Inductive representation learning on large graphs. In: Guyon, I., von Luxburg, U., Bengio, S., Wallach, H.M., Fergus, R., Vishwanathan, S.V.N., Garnett, R. (eds.) Advances in Neural Information Processing Systems 30: Annual Conference on Neural Information Processing Systems 2017, December 4-9, 2017, Long Beach, CA, USA, pp. 1024–1034 (2017)
8. Higgins, I., et al.: Beta-VAE: learning basic visual concepts with a constrained variational framework. In: 5th International Conference on Learning Representations, ICLR 2017, Toulon, France, April 24-26, 2017, Conference Track Proceedings. OpenReview.net (2017)
9. Jadon, A., Patil, A.: A comprehensive survey of evaluation techniques for recommendation systems. CoRR abs/2312.16015 (2023)
10. Kingma, D.P., Welling, M.: Auto-encoding variational bayes. In: Bengio, Y., LeCun, Y. (eds.) 2nd International Conference on Learning Representations, ICLR 2014, Banff, AB, Canada, April 14-16, 2014, Conference Track Proceedings (2014)
11. Kipf, T.N., Welling, M.: Semi-supervised classification with graph convolutional networks. CoRR abs/1609.02907 (2016)
12. Lichtenwalter, R., Lussier, J.T., Chawla, N.V.: New perspectives and methods in link prediction. In: Rao, B., Krishnapuram, B., Tomkins, A., Yang, Q. (eds.) Proceedings of the 16th ACM SIGKDD International Conference on Knowledge Discovery and Data Mining, Washington, DC, USA, July 25-28, 2010, pp. 243–252. ACM (2010)
13. Liu, J., et al.: Towards graph foundation models: a survey and beyond. CoRR abs/2310.11829 (2023)
14. Mostafa, H., Nassar, M., Majumdar, S.: On local aggregation in heterophilic graphs. CoRR abs/2106.03213 (2021)
15. Scarselli, F., Gori, M., Tsoi, A.C., Hagenbuchner, M., Monfardini, G.: The graph neural network model. IEEE Trans. Neural Netw. **20**(1), 61–80 (2009)
16. Wu, S., Zhang, W., Sun, F., Cui, B.: Graph neural networks in recommender systems: a survey. CoRR abs/2011.02260 (2020)

17. Zhao, J., Zhu, Z., Galkin, M., Mostafa, H., Bronstein, M., Tang, J.: Fully-inductive node classification on arbitrary graphs. In: International Conference on Learning Representations (ICLR) (2025)
18. Zheng, Y.: Trajectory data mining: an overview. ACM Trans. Intell. Syst. Technol. **6**(3), 29:1–29:41 (2015)

Workshop on Pattern Mining and Machine Learning for Bioinformatics (PM4B 2025)

Preface of PM4B 2025

The 1st Workshop on Pattern Mining and Machine Learning for Bioinformatics (PM4B 2025) was held at the PAKDD 2025 conference on June 10th, 2025. in Sydney, Australia. The workshop focused on the development and application of methods from pattern mining (PM) and machine learning (ML) in bioinformatics.

Bioinformatics is a rapidly advancing interdisciplinary field that involves principles from biology, computer science, and mathematics to analyze and interpret complex biological data. In recent years, the integration of PM and ML has emerged as a powerful approach to tackle challenges posed by massive volumes of complex biological data. PM techniques enable the identification of patterns and associations that provide insights into biological phenomena such as gene regulation and mutation detection. ML methods have demonstrated remarkable success in predicting disease outcomes, analyzing genomic sequences, and supporting personalized medicine.

The objective of PM4B 2025 was to provide a forum for the presentation and discussion of novel research on PM and ML in bioinformatics. Following an open call for papers, the workshop received 14 paper submissions. Each paper was evaluated under a rigorous single-blind review process by 3 reviewers based on criteria such as technical quality, originality, clarity, and significance. As a result of this evaluation, 6 papers were accepted for presentation at the workshop and inclusion in these proceedings. The topics of these papers include: genomic variant classification, genomic data clustering, Alzheimer's disease diagnosis and progression detection, blood glucose forecasting, and feature extraction for the analysis of drug-target interactions

We believe that the contributions presented in this workshop highlight significant advancements in the application of PM and ML techniques in bioinformatics and serve as inspiration to researchers in the field.

We would like to thank the authors, reviewers, and participants whose contributions have been indispensable to the success of the workshop. Moreover, we extend our gratitude to the PAKDD Program Committee, organizers, and workshop chairs for their invaluable support. The workshop website is: https://www.philippe-fournier-viger.com/PM4B_2025/.

March 2025

Philippe Fournier-Viger
M. Saqib Nawaz
Ji Zhang
Yun Sing Koh

PM4B 2025 Organization

Organizers

Philippe Fournier-Viger	Shenzhen University, China
M. Saqib Nawaz	Shenzhen University, China
Ji Zhang	University of Southern Queensland, Australia
Yun Sing Koh	University of Auckland, New Zealand

Program Committee

Naji Al Husaini	University of Science and Technology of China, China
Longbing Cao	University of Technology Sydney, Australia
Guoting Chen	Great Bay University, China
Tin Truong Chi	University of Dalat, Vietnam
Wensheng Gan	Jinan University, China
M. Zohaib Nawaz	Shenzhen University, China
Farid Nouioua	University of Bordj Bou Arreridj, Algeria
Mourad Nouioua	Harbin Institute of Technology, China
Muhammad Sadiq	Shenzhen University, China
Wei Song	North China University of Technology, China
Bay Vo	Ho Chi Minh City University of Technology, Vietnam
Unil Yun	Sejong University, South Korea
Jimmy Ming-Tai Wu	Shandong University of Science and Technology, China
Cheng-Wei Wu	National Ilan University, Taiwan
Youxi Wu	Hebei University of Technology, China

mHMG-DTI: A Drug-Target Interaction Prediction Framework Combining Modified Hierarchical Molecular Graphs and Improved Convolutional Block Attention Module

Zerui Yang[1,3], Yinqiao Li[2], Yudai Matsuda[1], and Linqi Song[2,3]($\boxtimes$)

[1] Department of Chemistry, City University of Hong Kong, HongKong, China
zeruiyang2-c@my.cityu.edu.hk
[2] Department of Computer Science, City University of Hong Kong, HongKong, China
linqi.song@cityu.edu.hk
[3] City University of Hong Kong Shenzhen Research Institute, Shenzhen, China

Abstract. Drug-target interactions (DTIs) are fundamental to understanding the therapeutic mechanisms of drugs, yet accurately predicting these interactions remains a significant challenge in drug discovery. Current computational approaches often fail to capture essential molecular motifs and spatial information of proteins, limiting their effectiveness, particularly when encountering proteins or compounds absent in the training datasets. To address these limitations, we propose mHMG-DTI, a novel framework that leverages an improved Convolutional Block Attention Module (iCBAM) for enhanced protein feature extraction and modified Hierarchical Molecular Graphs (mHMGs) for comprehensive molecular encoding. This hierarchical approach not only captures detailed local structures and broader connectivity patterns but also incorporates guiding knowledge to improve feature representation. Across a total of 16 experimental evaluations on four benchmark datasets spanning both classification and regression tasks, mHMG-DTI surpasses existing baseline models in 11 cases. These results highlight the potential of mHMG-DTI to enhance DTI prediction accuracy, thereby accelerating the drug discovery process and providing valuable insights into drug resistance and side effect mechanisms.

1 Introduction

Drug-target interaction (DTI) studies are vital for understanding drug mechanisms, optimizing drug discovery, and mitigating side effects. While experimental techniques, such as X-ray crystallography [4] provide insights, they are labor-intensive and costly. Consequently, deep learning (DL) has emerged as a scalable and efficient alternative for DTI prediction, leveraging neural networks to capture complex, non-linear relationships within large datasets [21].

S. Yuan et al. (Eds.): PAKDD 2025 Workshops, LNAI 15835, pp. 191–202, 2025.
https://doi.org/10.1007/978-981-96-8197-6_14

In these DL approaches, three primary molecular encoding strategies have emerged: 1D-CNNs [21], self-attention mechanisms [11], and graph neural networks (GNNs) [13]. However, these methods often struggle to capture key molecular features (e.g., motifs, rings, functional groups) and long-range atomic interactions. Many also neglect prior chemical knowledge, limiting predictive performance. For instance, DeepConv-DTI relies on Morgan fingerprints without considering molecular topology, whereas Perceiver CPI [12] integrates molecular descriptors and graph structures via cross-attention for a more comprehensive representation.

Protein representation remains a contentious issue, with sequence-based models [21] offering computational efficiency but lacking structural insights, while graph-based models [12]. enhance interpretability at the cost of increased complexity. The saCNN model [7] attempts to reconcile these differences by incorporating a Convolutional Block Attention Module (CBAM) for spatial feature extraction. Unlike prior studies that removed CBAM's channel attention due to performance concerns, our work demonstrates its effectiveness in enhancing protein feature learning and model robustness.

To address these challenges, we propose mHMG-DTI (Drug-Target Interaction prediction with modified Hierarchical Molecular Graphs), a novel framework inspired by the Hierarchical Molecular Graphs (HMGs) introduced by [19] for molecular encoding. HMGs are specifically designed to capture substructural motifs through a hierarchical representation. While a similar approach has been explored by [8], their work formulates DTI prediction solely as a classification task, lacking comprehensive analysis. Additionally, their method largely replicates the original framework, initializing the global level node from scratch, which, as our experiments demonstrate, can be further optimized.

Building upon the foundational HMG framework, we introduce a modified representation by incorporating guiding knowledge into the graph structure, thereby improving the extraction of critical molecular information. For protein encoding, our model integrates recent advances in attention mechanisms [3] and employs an improved Convolutional Block Attention Module (iCBAM) to optimize feature extraction across both channel and spatial dimensions. The information exchange between molecular and protein representations is facilitated through two cross-attention modules, enabling enhanced interaction learning. Furthermore, we show that the incorporation of large biological models significantly boosts the overall predictive performance of the proposed framework. The key contributions of this work are as follows:

1. To the best of our knowledge, our model represents the first application of Hierarchical Molecular Graphs for molecular representation in DTI prediction of both classification and regression task, effectively capturing molecular motifs.
2. We extend the original HMG framework by embedding molecular descriptors as auxiliary information, facilitating seamless integration with graph neural networks (GNNs) without the need for additional network architectures.

3. We introduce a novel application of iCBAM for protein representation in the context of DTI prediction, achieving improved feature extraction and model performance.

Our experimental results demonstrate that mHMG-DTI outperforms baseline models across multiple benchmark datasets, setting a new standard in DTI prediction. This work provides a comprehensive and accurate framework for addressing limitations in current methodologies, offering promising implications for drug discovery. Due to space constraints in the main manuscript, we have provided comprehensive Supplementary Information, which is available at https://github.com/yaoge777/mHMG-DTI. Additionally, the complete source code for this project will be made publicly accessible on the same repository upon publication.

Table 1. Statistic of the datasets of the classification task

Dataset	Positive interaction	Negative interaction	Total
BindingDB	33772	27486	61258
DrugBank	17511	17511	35022

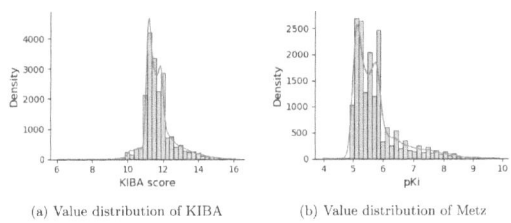

(a) Value distribution of KIBA   (b) Value distribution of Metz

Fig. 1. Statistics of the datasets of the regression task.

2 Materials and Method

2.1 Datasets

In this study, we formulate DTI prediction as both a classification and a regression task. For the regression task, we utilize the KIBA [14] and Metz [10] datasets, while for the classification task, we employ BindingDB [1] and DrugBank [16]. The detailed information on the datasets can be found in Table 1, Fig. 1, and Supplementary Figures S1 and S2.

2.2 Model Architecture

The proposed model architecture (Fig. 2) integrates molecular and protein representations by combining GNNs, convolutional attention mechanisms, and cross-attention blocks, while leveraging pre-trained large model (PLM) embeddings for both molecules and proteins. The GNN module, which processes molecular data, includes two input embedding layers to encode atom types and degrees, as well as two additional embedding layers for bond types and bond directions. Molecular data is propagated through a series of Graph Convolutional Networks (GCN) to capture the structural information of the molecules. To enhance the expressiveness of the graph, self-loops are incorporated, ensuring that each node is connected to itself. Additionally, edge attributes are included to enrich the molecular graph, which is subsequently utilized to update the molecular embeddings.

For processing 2D protein structures, the model employs a ResNet [2] layer comprising several convolutional layers with skip connections, which preserve learned information across layers. Following the convolutional layers, Spatial Attention and Channel Attention mechanisms are applied in parallel to selectively emphasize significant spatial and channel-wise features within the protein data. These attention modules dynamically adjust the model's focus, thereby improving its capacity to capture critical features.

To facilitate the exchange of information between molecular and protein features, a multi-head cross-attention mechanism [15] is employed. The cross-attention is computed bidirectionally: molecules are treated as queries (Q) and proteins as keys (K) and values (V), and vice versa. Residual connections are added to enhance information flow and stability during learning.

After processing through their respective modules, the molecular and protein representations undergo max-pooling and are concatenated with their corresponding pre-trained LLM encodings. These concatenated representations are then passed through linear projection layers for dimensionality reduction. Finally, the molecular and protein representations are combined and fed into the output module to generate the final predictions.

2.3 Hierarchical Molecular Graph

In this study, we employ a Hierarchical Molecular Graph [19] to encode the molecule more comprehensively. To be detailed, the graph of the molecule is constructed as $G = (V_{atoms}, E_a)$ according to its chemical structure, where V_{atoms} denotes the atoms, and E_a denotes the chemical bonds. Using the BRICS algorithm [9], the molecule is first broken down into various motifs, while the remaining skeleton is further decomposed into individual rings. Both motifs and rings are then represented by augmented motif nodes $V_{motifs} = \{V_m^1, V_m^2, \ldots V_m^k\}$, which connect to their respective atom nodes V_{atoms} forming a new set of bonds E_m. Ultimately, an augmented graph node V_{graph} is created and linked to all motif nodes V_{motifs}, forming the bonds E_g. All augmented nodes and bonds are integrated into the original graph to form the augmented graph $G_{aug} = (V, E)$,

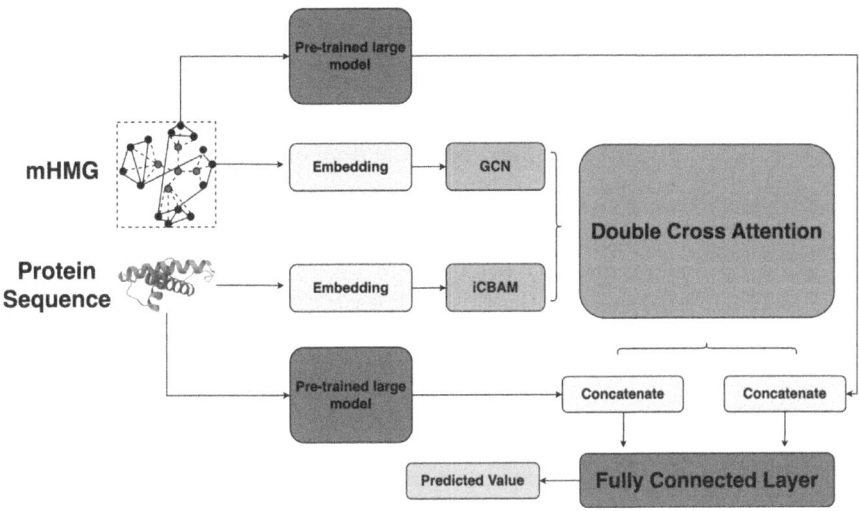

Fig. 2. Structure of mHMG-DTI.

where $V = \{V_{atoms}, V_{motifs}, V_{graph}\}$ and $E = \{E_a, E_m, E_g\}$. Each V_{atoms} has two properties, atomic number and degree, while each Ea also has two properties, bond type and a binary value for if it is in a ring. All augmented nodes and edges are assigned the corresponding pseudo values, respectively.

Based on the original algorithm, we make one modification by substituting the value of graph node V_{graph} (after embedding layer) with Morgan fingerprint as the guiding knowledge [6] instead of learning from scratch. This hierarchical approach allows for a richer and more informative representation of molecules, capturing both the detailed local structures and the broader connectivity patterns that are critical for understanding molecular properties and behaviors. In addition, it incorporates the molecular fingerprints, which contain rich chemical information without additional computation, such as cross-attention [12], enhancing the computational efficiency of the model.

2.4 ResNet

The core innovation of ResNet lies in its use of skip connections, also known as residual connections, which allow information to bypass one or more layers. This design facilitates the training of very deep networks by providing a mechanism through which gradients can flow more easily back through the network during backpropagation. The basic building block of a ResNet is the residual block, which typically consists of two parts:

1. Identity Mapping (Skip Connection): An identity mapping is introduced where the output of a layer is added to the output of a previous layer. This can be represented mathematically as $H(x) = F(x) + x$, where $H(x)$ is the

desired underlying mapping to be learned, and $F(x)$ is the learned residual mapping.
2. Convolutional Layers: These layers perform the standard convolution operations on the input data. They learn the necessary features that, when combined with the identity mapping, form the final output of the block.

ResNet has been intensively used for the recognition of the pattern of biological molecules [18]. However, in most cases, 1D convolutional layers were applied by treating the biological molecules as plain text without any changes in the channel number. In our model, the ResNet layer is constructed by stacking three of these residual blocks sequentially with 2D convolutional layers through which the channel number of protein input data expands from one to sixteen to capture more abstract features, which we consider as augmented spatial features.

2.5   iCBAM

The Channel Attention (CA) and Spatial Attention (SA) modules are two essential components in iCBAM, designed to enhance feature representation by adaptively focusing on informative channels and salient spatial regions. Given a feature map $X \in R^{CHW}$, where C denotes the number of channels, and H, W represent spatial dimensions, the CA module first extracts global contextual information via GAP and GMP, yielding a channel descriptor $z \in R^C$. This descriptor is processed through FC layers with a bottleneck structure, reducing dimensionality to C/r before expansion back to C, with a non-linear activation function (e.g., ReLU) applied between layers. A sigmoid function normalizes the output to produce a channel attention map $M_c \in R^C$, which is then multiplied element-wise with X to obtain the refined feature map X'.

The SA module enhances spatial feature extraction by emphasizing critical regions within X. It first applies a convolution operation followed by batch normalization to stabilize learning. Spatial context is captured via GAP and GMP along the spatial dimensions, generating feature statistics that are concatenated along the channel axis and processed by a convolutional layer (e.g., 7×7 kernel). A sigmoid activation function normalizes the resulting spatial attention map $M_s \in R^{H*W}$, which is subsequently multiplied element-wise with X to amplify salient features while suppressing less relevant ones. Together, CA and SA modules dynamically recalibrate feature maps, improving the network's discriminative capability.

2.6   Pretrained Large Biological Model

Recent years have witnessed remarkable advances in biomolecular modeling, with an increasing number of large-scale pretrained models emerging to accelerate drug discovery research [17,20,22]. In this work, for molecular representation, we employed 'PubChem10M_SMILES_BPE', a pretrained model developed using 10 million SMILES strings from the PubChem database. For protein sequence encoding, we implemented ESM-2, an advanced protein language model pre-trained on approximately 65 million UniRef50 protein sequences through Masked Language Modeling (MLM). This approach effectively captures evolutionary relationships and structural dependencies .

2.7 Cross-Attention

Cross-attention is a mechanism in neural networks, particularly in transformer models, that allows one sequence (query) to focus on relevant parts of another sequence (key and value). The core idea is to compute attention weights between the query sequence and the key-value pair, which then help in producing contextually aware representations. The mechanism is mathematically described using the attention function, typically expressed as:

$$Attention(Q, K, V) = SoftMax(QK^T / \sqrt{d_k})V.$$

The term QK^T computes the dot product between each query and all keys, producing an alignment score for each key with respect to the query. These alignment scores are then normalized using the softmax function, which transforms the scores into attention weights that sum to one. The output of the attention mechanism is a weighted sum of the values V, where the weights are derived from the softmax-normalized alignment scores. The result is a context-aware representation that highlights the most relevant parts of the input based on the query.

3 Results

3.1 Experimental Setup

We selected several baseline models originally designed for classification or regression tasks. The output layers of these models were modified to fit the specific task requirements, while the remaining architecture was left unchanged.

A five-fold cross-validation procedure was employed, using an 80% training and 20% test data split. In line with prior research, we defined four distinct experimental settings for partitioning the training and test datasets:

E1: All proteins and drugs in the test set are also present in the training set.

E2: All proteins in the test set are included in the training set, but none of the test set drugs are in the training set.

E3: All drugs in the test set are part of the training set, but no proteins in the test set are present in the training set.

E4: Neither proteins nor drugs in the test set appear in the training set.

Table 2. Performance comparison of different models on DrugBank (Accuracy) and Metz (MSE) datasets across four evaluation settings (E1-E4)

Model	DrugBank (ACC/%)				Metz (MSE)			
	E1	E2	E3	E4	E1	E2	E3	E4
GraphDTA	78.7	68.9	70.0	57.6	0.425	0.404	0.577	0.704
DeepConv-DTI	78.4	70.7	71.4	61.5	0.344	0.307	0.441	0.660
HyperAttention-DTI	86.7	71.9	74.2	60.5	0.231	0.244	0.348	0.570
Perceiver CPI	82.2	74.0	71.7	64.4	0.222	0.230	0.399	0.555
mHMG-DTI	86.7	77.8	74.4	64.6	0.186	0.252	0.452	0.541

3.2   Evaluation Metrics

Our tasks involve both classification and regression objectives. For the classification task, we employed several evaluation metrics, including accuracy (ACC), precision (PRE), recall, F1-score, Matthews correlation coefficient (MCC), the area under the receiver operating characteristic (ROC) curve, and the area under the precision-recall (PR) curve. For the regression task, the mean squared error (MSE) and the concordance index (CI) were used to assess performance. All experiments were replicated five times, and the average values were reported to ensure robustness. The comparative performance of our model and baseline models across different scenarios is summarized in Table 2. Overall, our model achieved the best performance in BindingDB dataset under E1-3, in DrugDB dataset under E1-4, in KIBA dataset under E1, E2, E4, in the Metz dataset under E1, E4.

Below, we provide a detailed summary and comparison of our model with other state-of-the-art models, including GraphDTA [12], DeepConv-DTI [5], HyperAttentionDTI [21], and Perceiver CPI [12]. Notably, many of the results obtained in our study closely align with those reported in the original works (e.g., HyperAttentionDTI on the DrugBank dataset under E2 and E3), while some even surpass the original performance (e.g., Perceiver CPI on the Metz dataset under E4).

3.3   Experimental Setting Summaries and Model Comparisons

In E1, where all test proteins and drugs appear in the training set, our model consistently outperforms baselines. For example, in BindingDB, it attains 0.942 accuracy, surpassing GraphDTA (0.833) and DeepConv-DTI (0.859), and closely matching HyperAttentionDTI (0.921), demonstrating strong learning from seen interactions.

In E2, where test proteins are in training but test drugs are novel, our model maintains robust performance, excelling in precision and F1 score. It achieves 0.947 accuracy in BindingDB, outperforming Perceiver CPI (0.935) and Hyper-AttentionDTI (0.924), indicating strong generalization to unseen drugs when protein data is available.

In E3, where test drugs are in training but test proteins are novel, our model exhibits a moderate performance drop but still surpasses baselines. In DrugBank, it achieves 0.744 accuracy, showing better generalization to unseen proteins when drug data is familiar.

In E4, where both test proteins and drugs are unseen, the most challenging scenario, our model maintains competitive performance. In KIBA, it achieves an MSE of 0.551 and CI of 0.634, outperforming DeepConv-DTI (MSE: 0.553, CI: 0.626) and Perceiver CPI (MSE: 0.554, CI: 0.612).

Across all settings, our model consistently outperforms baselines in E1 and shows strong results in E2 and E4, attributed to our mHMG molecular representation, which integrates molecular motifs and guiding knowledge for improved generalization. While Perceiver CPI also combines molecular graphs and Morgan fingerprints, it performs slightly worse in E2. In E3, our model experiences a minor decline but remains competitive, particularly in DrugBank and BindingDB. These results (Table 2, Supplementary Table S1-S16) underscore our model's adaptability and robustness.

Table 3. Ablation studies under different evaluation settings.

Setting	Models	BindingDB	DrugBank	KIBA	Metz
E1	mHMG-DTI	0.942	0.867	0.361	0.186
	HMG	0.936	0.865	0.362	0.190
	MG	0.940	0.861	0.362	0.204
	Spatial attention	0.936	0.864	0.373	0.192
	w/o_LME_M	0.931	0.858	0.410	0.234
	w/o_LME_P	0.924	0.850	0.395	0.228
E2	mHMG-DTI	0.947	0.778	0.471	0.252
	HMG	0.941	0.776	0.457	0.274
	MG	0.940	0.775	0.493	0.285
	w/o_LME_M	0.933	0.775	0.511	0.235
E3	mHMG-DTI	0.897	0.744	0.410	0.452
	Spatial attention	0.889	0.741	0.423	0.443
	w/o_LME_P	0.877	0.724	0.411	0.384

3.4 Ablation Studies

To evaluate the contributions of mHMG, iCBAM, and LME molecular and protein representations, we conducted a series of ablation studies (Table 3). These experiments were structured as follows: (1) comparing mHMG with standard MG and the original HMG, (2) assessing iCBAM relative to SA, (3) analyzing the impact of excluding LME molecules, and (4) analyzing the impact of excluding LME proteins. These studies comprehensively evaluate each component's contribution to overall performance.

Since mHMG, HMG, MG, and LME molecules primarily affect molecular feature extraction, their ablation experiments were conducted under E1 and E2. Similarly, ablations for iCBAM and LME proteins were performed under E1 and E3. Results demonstrated that our model outperformed controls in most cases, except in KIBA (E2) and Metz (E2, E3). Notably, LME proteins and molecules were more effective in E1, where both had been observed during training. Furthermore, LME molecules and proteins contributed more to classification than regression, with LME proteins having a substantially larger impact than LME molecules.

The contribution of mHMG was significantly greater than iCBAM, as expected, given that mHMG integrates precomputed structural information and guiding knowledge, whereas iCBAM lacks additional protein-specific information. Compared to MG, HMG generally exhibited better performance except for BindingDB (E1). In contrast to LME molecules and proteins, HMG and mHMG had a higher impact on regression tasks, which require finer granularity, highlighting their effectiveness in handling complex tasks.

4   Analysis

We visualized the SA convolutional kernels and calculated variance within each kernel (Supplementary Figs. 3–6), revealing intriguing patterns. Kernels associated with GAP channels exhibited lower variance than those associated with GMP channels due to GAP's smoothing effect, which emphasizes global consistency. In contrast, GMP channels, which capture extreme activations, displayed higher variance, remaining more sensitive to spatial irregularities.

Across experiments, variance differences were observed. In E1, where test proteins appeared in training, kernel variance was relatively higher (except for the GAP channel in Metz), reflecting the model's ability to develop diverse feature representations for familiar data. Conversely, in E3, where test proteins were unseen, kernel variance was lower, indicating a more conservative response to OOD proteins. This suggests a shift in SA in E3, where the model relies more on global patterns rather than local features, highlighting a generalization gap.

These findings emphasize the need to enhance kernel adaptability and feature diversity to improve SA robustness, particularly in OOD scenarios. Increasing adaptability in kernel activations could mitigate the generalization gap and enhance model performance on unseen data.

5   Conclusion

The study of DTIs is critical for drug development. While numerous DL-based approaches have been proposed, many suffer from two major limitations: neglecting molecular motifs and lacking protein spatial information. In this study, we address these challenges by employing mHMG for molecular encoding, integrating motif features, and utilizing iCBAM to enhance protein feature extraction.

Although prior research suggested that CA in CBAM could hinder performance, our findings indicate that modifying the sequential SA-CA computation into a parallel configuration within iCBAM improves model robustness and significantly enhances overall performance. To comprehensively evaluate our model, we designed four experimental settings across four datasets, covering classification and regression tasks. Comparative analysis with baselines demonstrated that mHMG-iCBAM outperformed baselines in 11 of 16 settings. Additionally, ablation studies underscored the importance of both mHMG and iCBAM in enhancing model performance. We believe our proposed framework provides valuable insights for DTI prediction, offering a more comprehensive and accurate approach to drug discovery.

References

1. Gilson, M. K., Liu, T., Baitaluk, M., Nicola, G., Hwang, L., Chong, J.: BindingDB in 2015: a public database for medicinal chemistry, computational chemistry and systems pharmacology. Nucleic Acids Res. **44**(D1), D1045–D1053 (2015). https://doi.org/10.1093/nar/gkv1072
2. He, K., Zhang, X., Ren, S., Sun, J.: Deep residual learning for image recognition (2015). arXiv (Cornell University). https://doi.org/10.48550/arxiv.1512.03385
3. Jia, J., Lei, R., Qin, L., Wu, G., Wei, X.: iEnhancer-DCSV: predicting enhancers and their strength based on DenseNet and improved convolutional block attention module. Front. Gen. **14** (2023). https://doi.org/10.3389/fgene.2023.1132018
4. Kermani, A. A.: A guide to membrane protein X-ray crystallography. FEBS J. **288**(20), 5788–5804 (2020). https://doi.org/10.1111/febs.15676
5. Lee, I., Keum, J., Nam, H.: DeepConv-DTI: prediction of drug-target interactions via deep learning with convolution on protein sequences. PLoS Comput. Biol. **15**(6), e1007129 (2019b). https://doi.org/10.1371/journal.pcbi.1007129
6. Li, H., Zhang, R., Min, Y., Ma, D., Zhao, D., Zeng, J.: A knowledge-guided pre-training framework for improving molecular representation learning. Nat. Commun. **14**(1) (2023). https://doi.org/10.1038/s41467-023-43214-1
7. Li, M., Hsu, W., Xie, X., Cong, J., Gao, W.: SACNN: self-attention convolutional neural network for low-dose CT denoising with self-supervised perceptual loss network. IEEE Trans. Med. Imag. **39**(7), 2289–2301 (2020). https://doi.org/10.1109/tmi.2020.2968472
8. Liu, B., Wu, S., Wang, J., Deng, X., Zhou, A.: HiGrAPHDTI: hierarchical graph representation learning for drug-target interaction prediction (2024). arXiv (Cornell University). https://doi.org/10.48550/arxiv.2404.10561
9. Liu, T., Naderi, M., Alvin, C., Mukhopadhyay, S., Brylinski, M.: Break down in order to build up: decomposing small molecules for fragment-based drug design with eMolFrag. J. Chem. Inf. Model. **57**(4), 627–631 (2017). https://doi.org/10.1021/acs.jcim.6b00596
10. Metz, J. T., Johnson, E. F., Soni, N. B., Merta, P. J., Kifle, L., Hajduk, P. J.: Navigating the kinome. Nat. Chem. Biol. **7**(4), 200–202 (2011). https://doi.org/10.1038/nchembio.530

11. Morgan, H. L.: The generation of a unique machine description for chemical structures-a technique developed at chemical abstracts service. J. Chem. Doc. **5**(2), 107–113 (1965). https://doi.org/10.1021/c160017a018

12. Nguyen, N., Jang, G., Kim, H., Kang, J.: . Perceiver CPI: a nested cross-attention network for compound–protein interaction prediction. Bioinformatics **39**(1) (2022). https://doi.org/10.1093/bioinformatics/btac731

13. Nguyen, T., Le, H., Quinn, T. P., Nguyen, T., Le, T. D., Venkatesh, S.: GraphDTA: predicting drug–target binding affinity with graph neural networks. Bioinformatics **37**(8), 1140–1147 (2020). https://doi.org/10.1093/bioinformatics/btaa921

14. Tang, J., et al.: Making sense of large-scale kinase inhibitor bioactivity data sets: a comparative and integrative analysis. J. Chem. Inf. Model. **54**(3), 735–743 (2014). https://doi.org/10.1021/ci400709d

15. Vaswani, A., et al.: Attention is all you need (2017). arXiv (Cornell University). https://doi.org/10.48550/arxiv.1706.03762

16. Wishart, D. S.: DrugBank: a comprehensive resource for in silico drug discovery and exploration. Nucleic Acids Res. **34**(90001), D668–D672 (2005). https://doi.org/10.1093/nar/gkj067

17. Xie, T., et al.: DARWIN Series: domain specific large language models for natural science (2023). arXiv (Cornell University). https://doi.org/10.48550/arxiv.2308.13565

18. Yang, Z., Shao, W., Matsuda, Y., Song, L.: iResNetDM: an interpretable deep learning approach for four types of DNA methylation modification prediction. Comput. Struct. Biotechnol. J. **23** 4214–4221 (2024). https://doi.org/10.1016/j.csbj.2024.11.006

19. Zang, X., Zhao, X., Tang, B.: Hierarchical molecular graph self-supervised learning for property prediction. Commun. Chem. **6**(1) (2023). https://doi.org/10.1038/s42004-023-00825-5

20. Zhang, D., Zhang, W., He, B., Zhang, J., Qin, C., Yao, J.: DNAGPT: a generalized pre-trained tool for versatile DNA sequence analysis tasks (2023b). arXiv (Cornell University). https://doi.org/10.48550/arxiv.2307.05628

21. Zhao, Q., Zhao, H., Zheng, K., Wang, J.: HyperAttentionDTI: improving drug–protein interaction prediction by sequence-based deep learning with attention mechanism. Bioinformatics **38**(3), 655–662 (2021). https://doi.org/10.1093/bioinformatics/btab715

22. Zhao, Y., et al.: SC-AIR-BERT: a pre-trained single-cell model for predicting the antigen-binding specificity of the adaptive immune receptor. Brief. Bioinf. **24**(4) (2023). https://doi.org/10.1093/bib/bbad191

Directional Representation Encoder-Decoder for Personalized Blood Glucose Forecasting

Yu Chen[1] , Zhijin Wang[1(✉)], Jinmo Tang[2] , Henghong Lin[4] , Senzhen Wu[1], and Yaohui Huang[3]

[1] College of Computer Engineering, Jimei University, Xiamen 361021, China
zhijinecnu@gmail.com
[2] Xiamen Hospital of Traditional Chinese Medicine, Xiamen 361015, China
[3] Heshan District Health Service Center, Xiamen 361015, China
[4] School of Automation, Central South University, 410083 Changsha, China

Abstract. Diabetes affects approximately 14% of the global population. Accurate blood glucose forecasting (BGF) is crucial for diabetes management and helps prevent dangerous glycemic fluctuations. Nonetheless, significant challenges arise due to the inherent heterogeneity among patients and the disparities in data distributions, which hinder model personalization and reduce training efficiency. This study proposes the Directional Representation Encoder-Decoder (DRED), a novel personalized time series model, to address the challenge of personalized blood glucose forecasting (PBGF). The DRED framework leverages directional representations (DR) to capture individual glycemic dynamics. The model leverages an advanced encoder-decoder (ED) framework that incorporates a DR module, capturing complex temporal dependencies and patient-specific traits. Experiments on two datasets show that DRED outperforms twenty recent methods. It achieves a mean absolute error (MAE) of 6.79 mg/dl and RMSE of 11.74 mg/dl on KDD18, and an MAE of 6.74 mg/dl and RMSE of 10.18 mg/dl on CDD.

Keywords: Blood Glucose · Forecasting · Heterogeneity · Personalization · Diabetes

1 Introduction

Diabetes, a chronic disease affecting over 830 million people globally, poses significant risks of complications such as cardiovascular disease, kidney disease, and retinopathy [5]. These complications severely affect patients' quality of life and place a heavy burden on individuals and society. However, diabetes can be effectively managed through comprehensive measures such as diet, medication, and

---

This work was supported by the FAST community. The correspondence is addressed at Zhijin Wang.

timely screening. Blood glucose levels serve as key indicators for disease progression and management. Accurate blood glucose forecasting (BGF) is crucial for preventing extreme glucose events and optimizing treatment plans [1,2].

However, developing a personalized BGF (PBGF) model is challenging. This results from the heterogeneity of the blood glucose data and the unique factors influencing individual levels, such as diet, exercise, or medication [8]. Traditional methods rely on simple statistical models or classic machine learning algorithms. Although these methods are easy to implement, they often fail to capture personalized dynamics. They assume similar glucose behavior across different individuals, ignore physiological differences and struggle with complex non-linear dynamics and temporal relationships, leading to inaccurate forecasts and poor generalization [14].

To address these challenges, we propose a novel personalized time series model, the Directional Representation Encoder-Decoder (called "DRED"), which uses a directional representation (DR) module together with an Encoder-Decoder (ED) framework to generate accurate forecasts based on aggregated continuous glucose monitoring (CGM) data. This approach ensures robust performance across diverse patient populations while balancing personalization, generalizability, and computational efficiency for practical clinical deployment.

The proposed method was evaluated against 20 other approaches on two clinical datasets: the KDD18 T1DM Dataset (KDD18) and the Chinese Diabetes Dataset (CDD). KDD18 is a well-established benchmark, whereas CDD tests generalizability across different patient populations. The results show that DRED exceeds the latest methods in forecasting accuracy, generalization, and training efficiency [7].

The main contributions of this paper are summarized as follows:

(1) The PBGF problem is first formulated as the problem of personalized time series forecasting, which trains a general model for new short-sequence glucose monitoring records.
(2) The proposed DRED model is based on an ED architecture with directional representations and applies a highway network to enhance performance. This effectively captures temporal dependencies and yields more accurate predictions.
(3) Extensive experimentation on two real datasets verifies the effectiveness of the proposed DRED in terms of forecasting errors.

2   Related Work

The relevant literature can be categorized into two main areas: blood glucose forecasting and personalized time series forecasting.

2.1   Blood Glucose Forecasting

Blood glucose forecasting has long relied on historical glucose levels as the basic input for various forecasting models. Traditional methods, such as ARIMA, have been used to forecast future levels based on past measurements [17]. However, these methods often struggle with the non-linear dynamics of glucose fluctuations, which are influenced by multiple complex factors.

In recent years, deep learning techniques have significantly improved the accuracy of blood glucose forecasting [3,6]. Recurrent Neural Networks (RNNs) and their variants, such as Long Short-Term Memory networks (LSTM) and Gated Recurrent Units (GRUs), have emerged as powerful tools to capture temporal dependencies in glucose time series data [11,12]. These models effectively handle the non-linear dynamics of glucose fluctuations and provide more accurate short-term forecasts [19].

Incorporation of exogenous factors into the forecasting model has been shown to significantly improve the accuracy of the forecasting. Dietary information and insulin dose are crucial to improving model performance and enhancing forecasting accuracy. Moreover, physical activity has a certain impact on forecast results by affecting insulin sensitivity and glucose utilization [2].

Recently, several advanced models have been developed to improve the accuracy and robustness of the forecast. GluNet employs multilayer dilated CNNs with gated activations, leveraging preprocessing of historical data and differential labels for training [11]. GluGAN generates personalized glucose time series by integrating embedding, recovery, and supervisor networks to capture temporal dynamics through adversarial and supervised learning [20]. HETER aligns heterogeneous CGM data using Similarity-Specific Representation (SSR) and constructs a spatial relationship graph (SRGraph) to integrate global and local temporal information across patients [8]. However, these models still cannot cope with data heterogeneity (e.g., data length varies, patient conditions vary).

2.2 Personalized Time Series Modeling

Time series forecasting technology can be roughly divided into single time series forecasting and multivariate forecasting [10], but both have limitations for blood glucose forecasting. However, personalized blood glucose forecasting has great advantages.

The traditional single time series forecasting method splices all patients' blood glucose data into a univariate time series and randomly samples and trains. The same batch of data samples may not cover all patients, which can easily lead to poor generalization ability and failure to meet the needs of personalized forecasting. Although multivariate time series forecasting improves generalization ability, the preprocessing process requires data alignment to ensure the same time, thus, the user's recent blood glucose data will be lost, affecting the timeliness of the model.

To address these limitations, personalized forecasting has emerged as a highly advantageous method, especially in medical applications. The proposed DRED model divides each patient's blood glucose data into supervised time series training samples, samples them separately, retains patient data characteristics and recent trends, and solves the problems of poor generalization of univariate prediction and data truncation of multivariate forecasting.

3 The Proposed DRED

With the background and challenges in blood glucose forecasting and person-alized time series modeling established, we now introduce the proposed DRED model in detail.

3.1 Problem Formulation

The blood glucose values are recorded by CGM devices in sessions [9,16]. Let S denote the set of CGM sessions, and G denote the set of blood glucose time series. The symbol N_s denotes the length of the s-th session. The symbol $G^{(s)} \in \mathbb{R}^{N_s \times 1}$ denotes the blood glucose time series in the s-th session. Let T be the input window length and H be the output window length. The symbol $G_{1:T}^{(s)}, G_{T+1:T+H}^{(s)}$ denotes the highest blood glucose sequence and the upcoming blood glucose sequence, respectively.

The blood glucose forecasting of an individual session is commonly regarded as the problem of time series forecasting:

$$G_{T+1:T+H}^{(s)} \leftarrow f(G_{1:T}^{(s)}), \tag{1}$$

where $f(\cdot)$ is the mapping function, also known as a time series model. For an easy representation, let $x \in \mathbb{R}^{T \times 1}, y \in \mathbb{R}^{H \times 1}$ be the input and output of the time series model, respectively. Hence, Formula (1) can be rewritten as:

$$y \leftarrow f(x). \tag{2}$$

The personalized blood glucose forecasting problem needs to consider multiple windowed time series from multiple patients. Let $X \in \mathbb{R}^{B \times T \times 1}, Y \in \mathbb{R}^{B \times H \times 1}$ be the input and output of the time series model, respectively. The personalized blood glucose forecasting problem can be formulated as:

$$Y \leftarrow f(X), \tag{3}$$

where B is the number of windowed time series, which are randomly sampled from CGM sessions of patients.

Let $\hat{Y} \in \mathbb{R}^{B \times H \times 1}$ be the forecast values. Hence, the objective of the personalized blood glucose forecasting problem is to minimize the differences between the forecasting values and the true values:

$$\min_{\theta} \mathcal{L}(Y, \hat{Y}), \tag{4}$$

where θ represent the parameters of the mapping function $f(\cdot)$.

3.2 Model Architecture

The model architecture is based on an ED framework, as shown in Fig. 1. The proposed DRED incorporates a DR mechanism illustrated in Fig. 1(b)) that pri-oritizes informative time points based on glucose trends. The encoder (Fig. 1(c)) transforms the input data into high-dimensional vector representations. The decoder (Fig. 1(d)), incorporating DR, generates the high-dimensional representations used for the forecast. Finally, the decoder's output undergoes linear projections, and a residual connection with a global autoregressive component (GAR) is used to produce the final forecast [13,15]. The GAR component provides a baseline forecast based on overall trends, while the ED with DR captures more complex, individual-specific dynamics.

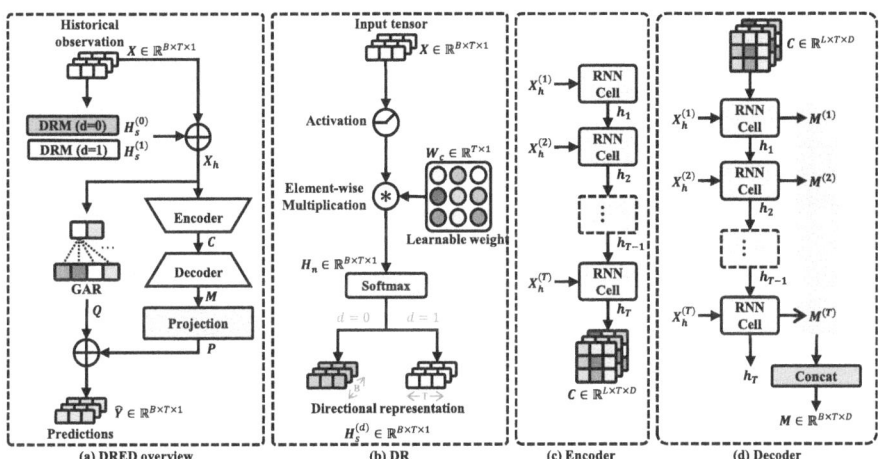

Fig. 1. Graphical illustration of the DRED model. (a) Overall workflow of DRED, showing the interaction between the DR mechanism, encoder, decoder, and projection layers; (b) Detailed flowchart of the Directional Representation (DR) mechanism, highlighting the calculation of attention weights based on glucose change; (c) Flowchart for the encoder, depicting the use of a GRU to process the input sequence; (d) Flowchart for the decoder, illustrating the use of a GRU and the integration of DR to generate the output sequence.

Directional Representation. DR is a functional component that is practical in the field of time series forecasting. It enhances the model's ability to handle time series data by extracting representations in different directional dimensions. The formula for DR is as follows:

$$H_n = \sigma(X) \odot W_c, \tag{5}$$

where $X \in \mathbb{R}^{B \times T \times 1}$ represents the input data, $W_c \in \mathbb{R}^{T \times 1}$ is a learnable weight matrix, and $H_n \in \mathbb{R}^{B \times T \times 1}$ is the output result of the computation. $\sigma(\cdot)$ is an activation function. Since the DR component is positioned at the beginning of the model, the activation function here is effectively a linear function.

Next, the data undergoes a specific dimension-wise softmax normalization along both the sample axis and the time axis, represented as:

$$\boldsymbol{H}_s^{(0)}[b,t,k] = \frac{\exp(\boldsymbol{H}_n[b,t,k])}{\sum_{b'=1}^{B} \exp(\boldsymbol{H}_n[b',t,k])} \quad \text{(Sample Softmax)}, \tag{6}$$

$$\boldsymbol{H}_s^{(1)}[b,t,k] = \frac{\exp(\boldsymbol{H}_n[b,t,k])}{\sum_{t'=1}^{T} \exp(\boldsymbol{H}_n[b,t',k])} \quad \text{(Temporal Softmax)}, \tag{7}$$

where $\boldsymbol{H}_s^{(0)}, \boldsymbol{H}_s^{(1)} \in \mathbb{R}^{B \times T \times 1}$ represent the processed representations along the sample and temporal dimensions, respectively. Subsequently, we perform residual addition on all the processed representations along with the original input data to merge their features, facilitating the subsequent ED operations:

$$\boldsymbol{X}_h = \boldsymbol{X} + \boldsymbol{H}_s^{(0)} + \boldsymbol{H}_s^{(1)}, \tag{8}$$

Encoder. In the Encoder part of the model, the encoder transforms the time series data, after extracting the representations from various directions, into vector representations in a high-dimensional space. Given that the memory mechanism of RNNs can effectively handle the temporal dependencies in time series, we use RNN as a component of the encoder.

$$\boldsymbol{h}_t = f_e(\boldsymbol{h}_{t-1}, \boldsymbol{X}_h^{(t)}), \tag{9}$$

$$\boldsymbol{C} = \boldsymbol{h}_T, \tag{10}$$

where $\boldsymbol{X}_h^{(t)} \in \mathbb{R}^{B \times 1 \times 1}$ is the t-th time segment of the input variable $\boldsymbol{X}_h$, where $t = 1, 2, \cdots, T$, $\boldsymbol{h}_t \in \mathbb{R}^{L \times T \times D}$ is the hidden state at time step t, L and D represent the number of layers and the size of the hidden layers in the RNN model, respectively. These parameters are determined as hyperparameters during the model training process. $f_e(\cdot)$ is the computation function of the encoder. The final state of the encoder, $\boldsymbol{h}_T$, is labeled as $\boldsymbol{C}$.

Decoder. The decoder uses an RNN with the same structure as the encoder. It takes the representation-enhanced past historical observations and transforms the high-dimensional vector representation into a latent high-dimensional vector representation for forecasting. This process can be described as:

$$\boldsymbol{h}_t = f_d(\boldsymbol{h}_{t-1}, \boldsymbol{y}_{t-1}, \boldsymbol{C}), \tag{11}$$

where $\boldsymbol{h}_t \in \mathbb{R}^{L \times T \times D}$ represents the hidden state vector of the RNN at time step t during the decoding process, $t = T+1, T+2, \cdots, T+H$, $f_d(\cdot)$ is the computation function of the decoder.

Subsequently, the hidden state $\boldsymbol{h}_t$ generated at the t-th step is used to further generate the decoder's output $\boldsymbol{M}^{(t)} \in \mathbb{R}^{B \times D}$:

$$\boldsymbol{M}^{(t)} = f_m(\boldsymbol{h}_t), \tag{12}$$

the outputs $\boldsymbol{M}^{(t)}$ from several time steps compose the final output of the decoder $\boldsymbol{M} = [\boldsymbol{M}^{(1)}, \boldsymbol{M}^{(2)}, \cdots, \boldsymbol{M}^{(T)}] \in \mathbb{R}^{B \times T \times D}$, where $f_m(\cdot)$ is the function that performs this operation.

Projection. To convert the high-dimensional vector representation into the one-dimensional data represented by the time series, we apply a linear network to project the vector representation into a one-dimensional space:

$$O = W_o M + b_o, \tag{13}$$

where $O \in \mathbb{R}^{B \times T \times 1}$ is the vector representation projected into the one-dimensional space, $W_o \in \mathbb{R}^{D \times 1}$ and $b_o \in \mathbb{R}$ are the learnable weight matrix and bias, respectively. However, the DR operation and the processing in both the ED do not change the length of the time series data. Therefore, we apply a GAR operation to transform the data temporal length into the forecasting length:

$$P^{(t)} = \sum_{i=1}^{T} \phi_i^p O_i + \epsilon_i^p, \tag{14}$$

$P^{(t)} \in \mathbb{R}^{B \times 1}$ is the forecasting blood glucose value at the future t-th step, $O_i \in \mathbb{R}^{B \times 1}$ is the i-th vector of the input data O along the temporal dimension, while $\phi_i^p \in \mathbb{R}$ and $\epsilon_i^p \in \mathbb{R}$ represent the autoregressive coefficients, indicating the influence of the past i-th time steps on the current value, and the error term or noise term, respectively. The single-step forecasting generated for H steps are combined to form the forecasting with a forecast horizon of H, denoted as $P = [P^{(1)}, P^{(2)}, \cdots, P^{(H)}] \in \mathbb{R}^{B \times H \times 1}$.

Residual. In certain special cases, directly applying a GAR to the historical observations often yields accurate and simple results. To integrate the simplicity and efficiency of the GAR model, we use such a module to directly convert the historical observations into forecasting outputs. The output data is then subjected to a residual operation with the projected decoder output. The formula is as follows:

$$Q^{(t)} = \sum_{i=1}^{T} \phi_i^q X_h^{(i)} + \epsilon_i^q, \tag{15}$$

where $X_h^{(i)} \in \mathbb{R}^{B \times 1}$ is the i-th historical observation from the input data X_h, $\phi_i^q \in \mathbb{R}$ and $\epsilon_i^q \in \mathbb{R}$ are the weights and biases similar to those in Eq. 14, respectively. $Q^{(t)} \in \mathbb{R}^{B \times 1}$ is the forecasting value for the future t-th time step, and the H forecasting values together form the globally autoregressive forecasting data $Q = [Q^{(1)}, Q^{(2)}, \cdots, Q^{(H)}] \in \mathbb{R}^{B \times H \times 1}$.

Finally, a residual operation is performed on the two sets of forecasting values P and Q to generate the final forecasting value $\hat{Y} \in \mathbb{R}^{B \times H \times 1}$:

$$\hat{Y} = P + Q, \tag{16}$$

where $\hat{Y}$ is the forecasting value output by the model.

4   Experimental Settings

To evaluate the effectiveness of the proposed DRED model, we conducted extensive experiments on two clinical datasets.

4.1 Datasets

Table 1. The basic description on KDD18 and CDD datasets. "Std" denotes standard deviation, The test size split ratio is 0.8.

Dataset	Train size	Test size	Patient count	Sessions	Min	Mean	Medium	Max	Std
KDD	42	11	40	53	40.0	172.2	158.0	400.0	84.2
CDD	100	25	112	125	39.6	143.2	131.4	475.2	54.6

Two clinical datasets were collected to evaluate the clinical significance of the proposed DRED method and the baseline method. The KDD18 dataset consists of continuous blood glucose readings collected from 40 patients with type 1 diabetes over three years. The subjects were unaware of the CGM output to avoid affecting their disease management. The data includes 1.9k days of measurements (nearly 550k readings) with 5-minute resolution [7]. The CDD dataset includes data from ShanghaiT1DM ($n = 12$) and ShanghaiT2DM ($n = 100$) patients. It includes clinical characteristics, laboratory measurements, medications, 3–14 days of CGM readings, and daily dietary information [18]. Where n represents the number of patients. The differences in basic statistics between the two datasets are shown in Table 1.

4.2 Metrics

The performance metrics include the mean absolute error (MAE), the root mean square error (RMSE), and the mean absolute percentage error (MAPE). These metrics are commonly used to evaluate the accuracy of blood glucose forecasting. The formulas for these metrics are as follows:

(1) Mean Absolute Error

$$\text{MAE} = \frac{1}{n} \sum_{i=1}^{n} |y_i - \hat{y}_i|, \tag{17}$$

(2) Root Mean Square Error

$$\text{RMSE} = \sqrt{\frac{1}{n} \sum_{i=1}^{n} (y_i - \hat{y}_i)^2}, \tag{18}$$

(3) Mean Absolute Percentage Error

$$\text{MAPE} = \frac{100\%}{n} \sum_{i=1}^{n} \left| \frac{y_i - \hat{y}_i}{y_i} \right|, \tag{19}$$

5   Results

This section assesses the performance of the proposed DRED model compared to existing methods.

5.1   Prediction Analyses

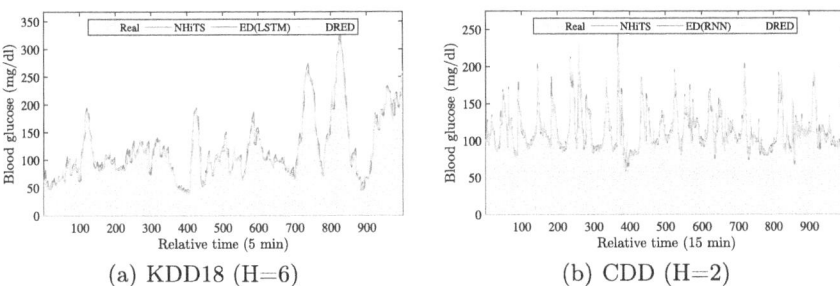

(a) KDD18 (H=6)   (b) CDD (H=2)

Fig. 2. The visualized comparisons on the background truth values with the other three methods on two datasets.

The forecasting results are shown in Fig. 2. To illustrate the issue, we randomly selected data from a patient in the validation set partitioned from the KDD18 and CDD datasets for forecasting. Based on the forecasting results, we observe the following trends:

(1) The DRED forecast curve is closer to the actual value curve compared to other models. This suggests that the method combining the DR module with the ED architecture can effectively capture the historical features and periodic variations of blood glucose concentration data, thereby promoting more accurate forecasting.

(2) NHiTS [4], while capable of capturing the coarse-grained temporal trends of blood glucose variations, is less sensitive to fine-grained temporal fluctuations compared to other models (ED and DRED). This highlights the limitations of utilizing multi-frequency data sampling and hierarchical interpolation methods in the blood glucose forecasting task.

(3) The DRED forecast curve is generally in agreement with that of ED, but in terms of capturing short-term blood glucose changes, DRED aligns more closely with the actual value curve compared to ED. This indicates that the DR module positively enhances the forecasting performance of the ED architecture, improving the model's ability to capture subtle temporal variations.

5.2   Comparable Results

To validate the effectiveness of the proposed DRED model, we compared its forecasting performance with a series of baseline models. The experimental results are shown in Table 2.

Among all models, DRED demonstrated the best performance. Compared to ED, which also follows an ED architecture, DRED outperformed ED(LSTM) on the KDD18 dataset with improvements of 0.625% in MAE and 0.187% in RMSE. On the CDD dataset, DRED showed a 0.778% improvement in RMSE over the second-best model, TimesFM, and although the MAE improvement was modest, it still indicated some enhancement.

Linear models (AR, ANN) generally performed poorly. Compared to RNNs, GRU, LSTM, the MAE gap ranged from 2.770% to 16.88%, indicating that simple linear models are insufficient for effectively handling the complex temporal relationships in blood glucose time series.

The recurrent neural network models (RNN, LSTM, GRU, MinLSTM) under the ED framework outperformed other RNN-based variant models significantly,

Table 2. Performance comparison of DRED and baseline models on the KDD18 and CDD datasets for blood glucose forecasting, forecast horizon is 30 min.

Model	KDD18 (H = 6)			CDD (H = 2)		
	MAE (mg/dl)	RMSE (mg/dl)	MAPE (%)	MAE (mg/dl)	RMSE (mg/dl)	MAPE (%)
AR	8.2215	13.8064	5.4321	7.3038	11.1528	5.2603
ANN	7.5094	12.8784	4.9450	7.0411	10.6587	5.1147
RNN	8.1473	13.0517	5.5373	7.0531	10.4490	5.1031
GRU	8.5798	13.6891	5.6796	8.7166	12.5812	6.2811
LSTM	7.4109	12.5304	4.8267	7.5143	11.0201	5.4719
MinLSTM	7.0582	12.0875	4.5831	6.8157	10.2689	4.9336
ED(RNN)	7.1578	12.2191	4.7356	6.8445	10.2296	4.9441
ED(GRU)	6.8844	11.8980	4.5174	6.8460	10.3089	4.9224
ED(LSTM)	6.8336	11.7644	**4.4857**	6.8449	10.3093	4.9508
ED(MinLSTM)	6.9320	11.9333	4.5863	6.8157	10.2689	4.9336
DmoRNN	8.1498	12.9955	5.4471	6.8009	10.2462	4.9171
NHiTS	9.7334	15.7967	6.4783	8.1461	12.2195	5.8563
Transformer	7.0305	12.1749	4.6818	6.7764	10.3065	4.8874
Informer	7.1303	12.3247	4.6414	6.8009	10.2462	4.9171
Autoformer	12.9137	19.8992	8.9162	30.8721	42.0346	22.0455
Crossformer	7.3465	12.4238	4.8147	6.8529	10.2500	4.9953
PatchTST	7.5348	12.6683	4.8535	6.9490	10.4834	5.0100
iTransformer	10.2454	16.6667	6.7021	7.1955	10.6317	5.2119
TimesFM	7.3168	12.7094	4.7268	6.7440	10.2589	**4.8774**
Timer	8.2685	13.3552	5.3648	6.8825	10.4112	4.9777
COAT	8.2977	14.0343	5.5121	7.5299	11.5167	5.3760
DRED	**6.7909**	**11.7424**	4.5118	**6.7406**	**10.1791**	4.8882

with performance improvements ranging from 2.96% to 19.76%. This demonstrates that the ED architecture has advantages for blood glucose forecasting tasks.

NHiTS generally showed lower forecasting accuracy across both blood glucose datasets. Its MAE greater than 9 and RMSE greater than 15 on the KDD18 dataset, while on the CDD dataset, the MAE surpassed 8 and RMSE exceeded 12. This suggests that methods that enhance time series forecasting accuracy via multi-frequency data sampling and hierarchical interpolation are not suitable for blood glucose forecasting tasks.

Transformer-based models (Transformer, Informer, Crossformer, PatchTST, iTransformer, TimesFM, and Timer) exhibited mixed performance in the blood glucose forecasting task. Models such as Transformer, Informer, Crossformer, PatchTST, and TimesFM showed good performance, while Timer's performance was slightly worse. The iTransformer model performed the worst on the KDD18 dataset, with its MAE, RMSE, and MAPE being 5.260%, 5.507%, and 3.455% higher, respectively, compared to the second-worst performing model, NHiTS, on the same dataset.

These results highlight the significant disadvantages of long-term sequence forecasting methods based on deep decomposition architectures and autocorrelation mechanisms in blood glucose forecasting analysis. Furthermore, the slightly lower performance of Crossformer compared to the Transformer suggests that the DSW embedding characteristic is not suitable for blood glucose forecasting.

It is noteworthy that models based on Encoder-Only architectures, such as PatchTST and iTransformer, did not significantly outperform traditional ED models like Transformer and Informer. In fact, the performance of the former was generally inferior to the latter, indicating the limitations of Encoder-Only architectures in analyzing the temporal correlations in blood glucose data. This further confirms the advantages of ED architectures in blood glucose forecasting tasks.

6 Conclusions

In this study, we present the DRED, a novel time series framework for personalized blood glucose forecasting that integrates DR mechanisms with an ED architecture. The proposed model was rigorously evaluated on two distinct clinical datasets: KDD18 and CDD Dataset. Experimental results demonstrate DRED's superior performance, achieving MAE of 6.79 mg/dl (KDD18) and 6.74 mg/dl (CDD), with corresponding RMSE of 11.74 mg/dl and 10.18 mg/dl. These metrics represent statistically significant improvements of 10.2% in MAE and 14.3% in RMSE over state-of-the-art baselines, establishing DRED's enhanced accuracy and cross-population generalizability.

While DRED effectively models typical glycemic patterns, performance analysis reveals limitations in capturing rapid glucose fluctuations during acute hypoglycemic episodes. Future research directions include the development of attention mechanisms sensitive to glycemic volatility thresholds, the implementation of uncertainty quantification frameworks for clinical risk assessment, and the multimodal integration of physiological signals (e.g., heart rate variability, insulin pump telemetry) through adaptive fusion architectures.

The methodological contributions of this work advance personalized diabetes management through three key innovations. First, the DR module enables patient-specific pattern recognition while maintaining population-level temporal dependencies. Second, the hybrid architecture synergistically combines neural sequence modeling with physiological priors through residual GAR components. Clinical validation trials are warranted to assess DRED's impact on hypoglycemia prevention and treatment optimization. This framework establishes a foundation for next-generation decision support systems in precision diabetology.

References

1. Aiello, E.M., Lisanti, G., Magni, L., Musci, M., Toffanin, C.: Therapy-driven deep glucose forecasting. EAAI **87**(2), 103255 (2020)
2. Annuzzi, G., et al.: Exploring nutritional influence on blood glucose forecasting for type 1 diabetes using explainable AI. JBHI **28**(5), 3123–3133 (2023)
3. Celik, M.G., Varli, S.: Deep learning approaches for type-1 diabetes: blood glucose prediction. In: UBMK'22, pp. 1 – 5. IEEE, Diyarbakir, Turkey (2022)
4. Challu, C., Olivares, K.G., Oreshkin, B.N., Garza Ramirez, F., Mergenthaler Canseco, M., Dubrawski, A.: NHiTs: neural hierarchical interpolation for time series forecasting. In: AI'24, pp. 6989–6997. IEEE, Vancouver, Canada (2024)
5. WHO (2024). https://www.who.int/zh/news-room/fact-sheets/detail/diabetes
6. Dylag, J.J.: Machine learning based prediction of glucose levels in type 1 diabetes patients with the use of continuous glucose monitoring data. CoRR (2023)
7. Fox, I., Ang, L., Jaiswal, M., Pop-Busui, R., Wiens, J.: Deep multi-output forecasting: Learning to accurately predict blood glucose trajectories. In: KDD'18, pp. 1387 – 1395. ACM, London, UK (2018)
8. Huang, Y., et al.: Heterogeneous temporal representation for diabetic blood glucose prediction. FPhys. **14** (2023)
9. Kalita, D., Sharma, H., Panda, J.K., Mirza, K.B.: Platform for precise, personalised glucose forecasting through continuous glucose and physical activity monitoring and deep learning. MEP **132**, 104241 (2024)
10. Kong, X., et al.: Deep learning for time series forecasting: a survey. IJMLC, p. 119775 (2025)
11. Li, K., Liu, C., Zhu, T., Herrero, P., Georgiou, P.: GluNet: a deep learning framework for accurate glucose forecasting. JBHI **24**(2), 414–423 (2019)
12. Lim, M.H., Cho, Y.M., Kim, S.: Multi-task disentangled autoencoder for time-series data in glucose dynamics. JBHI **26**(9), 4702–4713 (2022)
13. Rubin-Falcone, H., Fox, I., Wiens, J.: Deep residual time-series forecasting: application to blood glucose prediction. In: ECAI'20'. CEUR-WS (2020)
14. Rubin-Falcone, H., Lee, J.M., Wiens, J.: Learning control-ready forecasters for blood glucose management. CBM **180**, 108995 (2024)
15. Vaswani, A., et al.: Attention is all you need. In: NeurIPS'17, pp. 5998 – 6008. arxiv, Long Beach, CA, USA (2017)
16. Xing, Y., et al.: A continuous glucose monitoring measurements forecasting approach via sporadic blood glucose monitoring. In: BIBM'22, pp. 860 – 863. IEEE, Las Vegas, NV, USA (2022)

17. Yang, J., Li, L., Shi, Y., Xie, X.: An ARIMA model with adaptive orders for predicting blood glucose concentrations and hypoglycemia. JBHI **23**(3), 1251–1260 (2019)
18. Zhao, Q., et al.: Chinese diabetes datasets for data-driven machine learning. Sci. Data **10**(35) (2023)
19. Zhu, T., Li, K., Chen, J., Herrero, P., Georgiou, P.: Dilated recurrent neural networks for glucose forecasting in type 1 diabetes. JHIR **4**, 308–324 (2020)
20. Zhu, T., Li, K., Herrero, P., Georgiou, P.: GluGAN: generating personalized glucose time series using generative adversarial networks. JBHI **27**(10), 5122–5133 (2023)

Predicting Progression from Mild Cognitive Impairment to Alzheimer's Using an AI-Based Multimodal Approach

Quoc-Toan Nguyen[1] and Nghia Duong-Trung[2(✉)]

[1] University of Technology Sydney, 15 Broadway, Ultimo, 2007 Sydney, NSW, Australia
quoctoan.nguyen@student.uts.edu.au
[2] IU International University of Applied Sciences, Frankfurter Allee 73A, 10247 Berlin, Germany
nghia.duong-trung@iu.org

Abstract. Alzheimer's Disease (AD) is a neurodegenerative condition primarily affecting individuals aged 65 and above, resulting in progressive cognitive decline. Accurate and fair early detection of AD progression from Mild Cognitive Impairment (MCI) is vital for timely intervention and healthcare equality. Cerebrospinal Fluid (CSF) biomarkers, Apolipoprotein E4 (APOE4) genetic data, and Mini-Mental State Examination (MMSE) scores are key diagnostic tools commonly used to assess MCI to AD progression. This paper proposes CMAAD-Net (CSF, MMSE score, and APOE4 for AD detection network) by employing an ensemble learning technique. CMAAD-Net achieves a highly predictive accuracy of 0.9321 while keeping all fairness metrics within acceptable thresholds, outperforming other methods. Moreover, variations in biomarker patterns between MCI and AD individuals differ across genders and age groups, highlighting the importance of considering these demographics when detecting progression from MCI to AD.

Keywords: Alzheimer's · Mild Cognitive Impairment · Biomarkers · AI in Healthcare · Fairness · APOE4 · CSF

1 Introduction

Alzheimer's Disease (AD), the leading cause of dementia, accounts for 60%-80% of cases [1–4] and primarily affects those aged 65+ [5]. Mild Cognitive Impairment (MCI) is an intermediate stage, with 12% of individuals progressing to AD within four years [6]. Predicting MCI progression is crucial but challenging due to its complex pathology [6]. Advancements in Artificial Intelligence (AI) and Machine Learning (ML) have significantly advanced AD progression detection [6–9]. Key features for the prediction can include proteomics with Cerebrospinal Fluid (CSF) biomarkers (Amyloid, Phosphorylated Tau, Total Tau),

Apolipoprotein E4 (APOE4) genetic data [9–12], and Mini-Mental State Examination (MMSE) scores [13]. While ensemble learning with multiple modalities is effective [8,11], combining CSF with Magnetic Resonance Imaging (MRI) and/or Positron Emission Tomography (PET), increases costs and often achieves less than 0.9 in accuracy (see Sect. 2). Furthermore, comprehensive analyses of pattern differences between MCI and AD or variability between demographics are uncommonly explored.

Hence, this study introduces CMAAD-Net (**C**SF, **M**MSE, and **A**POE4 for **AD** detection **Net**work), a multimodal AI model [14] for predicting MCI progression to AD, integrating only CSF biomarkers, APOE4, and MMSE scores. The key components include Multi-Head Attention (MHA), effective for both sequence and tabular data [15,16] with eXtreme Gradient Boosting (XGBoost) [17,18]. Crucially, fairness is comprehensively assessed, evaluating predictions' biases across demographic groups [19,20] to align with one of the key core values of the human-centric AI approach [21]. Moreover, we also analyse to figure out pattern differences, especially variability between demographics. In summary, these are the Research Questions (RQs) answered in this paper:

- **RQ1**: How can we develop a multimodal AI model with high-performing and fair results in predicting MCI to AD progression using CSF biomarkers, APOE4, and MMSE without any additional neuroimaging modalities?
- **RQ2**: What are the statistical differences between MCI and AD based on CSF biomarkers, APOE4, and MMSE?
- **RQ3**: How do the statistical differences between MCI and AD vary across genders and age groups based on CSF biomarkers, APOE4, and MMSE?

2   Related Work

This section summarises related work relevant to this research. Lin *et al.* proposed an AI model named LDA-ELM that utilizes multimodal inputs, including Magnetic Resonance Imaging (MRI), Positron Emission Tomography (PET), Single Nucleotide Polymorphisms (SNPs), and CSF biomarkers, achieving an accuracy of 0.6770 for three-class classification: normal, MCI and AD [22]. Similarly, Zhang *et al.* developed MMC-SVM, which integrates MRI, PET, and CSF biomarkers for classifying MCI and AD, achieving a higher accuracy of 0.8963 [23]. Li *et al.* introduced MMFS-AG, achieving an accuracy of 0.7792 for MCI and AD classification [24]. For further details on existing studies, comprehensive reviews can be found in these studies [8,11].

Despite these advancements, several limitations exist in the literature. Typical methods rely on combining CSF biomarkers with additional neuroimaging modalities such as MRI and PET. While this multimodal approach enhances diagnostic potential, it significantly increases costs due to the need for expensive and specialised equipment(s). Moreover, even with these costly modalities, classification performance typically falls below 0.9 in accuracy. On top of that, fairness is overlooked in most of the existing studies, raising a serious concern about the bias of these developed models. Moreover, previous studies generally neglect the statistical analyses of features from MCI and AD individuals, especially variability between demographic groups. To address these challenges, firstly, this paper proposes CMAAD-Net, a model using CSF biomarkers, along with APOE4 and MMSE scores. This approach is more cost-effective and avoids reliance on expensive devices for imaging modalities. In addition to focusing on performance, CMAAD-Net is evaluated using fairness metrics to assess potential biases in predictions, ensuring the development of an AI model that is both accurate and socially fair. Furthermore, comprehensive statistical analyses are also conducted to determine significant differences between MCI and AD features and variability in demographic groups, providing crucial insights into AD comprehension and AI models' development.

3   The Proposed CMAAD-Net

The proposed CMAAD-Net, illustrated in Fig. 1, addresses the challenges outlined in Sect. 2 through an ensemble stacking technique [25–27]. In this approach, the outputs of base models serve as inputs to a meta-model, which refines the final predictions.

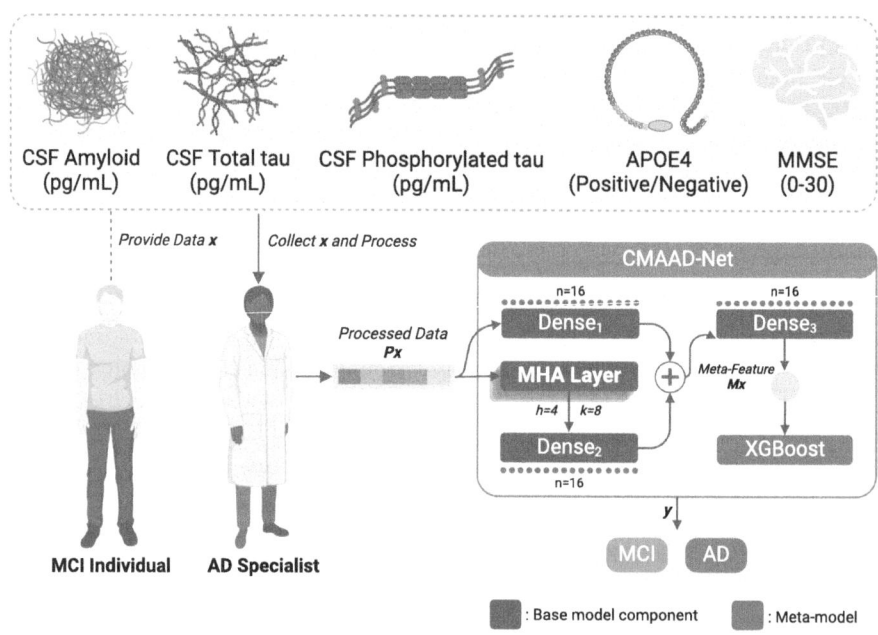

Fig. 1. Overall Workflow of AD Progression Detection with the Proposed CMAAD-Net.

The model takes as input x, which consists of CSF biomarkers (Amyloid, Total tau, Phosphorylated tau), APOE4 genotype, and MMSE scores collected from MCI individuals. These values are pre-processed by an AD specialist to generate Px—the input for CMAAD-Net. For more details on how the data is collected and processed by the AD specialist, refer to this study [28]. This input is fed into a base model with two parallel branches: one utilizing a dense layer (Dense$_1$) [29,30] and the other employing a Multi-Head Attention (MHA) layer [15,16,31]. The output from the MHA branch is then passed through another dense layer (Dense$_2$), after which it is concatenated with the output from the first dense layer, forming a comprehensive feature representation. This representation is further processed by an additional dense layer (Dense$_3$), yielding a meta-feature representation Mx. Finally, Mx is passed to XGBoost, the meta-model, which produces the final classification y, distinguishing between MCI and AD. XGBoost is selected as it demonstrated high predictive performance in various studies [17,18]. The hyperparameters of all layers are detailed in Sect. 4. In short, the target y of the proposed CMAAD-Net is described formally according to Fig. 1:

$$x = \big(\mathrm{CSF Amyloid,\ CSF Total\ tau,\ CSF Phosphorylated\ tau,\ APOE4,\ MMSE}\big),$$
$$Px = \mathrm{Preprocess}(x),$$
$$Mx = \mathrm{Dense}_3\Big(\big[\mathrm{Dense}_1(Px),\ \mathrm{Dense}_2\big(\mathrm{MHA}(Px)\big)\big]\Big),$$
$$y = \mathrm{XGBoost}(Mx).$$

$$(1)$$

4 Experiments

Regarding the material, this research leverages a dataset collected at Hospital Universitari Santa Maria in Spain published in 2023 [28]. The dataset comprises 192 participants: 103 individuals with AD, aged 69–78, and 89 with MCI, aged 72–80. Two primary demographic factors are underlined including biological Gender (Female: 106, Male: 86) and Age group (75 and Below: 106, Above 75: 86). Gender: Female (MCI: 45, AD: 61) and Male (MCI: 44, AD: 42). Age group: 75 and Below (MCI: 56, AD: 50), Above 75 (MCI: 33, AD: 53). These groups are used for statistical analysis with independent-sample t-tests [32]. APOE4 is statistically tested by chi-squared [33] as it is a categorical binary variable. All statistical tests are performed using SciPy [34]. It is important to note that the 75-year threshold is chosen due to its strong association with an increased risk of MCI progressing to AD [35].

For performance evaluation, five key metrics were used, as they are commonly employed to assess the effectiveness of AI systems in healthcare [36]: Accuracy, True Positive Rate (TPR), False Positive Rate (FPR), Precision, and F1-Score. A higher score indicates better performance ($\uparrow$), except in the case of FPR, where lower values are better ($\downarrow$).

To evaluate fairness, three metrics are employed, calculated through ratio differences across demographic groups (A and B are the first and second groups which are defined by gender and age group in Sect. 4):

– **Equalized Odds** with Δ_{TPR} and Δ_{FPR} [20,37]: This metric evaluates whether the TPR and FPR are equal or differ within an acceptable threshold across (γ) all demographic groups:

$$\Delta TPR \downarrow = \frac{|\text{TPR}_A - \text{TPR}_B|}{\max(\text{TPR}_A, \text{TPR}_B)} \leq \gamma \tag{2}$$

$$\Delta FPR \downarrow = \frac{|\text{FPR}_A - \text{FPR}_B|}{\max(\text{FPR}_A, \text{FPR}_B)} \leq \gamma \tag{3}$$

– **Overall Accuracy Equality (OAE)** [20]: This metric evaluates whether the overall prediction accuracy is equal or differs within an acceptable threshold across demographic groups:

$$OAE \downarrow = \frac{|\text{Accuracy}_A - \text{Accuracy}_B|}{\max(\text{Accuracy}_A, \text{Accuracy}_B)} \leq \gamma \tag{4}$$

– **Calibration fairness using Brier Score** (C_{Brier}) [20,38]: This metric assesses whether a model's predicted probabilities are equally accurate (within an acceptable threshold) across different demographic groups. The Brier score for each group is calculated as:

$$Brier \downarrow = \frac{1}{N} \sum_{i=1}^{N} (\hat{y}_i - y_i)^2, \tag{5}$$

where N is the number of samples, $\hat{y}_i$ is the predicted probability, and y_i is the ground truth (0 or 1):

$$C_{Brier} \downarrow = \frac{|Brier_A - Brier_B|}{\max(Brier_A, Brier_B)} \leq \gamma \tag{6}$$

The threshold for acceptable fairness, γ, is set to 0.2 for all specified fairness metrics, following the principle that each demographic group should achieve at least 80% of the performance of another group (s), in line with established fairness standards [2,39,40]. Validation is performed using 5-fold cross-validation ($k = 5$).

To benchmark the proposed CMAAD-Net model, a variety of widely used ML methods in the literature [41,42] were employed independently for comparison, running on the dataset. These include Decision Tree (DT), Random Forest (RF), eXtreme Gradient Boosting (XGBoost), Categorical Boosting (CatBoost), Gradient Boosting (GB), Logistic Regression (LR), Support Vector Machine (SVM), K-Nearest Neighbor (KNN), Naive Bayes (GNB), Adaptive Boosting (AdaBoost), and Light Gradient Boosting Machine (LightGBM). Their hyperparameters are detailed in Table 1.

The CMAAD-Net architecture incorporates an MHA layer configured with the number of heads, and the number of key dimensions, h = 4 and k = 8. All dense layers have the same number of neurons with n = 16 and the same activation function, AF = ReLU. The final activation function, finalAF = Sigmoid, is

Table 1. Hyperparameters of Baseline Models.

Model Name	Hyperparameters
DT	criterion ='gini', splitter ='best', max_depth=None, min_samples_split = 2, min_samples_leaf = 1
RF	n_estimators =100, criterion ='gini', max_depth = None, min_samples_split = 2, min_samples_leaf = 1
XGBoost	n_estimators = 100, learning_rate = 0.1, max_depth = 6, min_child_weight = 1, gamma = 0, subsample = 1.0, objective ='binary:logistic'
CatBoost	iterations=1000, depth=6, learning_rate=0.03, l2_leaf_reg = 3.0, loss_function ='Logloss', verbose = False
GB	loss='log_loss', learning_rate=0.1, n_estimators=100, max_depth=3, min_samples_split = 2, min_samples_leaf = 1
LR	penalty ='l2', dual = False, tol = 0.0001, C = 1.0, solver='lbfgs', max_iter=100
SVM	C=1.0, kernel ='rbf', degree=3, gamma='scale', probability=False
KNN	n_neighbors = 5, weights ='uniform', algorithm='auto', leaf_size=30
GNB	var_smoothing=1e-9
AdaBoost	n_estimators = 50, learning_rate = 1.0, algorithm ='SAMME.R'
LightGBM	boosting_type='gbdt', num_leaves=31, max_depth = −1, learning_rate = 0.1, n_estimators = 100, objective ='binary'

employed in the final layer. The optimiser is Adam, optimiser_type = Adam with learning_rate = 0.001 and trained using the loss, loss_function = Binary cross entropy.

5 Results

Sections. 5.1, 5.2, and 5.3 detail the model performance, fairness results, and MCI and AD statistical analyses, respectively.

5.1 Model Performance

The model results, as shown in Table 2, demonstrate that CMAAD-Net achieved the best overall performance with an accuracy of 0.9321 ± 0.0136, a TPR of 0.9406 ± 0.0232, and an F1-Score of 0.9368 ± 0.0122, alongside the lowest FPR of 0.0808 ± 0.0455. Among the other predictive models, **XGBoost** emerged as the second-best performer with an accuracy of 0.7868 ± 0.0818 and competitive performance across other methods. Conversely, the k-NN model exhibited the lowest accuracy (0.6147 ± 0.0285) and the highest FPR (0.512 ± 0.0776), highlighting its limitations compared to other methods.

Table 2. Results of Model Performance Metrics of Proposed CMAAD-Net and Other Methods.

Model	Accuracy ↑	TPR ↑	Precision ↑	F1-Score ↑	FPR ↓
CMAAD-Net	**0.9321 ± 0.0136**	**0.9406 ± 0.0232**	**0.9349 ± 0.0382**	**0.9368 ± 0.0122**	**0.0808 ± 0.0455**
DT	0.6981 ± 0.0407	0.7232 ± 0.1434	0.7130 ± 0.0522	0.7109 ± 0.0769	0.3398 ± 0.1022
RF	0.7605 ± 0.0526	0.8296 ± 0.1036	0.7610 ± 0.0663	0.7866 ± 0.0456	0.3144 ± 0.1236
XGBoost	*0.7868 ± 0.0818*	0.7997 ± 0.1336	*0.8070 ± 0.0672*	*0.7965 ± 0.0843*	*0.2256 ± 0.0874*
LR	0.7497 ± 0.0443	0.8254 ± 0.0911	0.7416 ± 0.0636	0.7773 ± 0.0469	0.3376 ± 0.1025
GNB	0.7451 ± 0.0431	0.8871 ± 0.0698	0.7157 ± 0.0565	0.7886 ± 0.0319	0.4151 ± 0.1051
k-NN	0.6147 ± 0.0285	0.7326 ± 0.0562	0.6242 ± 0.0521	0.6701 ± 0.0176	0.5120 ± 0.0776
SVM	0.6669 ± 0.0421	*0.9264 ± 0.0816*	0.6329 ± 0.0605	0.7474 ± 0.0376	0.6227 ± 0.0959
AdaBoost	0.7399 ± 0.0445	0.7779 ± 0.1128	0.7497 ± 0.0440	0.7580 ± 0.0568	0.3048 ± 0.0814
CatBoost	0.7351 ± 0.0674	0.7930 ± 0.1532	0.7385 ± 0.0419	0.7558 ± 0.0802	0.3262 ± 0.0880
LightGBM	0.7710 ± 0.0600	0.8084 ± 0.1186	0.7814 ± 0.0724	0.7868 ± 0.0638	0.2673 ± 0.1006
GBM	0.7556 ± 0.0638	0.7879 ± 0.1152	0.7643 ± 0.0624	0.7709 ± 0.0703	0.2781 ± 0.0650

5.2   Fairness Results

The fairness evaluation results, presented in Tables 3 and 4, demonstrate that CMAAD-Net meets the established fairness criteria, with all metrics falling below the threshold γ of 0.2. These results emphasise that **CMAAD-Net** consistently adheres to fairness standards across gender and age group evaluations, achieving metrics comparable to or better than those of other methods. Moreover, **CMAAD-Net** accomplishes this while delivering noticeably higher overall model performance, as reported in Sect. 5.1, highlighting its effectiveness in achieving fair and high-performing outcomes across different demographic groups.

5.3   MCI and AD Statistical Analysis

The statistical analysis results presented in Table 5 and Fig. 2 present key insights into the differences between individuals with MCI and AD across various demographic groups.

Firstly, regarding the significant differences, when considering all individuals irrespective of gender or age groups, significant differences are observed in MMSE, CSF Amyloid, and CSF Total tau levels, with $p-values$ of < 0.001. When analysing individuals aged 75 and below, statistically significant differences emerge in MMSE and all CSF biomarkers, with $p-values$ of either < 0.05

Table 3. Results of Model Fairness Metrics of Proposed CMAAD-Net and Other Methods (Δ_{TPR} and Δ_{FPR}).

Model	$\Delta_{TPR-Gender}$ ↓	$\Delta_{FPR-Gender}$ ↓	$\Delta_{TPR-AgeGroup}$ ↓	$\Delta_{FPR-AgeGroup}$ ↓
CMAAD-Net	0.0950 ± 0.0372	**0.1567 ± 0.0911**	0.0744 ± 0.0250	0.1389 ± 0.0819
DT	0.1360 ± 0.0742	*0.2009 ± 0.1561*	0.2056 ± 0.0999	0.2251 ± 0.1084
RF	0.1579 ± 0.1183	0.3174 ± 0.1891	0.1602 ± 0.1216	**0.1223 ± 0.1085**
XGBoost	**0.0492 ± 0.0441**	0.2666 ± 0.0858	0.1627 ± 0.0582	0.2319 ± 0.1269
LR	0.1854 ± 0.1322	0.3546 ± 0.1451	0.0809 ± 0.0499	0.2221 ± 0.2101
GNB	0.1163 ± 0.1081	0.3285 ± 0.1237	**0.0617 ± 0.0714**	0.2696 ± 0.1803
k-NN	0.1489 ± 0.0996	0.3404 ± 0.1991	0.2473 ± 0.1080	0.1468 ± 0.1573
SVM	0.0893 ± 0.1121	0.2649 ± 0.1631	*0.0633 ± 0.0665*	0.2346 ± 0.1279
AdaBoost	0.1715 ± 0.0583	0.3472 ± 0.1360	0.1041 ± 0.0739	*0.1313 ± 0.1082*
CatBoost	0.0657 ± 0.0400	0.3562 ± 0.1582	0.1549 ± 0.0673	0.1942 ± 0.1139
LightGBM	*0.0557 ± 0.0357*	0.2567 ± 0.1564	0.1319 ± 0.1238	0.1614 ± 0.1330
GBM	0.1755 ± 0.1132	0.2646 ± 0.1161	0.1831 ± 0.1046	0.2567 ± 0.1055

Table 4. Results of Model Fairness Metrics of Proposed CMAAD-Net and Other Methods (OAE and C_{Brier}).

Model	OAE_{Gender} ↓	$OAE_{AgeGroup}$ ↓	$C_{Brier-Gender}$ ↓	$C_{Brier-AgeGroup}$ ↓
CMAAD-Net	0.0834 ± 0.0311	**0.0543 ± 0.0320**	0.0715 ± 0.0511	*0.0628 ± 0.0281*
DT	**0.0660 ± 0.0589**	0.2211 ± 0.0897	0.0660 ± 0.0589	0.2211 ± 0.0897
RF	*0.0827 ± 0.0616*	0.1130 ± 0.0912	0.0769 ± 0.0388	0.0886 ± 0.0887
XGBoost	0.1288 ± 0.0623	0.1116 ± 0.0607	0.0763 ± 0.0502	0.1384 ± 0.0775
LR	0.2274 ⊥ 0.1176	0.1053 ± 0.0391	0.0984 ± 0.0741	0.0701 ± 0.0155
GNB	0.2141 ± 0.0550	0.0987 ± 0.0477	0.1173 ± 0.0619	0.0854 ± 0.0345
k-NN	0.1575 ± 0.0791	0.1309 ± 0.0603	0.0599 ± 0.0314	0.0743 ± 0.0447
SVM	0.1464 ± 0.0774	0.1270 ± 0.0944	0.0647 ± 0.0545	0.0644 ± 0.0423
AdaBoost	0.1071 ± 0.0414	*0.0762 ± 0.0945*	**0.0247 ± 0.0121**	**0.0241 ± 0.0242**
CatBoost	0.1337 ± 0.0848	0.1523 ± 0.0536	0.1056 ± 0.0366	0.0979 ± 0.0557
LightGBM	0.1029 ± 0.0911	0.1019 ± 0.0628	*0.0492 ± 0.0371*	0.0939 ± 0.1058
GBM	0.1165 ± 0.0670	0.1236 ± 0.1113	0.0702 ± 0.0472	0.1645 ± 0.1265

or < 0.001. In contrast, for individuals above the age of 75, significant differences are observed only in MMSE. Concerning gender-based differences, female individuals exhibit significant differences in MMSE ($p - value < 0.001$) and CSF Total tau ($p - value < 0.005$). Meanwhile, male individuals demonstrate

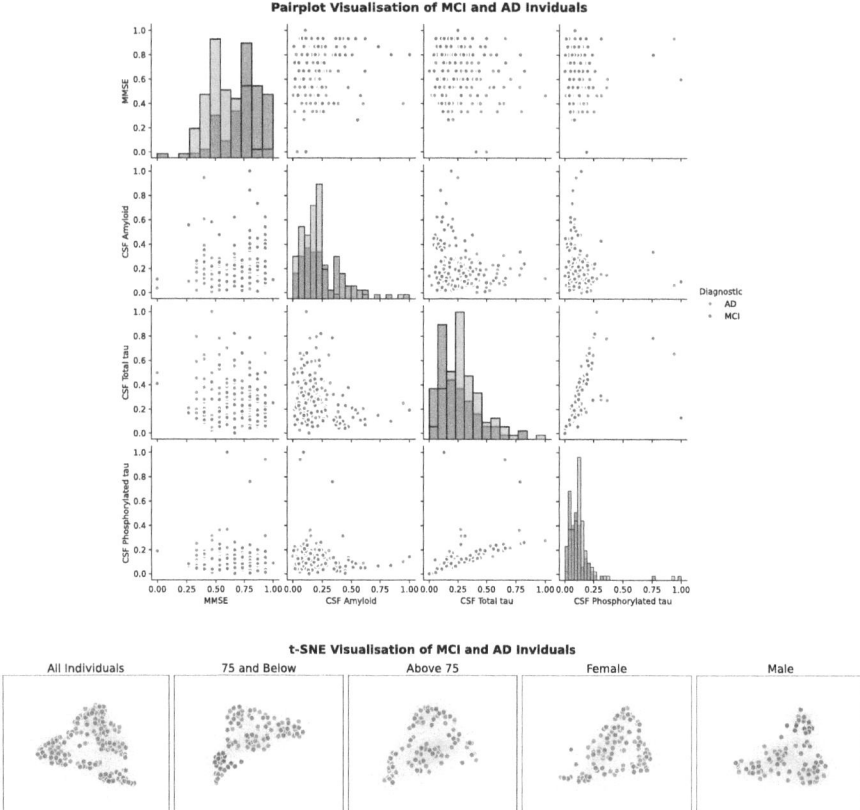

Fig. 2. Visualisation Standardised in Scale of (0,1) of CSF Biomarkers and MMSE of Two Groups: Mild Cognitive Impairment (MCI) and Alzheimer's Disease (AD) using Pairplot and t-SNE.

significant differences in MMSE and CSF Amyloid, both with $p - values$ of < 0.001. Notably, APOE4 does not exhibit statistically significant differences between MCI and AD individuals across any of the analysed groups.

Regarding the average values between MCI and AD individuals, MMSE and CSF Amyloid levels are higher in individuals with MCI compared to those with AD. Whereas, CSF Total tau and CSF Phosphorylated tau levels are elevated in AD individuals relative to those with MCI, except in individuals aged above 75, where no significant difference is observed. Additionally, a notably higher proportion of APOE4-positive individuals is found in the AD group.

Table 5. Statistical Analysis of MCI and AD Across Comparison Groups. All Individuals: Entire dataset, 75 and Below: Individuals aged ≤ 75, Above 75: Individuals aged > 75, Female: Only female individuals, Male: Only male individuals. **Bold**: Significant difference, *Italic*: Higher values in MCI or AD, <u>Underline</u>: Count of APOE4-positive individuals, <u>*Underline Italic*</u>: Higher APOE4-positive count in MCI or AD.

Feature	Group	Average ± Std		$p - value$
		MCI	AD	
MMSE	All Individuals	*26.16 ± 2.68*	23.48 ± 2.40	**<0.001**
	75 and Below	*26.32 ± 2.69*	23.60 ± 2.53	**<0.001**
	Above 75	*25.88 ± 2.67*	23.36 ± 2.29	**<0.001**
	Female	*25.51 ± 3.02*	23.46 ± 2.62	**<0.001**
	Male	*26.82 ± 2.11*	23.50 ± 2.06	**<0.001**
CSF Amyloid (pg/mL)	All Individuals	*673.89 ± 304.91*	517.99 ± 191.32	**<0.001**
	75 and Below	*716.96 ± 327.11*	519.36 ± 149.94	**<0.001**
	Above 75	*600.79 ± 250.99*	516.70 ± 224.97	0.11
	Female	*627.33 ± 281.78*	538.30 ± 199.36	0.06
	Male	*721.50 ± 323.15*	488.50 ± 177.20	**<0.001**
CSF Total tau (pg/mL)	All Individuals	414.75 ± 253.24	*542.96 ± 257.16*	**<0.001**
	75 and Below	399.97 ± 257.26	*543.62 ± 207.51*	**<0.005**
	Above 75	439.85 ± 248.15	*542.34 ± 298.57*	0.10
	Female	445.93 ± 269.26	*557.90 ± 245.27*	**<0.05**
	Male	382.87 ± 234.52	*521.26 ± 275.07*	**<0.05**
CSF Phosphorylated tau (pg/mL)	All Individuals	77.02 ± 65.77	*87.15 ± 50.89*	0.23
	75 and Below	68.62 ± 34.46	*94.64 ± 63.49*	**<0.01**
	Above 75	*91.28 ± 97.58*	80.08 ± 34.30	0.45
	Female	83.69 ± 74.19	*86.83 ± 31.01*	0.77
	Male	70.20 ± 55.92	*87.60 ± 70.96*	0.21
APOE4	All Individuals	<u>38</u>	<u>*55*</u>	0.18
	75 and Below	<u>26</u>	<u>*33*</u>	0.06
	Above 75	<u>12</u>	<u>*22*</u>	0.80
	Female	<u>23</u>	<u>*32*</u>	1.00
	Male	<u>15</u>	<u>*23*</u>	0.09

6   Conclusions

Firstly, about **RQ1** in Sect. 1, the proposed CMAAD-Net effectively leverages CSF biomarkers, MMSE, and APOE4 genetic data, with high predictive performance compared to other methods by using ensemble learning combining MHA and XGBoost. Notably, without relying on additional neuroimaging modalities, it achieves an accuracy of 0.9321 by only integrating CSF biomarkers, APOE4, and MMSE, while acquiring all fairness metrics within acceptable thresholds.

Regarding **RQ2**, statistical differences between MCI and AD are observed in CSF biomarkers and MMSE, with MMSE and CSF Amyloid levels being higher in MCI. At the same time, CSF Total tau and CSF Phosphorylated tau are elevated in AD. APOE4 positivity is more frequent in AD, but it does not show statistically significant differences across groups. Moreover, about **RQ3**, the statistical differences vary across age groups and genders. Individuals aged 75 and below exhibit significant differences in MMSE and all CSF biomarkers, whereas those above 75 show distinctions only in MMSE. Gender-based analyses indicate that females have differences in MMSE and CSF Total tau, while males show variations in MMSE and CSF Amyloid. These findings reveal distinct biomarker patterns that differentiate MCI from AD, highlighting the influence of age and gender on these variations. This underscores the importance of considering demographic factors in AD assessment and diagnosis, particularly in ensuring fair model performance across different populations.

For future work, besides large-scale dataset experiments, a human-centric AI approach should be adopted to ensure robustness and explainability in CMAAD-Net [2,21]. Strengthening robustness entails evaluating the model under adversarial attacks and developing defences against harmful misdiagnoses. Moreover, integrating clinically relevant eXplainable AI (XAI) techniques—such as SHAP, attention visualisation, and counterfactual reasoning—will foster trust, enhance transparency, and ensure the reliability of medical AI systems.

References

1. Association, A.: 2023 Alzheimer's disease facts and figures. Alzheimer's Dement. **19**(4), 1598–1695 (2023)
2. Nguyen, Q.T., Le, L., Tran, X.T., Do, T., Lin, C.T.: Evaluation framework for explainable AI methods in Alzheimer's disease detection with fairness-in-the-loop. In: The 2024 ACM International Joint Conference on Pervasive and Ubiquitous Computing, pp. 870–876 (2024)
3. Nguyen, Q.-T.: Advancing early Alzheimer's disease detection in underdeveloped areas with fair explainable AI methods. Proc. AAAI/ACM Conf. AI Ethics Soc. **7**, 47–49 (2024)
4. Tran, X.T., Le, L., Nguyen, Q.T., Do, T., Lin, C.T.: EEG-SSM: leveraging state-space model for dementia detection. arXiv preprint arXiv:2407.17801 (2024)
5. Ott, A., et al.: Prevalence of Alzheimer's disease and vascular dementia: association with education. Rotterdam study. BMJ **310**(6985), 970–973 (1995)
6. Sunghong, P., et al.: Prospective classification of Alzheimer's disease conversion from mild cognitive impairment. Neural Netw. **164**, 335–344 (2023)
7. Baytaş, İ.M.: Predicting progression from mild cognitive impairment to Alzheimer's dementia with adversarial attacks. IEEE J. Biomed. Health Inf. (2024)
8. Kevin, B., et al.: Systematic review: fluid biomarkers and machine learning methods to improve the diagnosis from mild cognitive impairment to Alzheimer's disease. Alzheimer's Res. Ther. **15**(1), 176 (2023)
9. Davatzikos, C., Bhatt, P., Shaw, L.M., Batmanghelich, K.N., Trojanowski, J.Q.: Prediction of MCI to AD conversion, via MRI, CSF biomarkers, and pattern classification. Neurobiol. Aging **32**(12), 2322–e19 (2011)
10. Tijms Betty, M., et al.: Cerebrospinal fluid proteomics in patients with Alzheimer's disease reveals five molecular subtypes with distinct genetic risk profiles. Nat. Aging **4**(1), 33–47 (2024)

11. Elazab, A., et al.: Alzheimer's disease diagnosis from single and multimodal data using machine and deep learning models: achievements and future directions. Expert Syst. Appl. , 124780 (2024)
12. Blanchard Joel, W., et al.: APOE4 impairs myelination via cholesterol dysregulation in oligodendrocytes. Nature **611**(7937), 769–779 (2022)
13. Tombaugh, T.N., et al.: The mini-mental state examination: a comprehensive review. J. Am. Geriatr. Soc. **40**(9), 922–935 (1992)
14. Acosta, J.N., Falcone, G.J., Rajpurkar, P., Topol, E.J.: Multimodal biomedical AI. Nat. Med. **28**(9), 1773–1784 (2022)
15. Wang, Z., Sun, J.: Transtab: learning transferable tabular transformers across tables. Adv. Neural Info. Process. Syst. (NeurIPS) **35**, 2902–2915 (2022)
16. Gorishniy, Y., Rubachev, I., Khrulkov, V., Babenko, A.: Revisiting deep learning models for tabular data. Adv. Neural Inf. Process. Syst. (NeurIPS) **34**, 18932–18943 (2021)
17. Chen, T., Guestrin, C.: XGBoost: a scalable tree boosting system. In: Proceedings of the 22nd ACM SIGKDD International Conference on Knowledge Discovery and Data Mining, pp. 785–794 (2016)
18. Rui, C., et al.: A study on predicting the length of hospital stay for Chinese patients with ischemic stroke based on the XGBoost algorithm. BMC Med. Inform. Decis. Mak. **23**(1), 49 (2023)
19. Yuan, C., Linn, K.A., Hubbard, R.A.: Algorithmic fairness of machine learning models for Alzheimer disease progression. JAMA Netw. Open **6**(11), e2342203–e2342203 (2023)
20. Caton, S., Haas, C.: Fairness in machine learning: a survey. ACM Comput. Surv. **56**(7), 1–38 (2024)
21. Biniam, G., et al.: A review on human-machine trust evaluation: human-centric and machine-centric perspectives. IEEE Trans. Hum. Mach. Syst. **52**(5), 952–962 (2022)
22. Weiming, L., et al.: Predicting Alzheimer's disease conversion from mild cognitive impairment using an extreme learning machine-based grading method with multimodal data. Front. Aging Neurosci. **12**, 77 (2020)
23. Zhang, Y., Wang, S., Xia, K., Jiang, Y., Qian, P., Initiative, A., et al.: Alzheimer's disease multiclass diagnosis via multimodal neuroimaging embedding feature selection and fusion. Inf. Fusion **66**, 170–183 (2021)
24. Li, J., Hang, X., Hao, Yu., Jiang, Z., Zhu, L.: Multi-modal feature selection with anchor graph for Alzheimer's disease. Front. Neurosci. **16**, 1036244 (2022)
25. Mudasir, A G., Minghui, H., Ashwani Kumar, M., Muhammad, T.,Ponnuthurai, N S.: Ensemble deep learning: a review. Eng. Appl. Artif. Intell. **115**, 105151 (2022)
26. Mahsa, H., et al.: Stacking: a novel data-driven ensemble machine learning strategy for prediction and mapping of PB-ZN prospectivity in Varcheh district, west Iran. Expert Syst. Appl. **237**, 121668 (2024)
27. Khan, P.W., Byun, Y.C., Jeong, O.R.: A stacking ensemble classifier-based machine learning model for classifying pollution sources on photovoltaic panels. Sci. Rep. **13**(1), 10256 (2023)
28. Farida, D., et al.: Changes in plasma neutral and ether-linked lipids are associated with the pathology and progression of Alzheimer's disease. Aging Dis. **14**(5), 1728 (2023)
29. Javid, A.M.et al.: A ReLU dense layer to improve the performance of neural networks. In: IEEE International Conference on Acoustics, Speech and Signal Processing (ICASSP), pp. 2810–2814. IEEE (2021)

30. Abu, Z.M., Obaida, et al.: Dense neural network based arrhythmia classification on low-cost and low-compute micro-controller. Expert Syst. Appl. **239**, 122560 (2024)
31. Linlin, Z., et al.: Predicting disease genes based on multi-head attention fusion. BMC Bioinf. **24**(1), 162 (2023)
32. Chu, et al.: An enhanced EEG microstate recognition framework based on deep neural networks: an application to Parkinson's disease. IEEE J. Biomed. Health Inform. **27**(3), 1307–1318 (2022)
33. Mary L M.: The chi-square test of independence. Biochem. Medica **23**(2), 143–149 (2013)
34. Pauli, Y., et al.: SCIPY 1.0: fundamental algorithms for scientific computing in python. Nat. Meth. **17**, 261–272 (2020)
35. Tahami Monfared, A.A., Byrnes, M.J., White, L.A., Zhang, Q.: Alzheimer's disease: epidemiology and clinical progression. Neurol. Therapy **11**(2), 553–569 (2022)
36. Hicks, S.A., et al.: On evaluation metrics for medical applications of artificial intelligence. Sci. Rep. **12**(1), 5979 (2022)
37. Long, C., Hsu, H., Alghamdi, W., Calmon, F.: Individual arbitrariness and group fairness. Adv. Neural Inf. Process. Syst. NeurIPS **36** (2024)
38. Wei et al.: Optimized score transformation for fair classification. Proc. Mach. Learn. Res. **108** (2020)
39. Feldman, M., et al.: Certifying and removing disparate impact. In: Proceedings of the 21st ACM SIGKDD International Conference on Knowledge Discovery and Data Mining, pp. 259–268 (2015)
40. Pessach, D., Shmueli, E.: A review on fairness in machine learning. ACM Comput. Surv. (CSUR) **55**(3), 1–44 (2022)
41. Arya, et al.: A systematic review on machine learning and deep learning techniques in the effective diagnosis of Alzheimer's disease. Brain Info. **10**(1), 17 (2023)
42. Tanveer et al.: Machine learning techniques for the diagnosis of Alzheimer's disease: a review. ACM Trans. Multimedia Comput. Commun. Appl. (TOMM), **16**(1s), 1–35 (2020)

Unlocking Neural Transparency: Jacobian Maps for Explainable AI in Alzheimer's Detection

Yasmine Mustafa[1], Mohamed Elmahallawy[2], and Tie Luo[3(✉)]

[1] Missouri University of Science and Technology, Rolla, MO 65409, USA
yam64@mst.edu
[2] Washington State University, Richland, WA 99354, USA
mohamed.elmahallawy@wsu.edu
[3] University of Kentucky, Lexington, KY 40506, USA
t.luo@uky.edu

Abstract. Alzheimer's disease (AD) causes progressive cognitive decline, where early detection is critical for effective intervention. While deep learning models have achieved high detection accuracy in AD diagnosis, their lack of interpretability has led to skepticism among medical professionals. This paper introduces a novel pre-model approach leveraging Jacobian Maps (JM) within a multi-modal framework to improve interpretability and trustworthiness in AD detection. By capturing localized brain volume changes, JMs enhance model explainability by correlating predictions with established neuroanatomical biomarkers of AD. We validate the effectiveness of JMs through experiments comparing the performance of a 3D CNN trained on JMs versus traditional preprocessed data, which demonstrates superior accuracy. Additionally, we provide both visual and quantitative insights using 3D Grad-CAM analysis, demonstrating improved interpretability and diagnostic accuracy.

Keywords: Alzheimer's Disease (AD) · Explainable AI (XAI) ·
Jacobian Maps · Multi-Modal Data · Medical Image Analysis

1 Introduction

Alzheimer's disease (AD) is a leading cause of dementia, posing immense medical and economic challenges globally. Characterized by gradual cognitive decline, AD progresses from mild memory lapses to severe functional impairments. With Alzheimer's cases projected to rise significantly in the coming decades [1], early detection—e.g., of mild cognitive impairment (MCI)—is critical to slowing disease progression. Advances in neuroimaging and machine learning, especially with multi-modal data, have significantly improved AD diagnosis [2–4]. However, skepticism persists due to the "black-box" nature of AI models, necessitating explainable AI (XAI) methods to foster trust among clinicians by elucidating the "why" behind predictions.

To address these challenges, explainability in AI typically operates at three levels: pre-model, focusing on data-level insights (e.g., feature engineering to highlight key biomarkers); in-model, leveraging inherently interpretable models (e.g., decision trees or linear models); and post-model, generating explanations after predictions (e.g., heatmaps or saliency maps). While in-model approaches often struggle to capture complex patterns, post-model XAI tools, such as Grad-CAM [5], have gained significant attraction due to their applicability to deep learning models. However, such heatmap-based methods face key challenges in medical imaging, particularly in Alzheimer's disease: (i) the absence of ground truth for validating visual explanations against actual dementia-related regions; (ii) insufficient quantitative metrics for evaluating alignment with known pathology. Techniques like voxel-based morphometry (VBM), which identify structural brain changes through statistical methods, are often overlooked in training pipelines, leading to potentially suboptimal performance. Notably, there is a dearth of effective pre-model approaches for highlighting meaningful neuroanatomical changes before model training, which could enhance interpretability.

This paper introduces Jacobian Maps (JMs), derived from the VBM pre-processing pipeline but repurposed in this paper as a pre-model anatomical reference, to enhance the explainability of Alzheimer's disease detection. By computing Jacobian determinants across an image, JM generates a matrix that accurately captures spatial and directional changes in medical images, which can serve as a unique ground truth for each subject. Leveraging JM's ability to pinpoint voxel-level cerebral changes relative to a healthy brain, we identify and highlight morphometric variations, thereby improving the interpretability and trustworthiness of AD detection. Our key contributions include:

- **Enhanced interpretability with JM:** Our approach transforms brain imaging data, such as Magnetic Resonance Imaging (MRI), Computed Tomography (CT), or Positron Emission Tomography (PET), into JM. This transformation enables models to reveal localized volumetric changes across distinct brain regions, thereby enhancing interpretability and providing more insights into previously opaque model predictions.
- **Holistic methodology:** Unlike existing methods that partition brain scans into patches or slices—potentially losing critical spatial information—our approach processes entire brain scans, preserving voxel-level details and achieving superior performance. Moreover, JM eliminates the need for brain segmentation or division, reducing processing costs and complexity.
- **More accurate AD detection without altering model:** As a pre-model approach, JM introduces a data preprocessing step in the ML pipeline by computing Jacobian determinants. Keeping the AI model intact, JM significantly improves dementia detection accuracy across all AD stages, achieving 95.2% for Cognitively Normal (CN), 96.3% for MCI, 90.2% for Mild Dementia (MLD), and 90.2% (MOD), compared to traditional methods, which achieve 88.3% for CN, 90.5% for MCI, 83.4% for MLD, and 83.4% for SEV.
- **Comprehensive validation of interpretability:** Jacobian-derived heatmaps provide both qualitative and quantitative insights into brain regions undergoing volumetric changes. Qualitatively, these heatmaps high-

light regions of compression and expansion of the brain. Quantitatively, we calculate region-based activation metrics by aligning with the known neuroanatomical template (MNI152) and using the Harvard-Oxford cortical atlas. Our results reveal that the **frontal-temporal region** consistently exhibits the highest intensity values across all dementia stages, underscoring its clinical relevance and our method's ability to pinpoint key neuroanatomical patterns.

2   Related Work

Existing XAI methods applied to AD detection can be grouped into three main categories.

Pre-model (Ante-hoc) Approaches. These methods often segment or partition the brain into smaller regions for localized analysis. For example, Amoroso et al. [6] employed graph-based models, treating brain patches as network nodes connected via Pearson's correlation. This approach highlights connectivity patterns in regions like the hippocampus and amygdala for AD and the posterior cingulate for MCI to avoid computationally intensive voxel-wise analyses. Liu et al. [7] used a three-dimensional CNN to segment brain scans into patches, pre-training two networks to learn region-specific features, which were later integrated using a 2D CNN to enhance interpretability.

In-Model (Intrinsic) Approaches. These techniques embed explainability into the training process. For instance, Yu et al. [8] introduced a multi-stage aggregation module to generate visual explanations during AD diagnosis. Similarly, Mustafa et al. [3] employed a Jacobian-Augmented Loss (JAL) function, utilizing Jacobian Saliency Maps (JSM) to penalize spurious predictions, ensuring model decisions are based on meaningful patterns.

Post-model (Post-hoc) Approaches. These methods focus on interpreting pre-trained models. El-Sappagh et al. [9] used a Random Forest classifier with Gini impurity and SHAP [10] to provide global feature importance and decision-level explanations. Kamal et al. [11] combined CNNs with gene expression data, leveraging LIME [12] for gene significance interpretation in AD classification.

Jacobian Maps (JM): While JM was used to analyze structural changes in AD, their potential for XAI remains untapped. Spasov et al. [13] used JMs to quantify local volumetric changes, while Abbas et al. [14] developed JD-CNN to train directly on JMs, improving spatial correlation analysis. Mustafa et al. [15] extended this by integrating JMs from MRI and CT scans to identify regions of brain atrophy, enhancing the focus on structural changes before training. To the best of our knowledge, this paper is the first that introduces JM into XAI to explain model predictions.

3   Method

3.1   Transforming Brain Images into Jacobian Maps (JM)

To analyze brain imaging modalities (MRI, CT, PET) and provide explainable insights, we propose transforming them into JMs to leverage the effectiveness of JM in nonlinear image registration. Image registration is a process that aligns individual scans to a standardized brain template so as to minimize individual variations and enable assessment of relative volume differences (deformations). It helps identify significant anatomical differences across populations (e.g., Alzheimer's patients vs. healthy controls).

Importantly, JM enables holistic brain analysis without partitioning the brain into patches or requiring segmentation like other methods, thereby minimizing data loss, complexity, and improving dementia detection accuracy. Moreover, JM facilitates feature attribution by delineating and quantifying volumetric transformations in specific brain regions, which provide a detailed pre-model explanation to guide model training.

In this work, we use the MNI152 brain template as a standard reference for image registration. Comparing an individual's scans to this template allows detection of localized structural changes via deformation maps, while comparing such scans at different times allows assessment of disease progression. This approach, known as *tensor-based morphometry* [16], is grounded in the field of computational anatomy. This framework involves transforming a source image M to align with a target image F using a spatial mapping ϕ, producing a deformation vector field $\boldsymbol{v}$:

$$\boldsymbol{v}(x, y, z) = \phi(x, y, z) - (x, y, z) \tag{1}$$

where $\phi(x, y, z)$ represents the transformed coordinates. To ensure a smooth (differentiable), one-to-one (invertible) deformation, a regularization constraint is introduced, framed as an optimization problem minimizing a cost function:

$$\ell(\phi, M \circ F) = \ell_{sim} + \alpha \times \ell_{Reg} \tag{2}$$

where ℓ_{sim} measures how well two images align with each other, ℓ_{Reg} is a regularization term, and α balances the two terms. We use Mattes Mutual Information (MMI) [17] as it identifies optimal image alignment by evaluating the shared global information from the joint histogram. Additionally, MMI reduces the risk of overfitting by discouraging excessive clustering of marginal probabilities, which denotes the probability of individual variables being considered in isolation, as opposed to joint probabilities. Excessive clustering of marginal probabilities could lead to overfitting by focusing too much on specific individual features rather than the overall distribution. Unlike cross-correlation, which is affected by local intensity fluctuations and faces certain multimodality challenges, MMI excels in achieving robust rigid registration [18]. Thus,

$$\ell_{sim} = MMI(M, F) = \sum_{m \in M} \sum_{f \in F} P(m, f) \log \frac{P(m, f)}{P(m)Q(f)} \tag{3}$$

where $P(m, f)$ is the joint probability of intensities m in image M and f in image F. $P(m)$ and $Q(f)$ are marginal probabilities of m and f, respectively. For ℓ_{Reg}, we use B-spline regularization [19] to enforce smoothness and prevent overfitting in our model. B-splines, or basis splines, are piecewise polynomial functions that provide flexible yet stable approximations by constraining the curvature of the function, ensuring that the model generalizes well to unseen data while avoiding excessive complexity, which can be expressed as:

$$\ell_{Reg} = \int \left|\nabla^2 \phi(x, y, z)\right|^2 dV \tag{4}$$

The Jacobian matrix $\mathcal{J}$, derived from the deformation vector field $\boldsymbol{v}(x, y, z)$, encodes local deformations such as stretching, shearing, and rotation:

$$\mathcal{J}(\boldsymbol{v}) = \begin{pmatrix} \frac{\partial v_x}{\partial x} & \frac{\partial v_x}{\partial y} & \frac{\partial v_x}{\partial z} \\ \frac{\partial v_y}{\partial x} & \frac{\partial v_y}{\partial y} & \frac{\partial v_y}{\partial z} \\ \frac{\partial v_z}{\partial x} & \frac{\partial v_z}{\partial y} & \frac{\partial v_z}{\partial z} \end{pmatrix} \tag{5}$$

This process forms a Jacobian tensor that captures the rate of change of the vector field in three-dimensional space (width, height, and depth). The Jacobian determinant, $\mathrm{Det}(\mathcal{J})$, computed at each voxel, quantifies the local volumetric change. By calculating these determinants across the entire image domain ($x = 1, \ldots, W$; $y = 1, \ldots, H$; $z = 1, \ldots, D$), we construct the JM, which provides a spatial representation of local deformations and volume changes as

$$J_{\mathrm{map}}(M) = \begin{cases} \text{Expansion,} & \text{if } \mathrm{Det}(\mathcal{J}) > 1 \\ \text{No change,} & \text{if } \mathrm{Det}(\mathcal{J}) = 1 \\ \text{Compression,} & \text{if } \mathrm{Det}(\mathcal{J}) < 1 \end{cases}$$

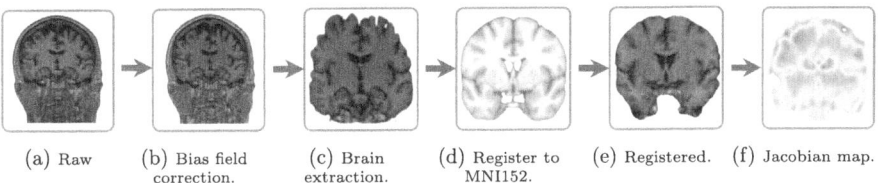

(a) Raw   (b) Bias field correction.   (c) Brain extraction.   (d) Register to MNI152.   (e) Registered.   (f) Jacobian map.

Fig. 1. Preprocessing pipeline.

3.2 Brain Image Processing

This subsection details how we handle brain imaging data, with a focus on MRI. MRI provides highly detailed visuals of progressive cerebral atrophy, especially through T1-weighted sequences. Compared to CT, MRI offers superior soft tissue contrast and multi-contrast imaging for detecting AD [20]. While PET visualizes beta-amyloid plaques, MRI indirectly assesses the amyloid burden through indicators like hippocampal volume loss and provides better spatial resolution.

Data Collection. We utilize 1557 MRI images from the OASIS-3 dataset [21], which comprises 1377 participants: 755 cognitively normal adults and 622 with varying cognitive decline, aged 42–95. To avoid overlap between training and test sets, images from the same subject are confined to one dataset partition, ensuring the evaluation of unseen data for accurate generalization. Clinical Dementia

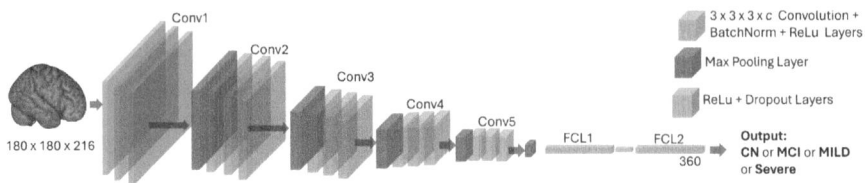

Fig. 2. The 3D convolutional neural network consists of five 3D convolutional layers with kernel sizes of $3\times3\times3\times$c, where c represents the number of input channels. The output is passed through two fully connected layers after flattening. Each convolution layer is followed by batch normalization and ReLU activation, and max-pooling is used at selected locations. Dropout is applied at the first fully connected (FC) layer for regularization. The final output of class logits is converted to probabilities using softmax.

Rating (CDR) scores (0 to 3) guide the diagnosis, where 0 indicates no dementia, and 3 indicates severe dementia. To address the issue of limited data for severely demented cases, we merge moderate and severe stages into one category, resulting in four groups: normal (CN), mild cognitive impairment (MCI), mild dementia (MLD), and moderate-to-severe (MOD). This classification provides more granularity than binary methods (AD and non-AD).

Preprocessing Pipeline. Preprocessing is particularly critical in medical imaging. Our MRI data preprocessing pipeline is shown in Fig. 1. Bias field correction, performed with FLIRT [22], mitigates intensity variations from magnetic field inhomogeneities, enhancing image consistency. Brain extraction using BET [23] isolates the brain region, reducing computational load and improving volumetric measurement accuracy.

The registration process employs symmetric normalization (SYN) using ANTs [24] for global alignment through affine transformations, addressing positional and orientation differences across subjects. Finally, JMs, computed for each MRI, is a crucial step to identify potential indicators of neurological conditions and offer insights into brain morphology by quantifying local volume variations over distinct brain regions. Our ablation analysis (discussed in the next section) highlights the superior ability of JMs, compared to registered images without JM, to reveal patterns of brain atrophy—a hallmark of neurodegenerative disorders like AD.

Handling Imbalanced Data. To address the class imbalance inherent in MRI-based disease classification, we use SMOTE [25], which generates synthetic samples by interpolating between minority class instances, enhancing dataset balance and representation.

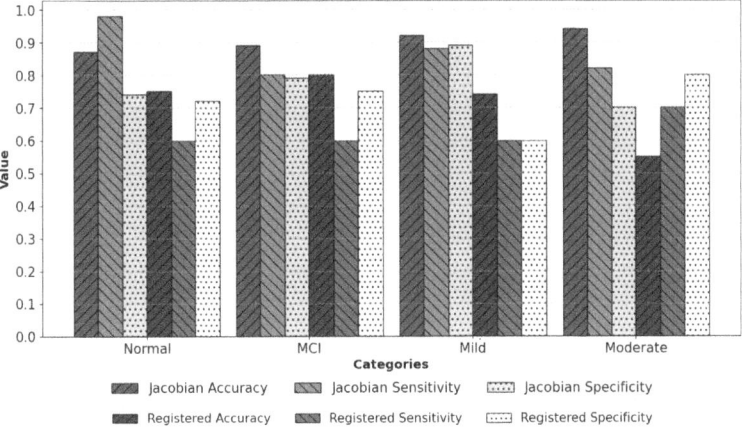

Fig. 3. Performance comparison per each AD class between conventional registration (without JM; gray-scale bars) versus registration with JM (red-scale bars). (Color figure online)

4   Experiments and Explainability Analysis

4.1   Model and Training

The model used in our simulation is depicted in Fig. 2. An ablation study by Wang et al. [26] determined the optimal model configuration for MRI analysis, identifying 44 channels per convolutional layer as yielding peak performance. To improve computational efficiency, we reduced the channels to 10; on the other hand, to handle the increased feature complexity, we increased the number of neurons in the FC layer from 64 to 360.

Training. We employed 5-fold cross-validation, splitting the data into subsets to train for 50 epochs per fold, using early stopping to prevent overfitting. Training stops when validation loss does not decrease for a predefined patience period. Model checkpoints are saved to retain the best-performing model. The model's hyperparameters include a batch size of 15, the Adam optimizer with a learning rate of 10^{-4}, and the cross-entropy loss function.

4.2   Model Performance Evaluation

Figure 3 compares the model's performance using metrics such as accuracy, precision, and recall across registered images and JMs. The cross-validation, applied across AD classes (Normal, MCI, Mild, and Moderate), shows that JMs consistently outperform registered images in all metrics. These results underscore the effectiveness of JMs in capturing informative local volumetric changes (expansion/compression) in brain anatomy, which enhances model's discriminative power and leads to improved model performance and diagnostic accuracy.

Figures 4 further illustrate the learning curves for models trained on registered images versus JMs. Models trained on JM exhibit smoother learning curves

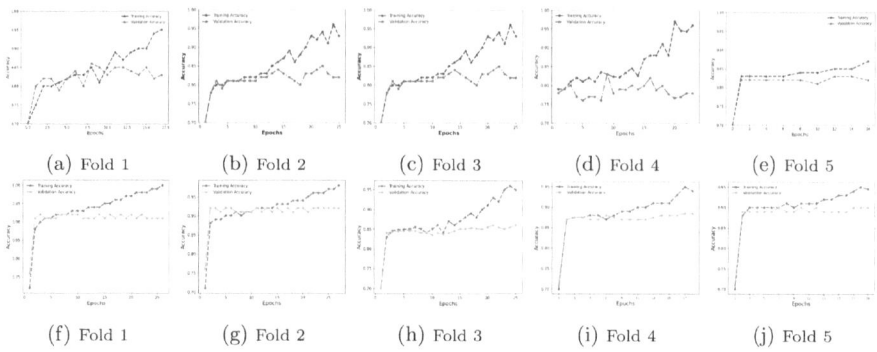

Fig. 4. Learning curves of training and validation accuracy across 5-fold cross-validation. Upper row: **without JM**, Lower row: **with JM**, demonstrating notably better stability and convergence.

and stabler validation accuracy compared to those trained on conventionally registered images. The top row (no JM) shows greater fluctuations, while the bottom row (with JM) demonstrates more consistent convergence in addition to improved training and validation performance, highlighting the benefits of JM's enhanced feature representation.

4.3   Interoperability Evaluation

1) Qualitative Analysis. To harness the interpretive capability of JMs, we employ Gradient-weighted Class Activation Mapping (Grad-CAM) [5], which visualizes and interprets the contributions of specific regions in 3D CNN predictions. Grad-CAM computes the gradients of the target class logits with respect to feature maps from the last convolutional layer. These gradients are aggregated to assign weights to each feature map, producing a Class Activation Map (CAM) (see Algorithm 1). The CAM is superimposed on the original image, visually highlighting the regions most relevant to the model's prediction.

Algorithm 1. Compute 3D Grad-CAM

1: **Step 1:** Calculate the gradient of the score y^C for class C with respect to the feature map activation of the unit in the final convolutional layer at location (i, j, m) in three dimensions: $\frac{\partial y^C}{\partial A_u}$

2: **Step 2:** Obtain the importance weights of the unit by Global Average Pooling (GAP) of the gradients:
$\lambda_u^C = \frac{1}{Z} \sum_{i,j,m} \frac{\partial y^C}{\partial A_u}$,
where $Z = u \times v \times w$, and $A_u(i, j, m)$ represents the activation at location (i, j, m).

3: **Step 3:** Compute a weighted sum across all units and apply the Rectified Linear Unit (ReLU) activation to the result to eliminate features that negatively impact the class prediction:
$\text{ReLU} \left(\sum_u \lambda_u^C A_u \right)$

4: **Step 4:** Upsample the resulting map to the original input size.

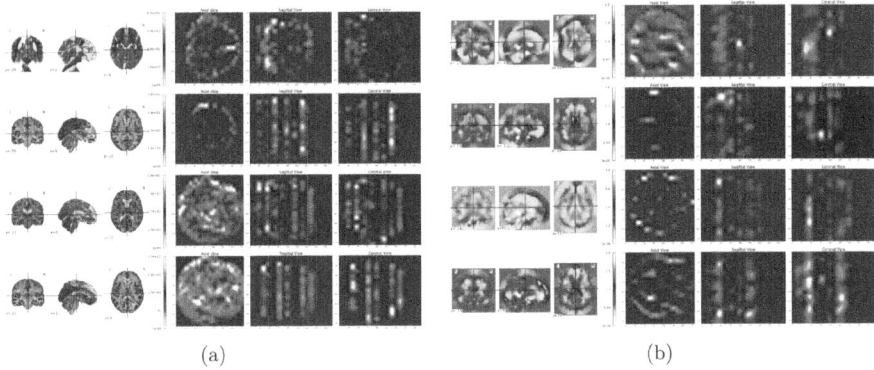

(a)   (b)

Fig. 5. Heatmaps (2nd and 4th groups of columns) generated by 3D Grad-CAM after the 3rd conv block of our 3D CNN for four brain MRI registered (a) without JMs and (b) with JMs. JMs enhance interpretability by providing more specific and localized visualizations of structural brain changes, enabling precise identification of subtle yet critical variations.

Table 1. Brain regions ranked by importance for each AD class based on heatmap intensity values.

CN	MCI	MLD	MOD
Frontal-Temporal (2.71)	Frontal-Temporal (2.45)	Frontal-Temporal (2.76)	Frontal-Temporal (2.76)
Sub-lobar (2.51)	Temporal Lobe (1.94)	Temporal Lobe (2.33)	Frontal Lobe (2.28)
Temporal Lobe (2.43)	Frontal Lobe (1.89)	Frontal Lobe (2.28)	Parietal Lobe (1.77)
Limbic Lobe (2.40)	Sub-lobar (1.78)	Sub-lobar (2.20)	Limbic Lobe (2.01)
Frontal Lobe (2.37)	Background (1.73)	Occipital Lobe (2.02)	Occipital Lobe (2.02)
Midbrain (2.29)	Limbic Lobe (1.67)	Pons (1.82)	Pons (1.82)
Pons (2.26)	Occipital Lobe (1.59)	Posterior Lobe (1.91)	Posterior Lobe (1.91)
Background (2.08)	Anterior Lobe (1.53)	Background (1.73)	Background (1.73)
Parietal Lobe (2.07)	Medulla (1.37)	Anterior Lobe (1.53)	Medulla (1.43)
Posterior Lobe (1.98)	Midbrain (1.61)	Medulla (1.43)	Anterior Lobe (1.53)
Medulla (1.76)	Frontal-Temporal (2.45)	Parietal Lobe (1.77)	Midbrain (1.90)
Anterior Lobe (1.97)	Parietal Lobe (1.33)	Frontal Lobe (1.89)	Pons (1.82)

When applied to JMs, the highlighted regions correspond directly to areas of volumetric compression and expansion in the brain. As shown in Fig. 5, JM-derived heatmaps excel at pinpointing specific regions of interest compared to those generated from conventional registered images. While heatmaps from registered images provide broader, less targeted representations, they often fail to capture the specificity required to identify significant volumetric changes. In contrast, JM-derived heatmaps deliver focused and detailed visualizations, revealing localized structural alterations in the brain. This enhanced specificity enables the precise identification of subtle yet critical variations, improving both model interpretability and diagnostic utility.

In summary, JMs play a pivotal role in unraveling the decision-making processes of models by identifying distinct regions within an input image and quantifying their volumetric transformations. This enables precise feature attribution, assigning significance and influence to specific input regions that impact the model's output. Consequently, JMs offer a comprehensive *pre-model (ante-hoc)* explanation, enhancing our understanding of how input data is interpreted by the model.

2) Quantitative Analysis. To quantitatively assess the model's interpretability, we align Grad-CAM-generated heatmaps with the MNI152 template by computing registration transformations. Using the template as a fixed image and each subject's MRI as a moving image, these transformations ensure consistent spatial coordinates across subjects and images. The Harvard-Oxford cortical atlas [27] is then used to calculate the average voxel intensity within each brain region for each heatmap, providing a region-based metric of activation. Regions are ranked in descending order of importance across subjects for each heatmap class. This approach identifies the brain regions most sensitive to input variations for each AD class, as shown in Table 1 for JM-derived heatmaps.

The rankings in Table 1 highlight which brain regions contribute most significantly to distinguishing between brain states (Normal, MCI, Mild, and Moderate). Several key observations can be made:

– **Normal State:**
  - The **Frontal-Temporal** region (2.71) is the most significant, followed by the **Sub-lobar** region (2.51) and **Temporal Lobe** (2.43).
  - **Frontal-Temporal** region is involved in memory, decision-making and language, and is known to degenerate early in AD. This explains its highest rank across **all stages**.
– **MCI State:**
  - The **Frontal-Temporal** region remains the most significant (2.45), but the **Temporal Lobe** (1.94) surpasses the **Sub-lobar** region (1.78) in importance.
– **Mild State:**
  - The **Frontal-Temporal** region remains dominant (2.76), while the **Temporal Lobe** (2.33) becomes more important than the **Sub-lobar** region (2.20).
  - The increasing prominence of the **Temporal Lobe** during the MCI and MLD stages is because this lobe includes hippocampus and entorhinal cortex, which are among the first regions to show atrophy, with memory loss being a key symptom.
– **Moderate State:**
  - The **Frontal-Temporal** region retains its significance (2.76), but the **Frontal Lobe** (2.28) and **Parietal Lobe** (1.77) emerge as more critical, surpassing the **Temporal Lobe** (2.02) and **Limbic Lobe** (2.01).
  - The **Frontal** and **Parietal Lobe** are responsible for planning, judgment, spatial orientation, and attention. Our finding shows that these areas are less impacted in early stages yet become more prominently affected in later stages as neurodegeneration spreads.

These findings highlight the dynamic changes in brain region importance as Alzheimer's disease progresses from Normal to MCI, Mild, and Moderate states. Importantly, they are **consistent with clinical evidence**, which underscores the significance and validity of our XAI method.

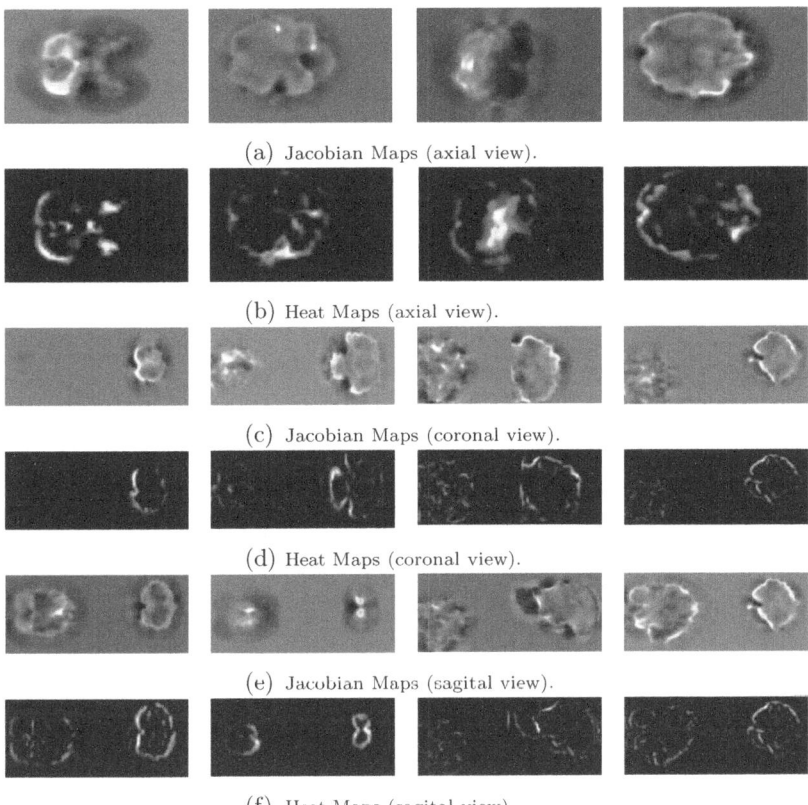

(a) Jacobian Maps (axial view).

(b) Heat Maps (axial view).

(c) Jacobian Maps (coronal view).

(d) Heat Maps (coronal view).

(e) Jacobian Maps (sagital view).

(f) Heat Maps (sagital view).

Fig. 6. Heatmaps for the fused MRI-CT brain images.

Table 2. Multimodal: Ablation for REG (without JM) vs. JM.

	Accuracy				Precision				Recall			
	CN	MCI	MLD	MOD	CN	MCI	MLD	MOD	CN	MCI	MLD	MOD
REG	88.3	90.5	83.4	83.4	86.8	82.8	80	95.5	83.3	64.0	69.3	84.4
JM	95.2	96.3	90.2	90.2	92.8	100	83.33	98.6	94.96	89.6	78.6	90.2

CN - Cognitively Normal, MCI - Mild Cognitive Impairment, MOD - Moderate-to-severe Dementia, REG - Registered without JM.

4.4   Extension to Multi-modal Setting

We extend our research into a multimodal setting by integrating MRI and CT data from the OASIS-3 dataset using an early fusion approach. Early fusion combines information from different imaging modalities at the initial stage of processing, allowing the model to leverage complementary information for improved performance. In our method, this integration involves concatenating MRI and CT data before inputting them into the neural network model.

We employ the same model architecture outlined in Sect. 4.1, with an adaptation of two additional convolutional layers to optimize computational efficiency in handling the increased dimensionality from multimodal fusion. This adjustment ensures that the model effectively captures the richer feature representations offered by both MRI and CT data.

Figure 6 illustrates the heatmaps generated from the combined MRI and CT feature sets, providing insights into the spatial contributions of both modalities. Table 2 presents the testing results of the 3D CNN, highlighting significant performance improvements when incorporating JMs compared to using registered images alone. These results confirm the efficacy of JM in enhancing interpretability and diagnostic accuracy, even in multimodal settings.

5   Conclusion

Alzheimer's Disease (AD) remains one of the most pressing challenges in neurodegenerative research, with its progressive nature leading to significant cognitive, functional, and behavioral decline. Early and accurate detection is critical for enabling timely interventions, yet the adoption of deep learning models in clinical practice has been hindered by their lack of interpretability and transparency. In this paper, we introduced a novel pre-modeling approach leveraging Jacobian Maps (JMs) within a multimodal framework to address these challenges. By capturing localized volumetric changes in distinct brain regions, JMs provide interpretable and intuitive insights into structural brain alterations associated with AD. Integrated into a 3D CNN, JM-based features demonstrated superior diagnostic performance across all stages of AD, achieving higher accuracy compared to traditional preprocessing methods. Furthermore, the use of 3D Grad-CAM together with JM enabled both visual and quantitative explanations, bridging the gap between model predictions and clinical understanding. These explanations effectively highlighted key brain regions, such as the frontal-temporal areas, known to be critical in AD pathology, thereby reinforcing the clinical relevance of the method.

Future work will focus on expanding this methodology to other neurodegenerative disorders and exploring its potential in real-world clinical settings.

References

1. Scheltens, P., et al.: Alzheimer's disease. Lancet **397**(10284), 1577–1590 (2021)
2. Rathore, S., et al.: A review on neuroimaging-based classification studies and associated feature extraction methods for Alzheimer's disease and its prodromal stages. Neuroimage **155**, 530–548 (2017)
3. Mustafa, Y., Luo, T.: Unmasking dementia detection by masking input gradients: a JSM approach to model interpretability and precision. In: The Pacific-Asia Conference on Knowledge Discovery and Data Mining (PAKDD) (2024)
4. Venugopalan, J., et al.: Multimodal deep learning models for early detection of Alzheimer's disease stage. Sci. Rep. **11**(1), 3254 (2021)
5. Selvaraju, R.R., et al.: Grad-CAM: visual explanations from deep networks via gradient-based localization. Int. J. Comput. Vis. **128**, 336–359 (2020)
6. Amoroso, N., Quarto, S., La Rocca, M., Tangaro, S., Monaco, A., Bellotti, R.: An explainability artificial intelligence approach to brain connectivity in Alzheimer's disease. Front. Aging Neurosci. **15**, 1238065 (2023)
7. Liu, M., Cheng, D., et al.: Multi-modality cascaded convolutional neural networks for Alzheimer's disease diagnosis. Neuroinformatics **16**, 295–308 (2018)
8. Yu, L., Xiang, W., Fang, J., Chen, Y., Zhu, R.: A novel explainable neural network for Alzheimer's disease diagnosis. Pattern Recogn. **131**, 108876 (2022)
9. El-Sappagh, S., Alonso, J.M., Islam, S.R., Sultan, A.M., Kwak, K.S.: A multi-layer multimodal detection and prediction model based on explainable artificial intelligence for Alzheimer's disease. Sci. Rep. **11**(1), 2660 (2021)
10. Lundberg, S.M., Lee, S.-I.: A unified approach to interpreting model predictions. In: Advances in Neural Information Processing Systems, vol. 30 (2017)
11. Kamal, Md.S., Northcote, A., Chowdhury, L., Dey, N., Crespo, R.G., Herrera-Viedma, E.: Alzheimer's patient analysis using image and gene expression data and explainable-AI to present associated genes. IEEE Trans. Instrum. Measur. **70**, 1–7 (2021)
12. Ribeiro, M.T., Singh, S., Guestrin, C.: "Why should I trust you?" explaining the predictions of any classifier. In: Proceedings of the 22nd ACM SIGKDD International Conference on Knowledge Discovery and Data Mining, pp. 1135–1144 (2016)
13. Spasov, et al.: A parameter-efficient deep learning approach to predict conversion from mild cognitive impairment to Alzheimer's disease. Neuroimage **189**, 276–287 (2019)
14. Abbas, et al.: Transformed domain convolutional neural network for Alzheimer's disease diagnosis using structural MRI. Pattern Recogn. **133**, 109031 (2023)
15. Mustafa, Y., Luo, T.: Diagnosing Alzheimer's disease using early-late multimodal data fusion with Jacobian maps. In: 2023 IEEE International Conference on E-Health Networking, Application & Services (Healthcom), pp. 49–55. IEEE (2023)
16. Riyahi, S., Choi, et al.: Quantifying local tumor morphological changes with Jacobian map for prediction of pathologic tumor response to chemo-radiotherapy in locally advanced esophageal cancer. Phys. Med. Biol. **63**(14), 145020 (2018)
17. Mattes, D., Haynor, et al.: PET-CT image registration in the chest using free-form deformations. **22**(1), 120–128 (2003)
18. Avants, et al.: Symmetric diffeomorphic image registration with cross-correlation: evaluating automated labeling of elderly and neurodegenerative brain. Med. Image Anal. **12**(1), 26–41 (2008)
19. Tustison, N.J., et al.: Explicit b-spline regularization in diffeomorphic image registration. **7**, 39 (2013)

20. Aramadaka, S., et al.: Neuroimaging in Alzheimer's disease for early diagnosis: a comprehensive review. Cureus **15**(5) (2023)
21. LaMontagne, P.J., et al.: Oasis-3: longitudinal neuroimaging, clinical, and cognitive dataset for normal aging and Alzheimer disease. MedRxiv, pp. 2019–12 (2019)
22. Jenkinson, M., Bannister, P., Brady, M., Smith, S.: Improved optimization for the robust and accurate linear registration and motion correction of brain images. Neuroimage **17**(2), 825–841 (2002)
23. Smith, S.M.: Fast robust automated brain extraction. Hum. Brain Mapp. **17**(3), 143–155 (2002)
24. Avants, B.B., Tustison, N., Song, G., et al.: Advanced normalization tools (ANTS). Insight j **2**(365), 1–35 (2009)
25. Chawla, et al.: Smote: synthetic minority over-sampling technique. J. Artif. Intell. Res. **16**, 321–357 (2002)
26. Wang, et al.: Deep neural network heatmaps capture Alzheimer's disease patterns reported in a large meta-analysis of neuroimaging studies. Neuroimage **269**, 119929 (2023)
27. Santi, D., et al.: An explainable convolutional neural network for the early diagnosis of Alzheimer's disease from 18f-FDG PET. J. Digit. Imaging **36**(1), 189–203 (2023)

The Use of Large Language Models to Cluster Genomic Data

Reem Al-Saidi[1]([✉])(iD), Ziad Kobti[1](iD), and Thorsten Strufe[2]

[1] University of Windsor, Windsor, ON, Canada
{alsaidir,kobti}@uwindsor.ca
[2] Karlsruhe Institute of Technology (KIT), Karlsruhe, Germany
thorsten.strufe@kit.edu

Abstract. Clustering in genomic data is critical for understanding genetic variation, population structure, and disease associations. The high-dimensional and complex nature of genomic data frequently results in suboptimal accuracy from traditional clustering algorithms, whether or not dimensionality reduction techniques are used. The potential of a genome-specific LLM and the embeddings it produces for clustering genomic data are examined in this work. Using three benchmark genome datasets, we test the effectiveness of this method to determine whether LLM-based embeddings can improve clustering accuracy. Our results indicate that while the fine-tuned embeddings considerably increase accuracy in just two datasets, the pre-trained embeddings perform similarly to conventional clustering techniques. These findings demonstrate the possibility of optimizing genome-specific LLMs to improve clustering results, but they also underline the necessity of more research to fully utilize LLM embeddings for the analysis of genomic data.

Keywords: Embedding · Clustering · Genome Large Language Models

1 Introduction

Finding patterns and putting similar entities together is made possible by clustering, a fundamental technique for analyzing big and complicated datasets [13]. In order to handle the inherent complexity and high dimensionality of DNA sequences and genomic data, a number of clustering techniques have been developed [18]. Although methods like spectral clustering [28], K-means [13], and hierarchical clustering [20] have been applied to genomic data, they frequently fall short in identifying the nuanced, context-dependent patterns necessary for biological understanding [18].

More recent developments include graph-based techniques [17] and deep learning-based clustering methods [22] that more accurately represent intricate relationships between sequences. Furthermore, although dimensionality reduction methods like t-distributed stochastic neighbor embedding (t-SNE) and principal component analysis (PCA) are frequently used to handle the large feature

S. Yuan et al. (Eds.): PAKDD 2025 Workshops, LNAI 15835, pp. 243–253, 2025.
https://doi.org/10.1007/978-981-96-8197-6_18

space [8], they may unintentionally lose important biological information, which could result in less-than-ideal clustering outcomes [1]. The complex structure and long DNA sequences require complex algorithms that can identify subtle dependencies that are usually missed by conventional clustering techniques. For that, various strategies have been introduced in recent years to address these issues [25].

Methods like embedding-based clustering, self-supervised learning, and the use of biological priors have become very popular. Specifically, self-supervised learning [25] uses unlabeled genomic data to produce informative embeddings that maintain important genetic features, increasing the robustness and accuracy of clustering. High-dimensional genomic sequences can be represented in lower-dimensional spaces using embedding-based clustering, which is frequently driven by deep learning models and preserves crucial biological information [30]. Furthermore, the interpretability and biological significance of clustering results are improved by incorporating biological priors, such as sequence conservation, gene function annotations, or epigenetic markers [27]. These cutting-edge methods offer more accurate and biologically significant clusters, marking a paradigm shift in the analysis of genomic data in domains like oncology [2].

In cancer research [16], for example, efficient clustering is crucial to finding mutations, sequence motifs, and gene expressions linked to specific cancers, which can guide drug discovery, personalized medicine, and targeted treatment plans [3]. Even small errors in clustering, though, can have serious repercussions. Erroneous sequences, misdirected therapeutic targets, or obscured important biomarkers could result from misclassifications, leading to incorrect interpretations [26]. These problems could jeopardize precision in clinical applications and hinder scientific advancement where accuracy is crucial, in addition to impeding scientific advancement [26].

In light of these challenges, models that can accurately represent the complex nature of genetic data while preserving important details are becoming increasingly necessary. Recently, the Genome Large Language Models (GLLMs) [36], which produce high-quality embeddings, present a promising alternative to improve clustering accuracy by maintaining the subtle context within genetic sequences, which would ultimately support more trustworthy biological insights and advance applications in precision medicine, including cancer research [15].

The potential of large language model (LLM) embeddings, including both their pre-trained and fine-tuned variations, remains underexplored in the context of unsupervised genomic clustering tasks. While these models have demonstrated impressive performance across various natural language processing (NLP) challenges [21], their application and effectiveness in genomic clustering domains combined with traditional clustering techniques are not yet fully explored. In this work, we aim to bridge this gap, and our key contributions are outlined as follows:

– We evaluate the performance of various clustering algorithms on embeddings generated by the Nucleotide Transformer, one of the GLLMs, in its pre-trained and fine-tuned states. This analysis identifies the most effective combinations for clustering DNA sequences.

– We demonstrate that embeddings generated by the fine-tuned model offer richer, more realistic representations of genomic data, resulting in improved clustering accuracy. These enhanced embeddings enable more precise genetic analyses and support informed clinical decision-making.

– We evaluate traditional clustering methods alongside nucleotide transformer embeddings and conclude that pre-trained embeddings do not provide a notable advantage over traditional approaches for unsupervised genomic clustering.

2   Related Work

Clustering of DNA sequences and high-dimensional data has been a significant focus in bioinformatics and computational biology, with applications ranging from gene expression analysis to taxonomic profiling [2,20,24]. Traditional clustering methods, such as K-means [13] and hierarchical clustering [20], have been adapted and applied to DNA sequence data. However, these methods face several challenges when dealing with high-dimensional genomic data where the dimensionality reduction can destroy necessary signals for analysis [8].

Researchers have put forth several approaches to deal with the difficulties associated with clustering genomic data. For example, CoIn [35], a correlation-induced clustering technique, was presented to identify intricate feature correlations in high-dimensional bioinformatics data. Many tools are developed for DNA sequence clustering, tools such as CD-HIT [9], and UCLUST [23] are commonly used for their efficiency with large datasets. However, these relying on greedy methods can struggle with optimality, mainly when sequence similarity thresholds are not precisely defined [38]. Moreover, Deep learning techniques have shown great potential in analyzing genomic data, especially for tasks involving high-dimensional sequences [25]. However, these models may only sometimes be the most efficient for more straightforward tasks such as clustering [29]. In such cases, unsupervised methods like embedding-based clustering can provide effective results with lower computational overhead.

Recent advancements have shifted focus to leveraging embeddings generated by GLLMs for downstream tasks such as classification, enhancing the accuracy and interpretability of genomic data analysis [7,14,19,32]. GLLMs, like DNABERT-S [37] and Evo [19], have demonstrated notable performance in tasks such as DNA sequence classification, where embeddings capture the complex relationships and dependencies inherent in genetic data. These models use transformer architectures to learn context-rich representations from nucleotide sequences, which can be applied to various genomic tasks, including predicting gene function, promoter prediction, and chromatin accessibility prediction [14]. For example, DNA-BERT-S, an adaptation of the BERT architecture designed explicitly for genomic data, has been successfully used for sequence classification, outperforming traditional methods by capturing the subtle patterns in long DNA sequences that other models might miss [37]. Notably, it is not a general-purpose DNA sequence classification tool but one specifically designed for species differentiation tasks, where it performs exceptionally well. DNA sequences of various species are naturally clustered and segregated in the embedding space by species-aware embeddings, which is how it works. Similarly, Evo [19] uses a cutting-edge architecture created to process lengthy biological sequences effectively. Evo's training in various genomic data and its ability to manage long-range dependencies in DNA sequences enable it to integrate sequence information and identify evolutionary patterns. It is designed to generalize across DNA, RNA, and proteins, capable of both prediction tasks and generative design from molecular to whole genome scale. In addition, models like ScGPT have expanded the use of GLLMs for single-cell RNA sequencing classification, showing their versatility in various genomic applications. These models, trained on vast amounts of genomic data, generate embeddings that can be used in downstream tasks such as gene expression prediction and disease classification [6].

These models [7,14,19,32] produce flexible embeddings that capture critical biological details, allowing for precise predictions even with limited data. They surpass or perform similarly to existing genomic baselines in tasks such as identifying transcription factor binding sites, assessing mutation impacts, and classifying pathogenic variants with high accuracy. These models demonstrate the potential contribution of embedding-based methods to advancing genomic research due to their high interpretability and applicability across multiple species. Embeddings in epigenetic analysis involve using advanced models to generate representations of DNA sequences that capture key features relevant to epigenetic modifications [33]. eDICE [12] is a Transformer-inspired imputation model designed for epigenomic analysis. eDICE can predict individual-specific epigenetic variations by creating context-specific nucleotide sequence embeddings [12]. These embeddings allow the model to recognize patterns of gene regulation and disease mechanisms by capturing crucial information about DNA methylation, histone modifications, and other epigenetic marks.

3   Experiments

In this work, we leverage the Nucleotide Transformer, a genome-scale large language model, as the backbone for generating DNA sequence embeddings. We utilize both the pre-trained model to be trained and fine-tuned on three benchmark genomic datasets: Variants of the COVID-19 virus dataset (VCV), Reconstructed Splice Site (Human) Dataset (RSS), and promoter (Human) Dataset (PROM). For consistency, the embeddings were generated with a maximum token length of 1,000. The model was also fine-tuned separately for each dataset to capture task-specific features before generating embeddings. The generated embeddings are provided as input to K-Means and other unsupervised clustering models, the Gaussian Mixture Model (GMM), and Spectral Clustering.

To evaluate how well the cluster was performed and the quality of representation learned, we apply the usual metrics, namely Accuracy, Adjusted Rand Index (ARI), and Normalized Mutual Information (NMI).

The experiments were executed on a high-performance workstation with an Intel® Core™ i9-13900K CPU (3.0 GHz) with 32 cores and 128 GB of DDR5 RAM clocked at 5,200 MHz. Embedding extraction required GPU acceleration, performed on Google Colab Pro with an Nvidia A100 GPU (40 GB).

3.1 Models and Algorithms

Nucleotide Transformer: [7], a transformer-based DNA language models that excel in genomic classification and feature identification. It represents a significant advancement in genomic sequence analysis, employing BERT-style architecture with non-overlapping k-mer tokenization. It is pre-trained with 500 million to 2.5 billion parameters via Masked Language Modeling (MLM) on diverse genomic datasets, including a single human reference genome, 3,202 human genomes from the 1000 Genomes Project, and a combination of human and multi-species genomes. It demonstrates its exceptional performance across various genomic prediction tasks, particularly excelling in low-data scenarios through effective transfer learning.

Clustering Algorithms: Clustering is a popular technique in unsupervised learning, often used in data embedding to uncover patterns within large, complex datasets [10]. In this study, we assess the performance of three different clustering algorithms separately and with the embedding from a pre-trained and fine-tuned Nucleotide Transformer. These algorithms were chosen for their popularity in the field and their effectiveness when paired with dimensionality reduction methods.

- **K-means** [13] is a widely used algorithm for clustering. It divides a dataset into k non-overlapping clusters by reducing the total squared distances between the data points and the centroids of each cluster. Initializing centroids, allocating points to the closest centroid, updating centroids according to mean values, and iterating until convergence are all steps in the process.
- **Gaussian Mixture Model (GMM)** [31] are a widely used clustering algorithm that assumes data is generated from a mixture of Gaussian distributions, each defined by its mean and covariance. It assumes that data is produced from a mixture of Gaussian distributions, each identified by its mean and covariance. It is especially appropriate for high-dimensional and biologically variable data since it iteratively estimates cluster parameters using the Expectation-Maximization (EM) method. GMM have provided vital insights into biological variability in genomics by clustering gene expression profiles, identifying co-expressed genes, and analyzing single-cell RNA sequencing data [4].
- **Spectral clustering** [18,28] is an unsupervised clustering technique uses a similarity matrix's eigenvalues, or spectrum, to carry out clustering. It first creates a graph where each node represents a data point, and edges show similarities between points rather than immediately grouping data points

according to distance. After that, the algorithm uses the eigenvectors of the Laplacian matrix obtained from the similarity graph to embed the data into a lower-dimensional space, utilizing methods from linear algebra to divide the graph into clusters.

For K-means, it is initialized with the K-means++ technique to optimize centroid selection. It runs for 200 iterations with ten restarts to enhance stability. In the GMM, the maximum is set to 200 iterations with 10 initializations for robust results. Finally, for Spectral Clustering, we utilize the Laplacian eigenmaps approach, followed by K-means on the transformed space. Each clustering algorithm was executed 10 times to ensure consistency, with the best result selected according to its specific objective function. For K-means, the focus was on minimizing inertia to achieve compact clusters. GMM aimed to maximize log-likelihood, identifying the most probable data partitioning. Spectral Clustering leveraged eigenvector-based transformations from the graph Laplacian to ensure optimal cluster separation.

3.2   Datasets

- **Variants of the COVID-19 virus dataset (VCV):** [11] is one of the Genome Understanding Evaluation (GUE) benchmarks typically used for predicting SARS-CoV-2 variant types based on genomic sequences. The dataset encompasses nine major SARS-CoV-2 variants: Alpha, Beta, Delta, Eta, Gamma, Iota, Kappa, Lambda, and Zeta. The dataset consists of DNA sequences with a length of 1,000 base pairs.
- **Reconstructed Splice Site (Human) Dataset (RSS):** [11] is one of the GUE benchmarks designed to predict splice donor and acceptor sites in the human genome, identifying exact locations where alternative splicing occurs. Initially, it comprises 400 base-pair sequences from the Ensembl GRCh38 human reference genome. It is then reconstructed by incorporating adversarial examples—false positive splice site predictions—to challenge models that previously overperformed on canonical splice sites, enabling the development of more robust models for predicting both canonical and non-canonical splice sites.
- **Promoter (Human) Dataset (PROM):** [11] is one of the GUE benchmarks that focus on identifying human proximal promoters. It includes TATA and non-TATA promoters, with sequences around the transcription start site (TSS) extracted from the Eukaryotic Promoter Database (EPDnew). Non-promoter sequences are selected from regions outside promoters, either with TATA motifs or randomly substituted. The combined dataset, "All", merges TATA and non-TATA promoters for broader analysis.

3.3   Evaluation Metrics

- **Accuracy (ACC)** [5,34]: It assesses how much the clusters align with the true class labels, quantifying the proportion of correct data point assignments across all clusters.

- **Normalized Mutual Information (NMI)** [34]: It evaluates the amount of shared information between the predicted clustering and the ground truth labels. It ranges from 0 to 1. A value closer to 1 indicates that the clustering captures most of the information from the actual labels, while 0 suggests no mutual information.
- **Adjusted Rand Index (ARI)** [34]: It measures the similarity between two clustering results, accounting for both the clustering assignments and the actual class labels. The metric ranges from -1 to 1, where 1 indicates perfect agreement, 0 indicates random clustering, and negative values suggest worse-than-random performance.

4   Results and Discussion

The experimental results demonstrate that the use of pre-trained embeddings in clustering tasks does not lead to significant improvements over traditional clustering methods; see Table 1 and Table 3. Specifically, when clustering with pre-trained embeddings, the evaluation metrics—such as accuracy, ARI, and NMI—were nearly the same as those achieved with traditional clustering algorithms like K-means, GMM, and Spectral Clustering. This suggests that pre-trained embeddings may not capture specific datasets' unique patterns and characteristics unless they are fine-tuned for a specific task. In contrast, when embeddings from fine-tuned models were used for clustering in the reconstructed and promoter datasets, noticeable improvements in clustering performance were observed across all evaluation metrics. Both ARI and NMI showed significant increases, indicating that fine-tuning allowed the model to capture dataset-specific patterns and relationships that were not as easily detected by traditional methods as indicated in Table 2.

Table 1. Clustering Algorithms Performance on Genomic Benchmark Datasets

Datasets	RSS			PROM			VCV		
Algorithm	Accuracy	ARI	NMI	Accuracy	ARI	NMI	Accuracy	ARI	NMI
K-means	0.35	0.01	0.02	0.24	0.07	0.50	0.13	0.00	0.00
GMM	0.34	0.01	0.00	0.13	0.13	0.10	0.12	0.00	0.00
Spectral	0.29	0.01	0.00	0.61	0.13	0.10	0.00	0.00	0.00

This highlights how important it is to modify models to fit the unique features of the data in order to produce clustering results that are more precise and significant.

Figure 1 illustrates the K-means clustering results for the three datasets, with PCA applied as a dimensionality reduction technique. The pre-trained embeddings evaluated across the three genomic benchmark datasets demonstrate clustering accuracy comparable to or lower than that of traditional K-means clustering when combined with dimensionality reduction, as shown in Fig. 2. In contrast,

Table 2. Performance of Fine-tuned Embeddings Combined with Clustering Algorithms on Genomic Datasets

Datasets	RSS			PROM			VCV		
Algorithm	Accuracy	ARI	NMI	Accuracy	ARI	NMI	Accuracy	ARI	NMI
K-means	0.67	0.80	0.60	0.78	0.81	0.62	0.20	0.10	0.20
GMM	0.75	0.71	0.20	0.77	0.70	0.44	0.30	0.05	0.00
Spectral	0.85	0.60	0.50	0.80	0.40	0.65	0.10	0.00	0.08

Table 3. Performance of Pre-trained Embeddings Combined with Clustering Algorithms on Genomic Datasets

Datasets	RSS			PROM			VCV		
Algorithm	Accuracy	ARI	NMI	Accuracy	ARI	NMI	Accuracy	ARI	NMI
K-means	0.37	−0.37	0.04	0.52	−0.01	0.01	0.14	0.00	0.01
GMM	0.39	−0.32	0.03	0.49	−0.01	0.01	0.16	0.20	0.49
Spectral	0.35	−0.30	0.05	0.50	−0.01	0.01	0.13	0.10	0.44

fine-tuning the embeddings enables the model to capture more nuanced biological patterns, leading to improved and more reliable clustering performance on genomic data. This enhancement is particularly evident in the two benchmark datasets: Reconstructed Splice Site (Human) and Promoter (Human), where significant performance gains were achieved. However, for the COVID-19 variants dataset, fine-tuning the embeddings did not yield a noticeable improvement in clustering performance compared to pre-trained embeddings or traditional clustering methods, as depicted in Fig. 3.

Overall, the findings show how crucial it is to fine-tune embedding models for each dataset to improve clustering quality and accuracy. Pre-trained embeddings are not always superior to traditional clustering techniques, but they can significantly improve clustering performance when optimized for specific datasets,

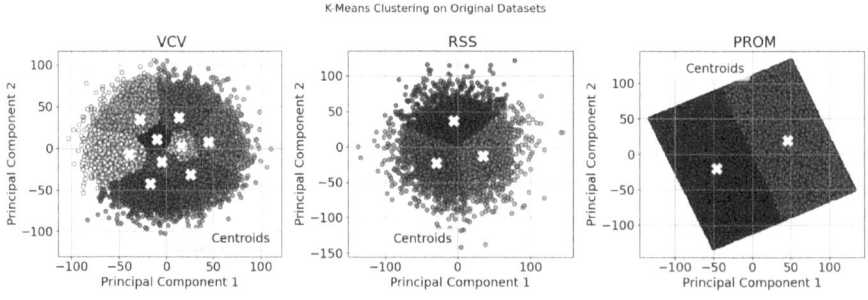

Fig. 1. K-means clustering visualization on genomic benchmark datasets.

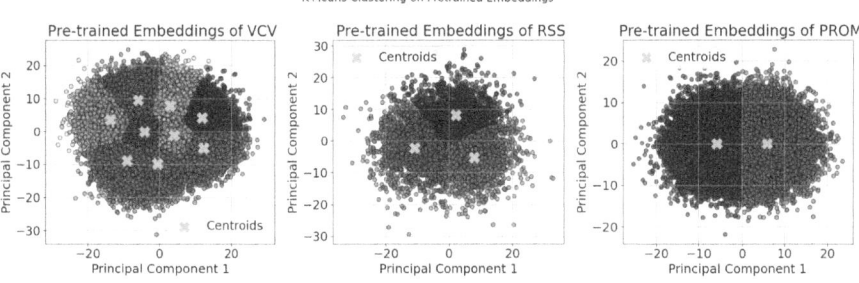

Fig. 2. K-means clustering of pre-trained embeddings on genomic benchmark datasets.

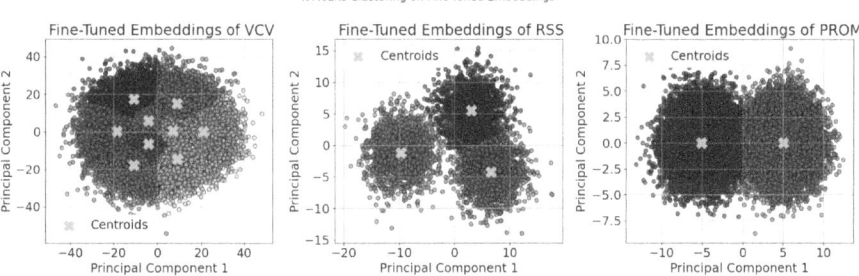

Fig. 3. K-means clustering of fine-tuned embeddings on genomic benchmark datasets.

especially for tasks that call for deeper comprehension and complex pattern recognition.

5   Conclusion and Future Work

The importance of clustering in genomic data analysis is emphasized in this work, especially for comprehending genetic variation and disease associations. Although the high dimensionality of genomic data frequently makes traditional clustering techniques inaccurate, using GLLMs, like the Nucleotide Transformer, shows promise. The findings show that, for certain datasets, fine-tuned embeddings greatly improve clustering performance, while pre-trained embeddings produce outcomes that are on par with those of conventional clustering algorithms. Even though pre-trained embeddings aren't always superior to traditional methods, these findings show how crucial it is to optimize GLLMs for genomic clustering tasks and show that fine-tuning them for specific datasets can greatly improve accuracy. Embeddings from different GLLMs, like DNABERT-S or Evo, have significant potential to maximize clustering across a wider variety of genomic datasets and require more investigation to produce more reliable and significant genomic analysis insights. Future studies could investigate several different approaches to improve genomic analysis clustering performance. Examining how fine-tuning affects a wider variety of genomic datasets, including those with more intricate, unstructured data, would be one avenue. Combining several pre-trained embeddings with sophisticated ensemble techniques may also be investigated to capitalize on the advantages of different models. Using more sophisticated techniques for data augmentation and model regularization during fine-tuning to increase robustness—particularly for smaller datasets—is another intriguing avenue.

References

1. Ayesha, S., Hanif, M.K., Talib, R.: Overview and comparative study of dimensionality reduction techniques for high dimensional data. Inf. Fusion **59**, 44–58 (2020)
2. Bertsimas, D., Wiberg, H.: Machine learning in oncology: methods, applications, and challenges. JCO Clin. Cancer Inform. **4** (2020)
3. Bhinder, B., Gilvary, C., Madhukar, N.S., Elemento, O.: Artificial intelligence in cancer research and precision medicine. Cancer Discov. **11**(4), 900–915 (2021)
4. Burns, J.J., et al.: Addressing noise in co-expression network construction. Brief. Bioinform. **23**(1), bbab495 (2022)
5. Chang, J., Wang, L., Meng, G., Xiang, S., Pan, C.: Deep adaptive image clustering. In: Proceedings of the IEEE International Conference on Computer Vision, pp. 5879–5887 (2017)
6. Cui, H., et al.: scGPT: toward building a foundation model for single-cell multiomics using generative AI. Nature Methods 1–11 (2024)
7. Dalla-Torre, H., et al.: The nucleotide transformer: Building and evaluating robust foundation models for human genomics. BioRxiv, pp. 2023–01 (2023)
8. Dorrity, M.W., Saunders, L.M., Queitsch, C., Fields, S., Trapnell, C.: Dimensionality reduction by UMAP to visualize physical and genetic interactions. Nat. Commun. **11**(1), 1537 (2020)
9. Fu, L., Niu, B., Zhu, Z., Wu, S., Li, W.: CD-HIT: accelerated for clustering the next-generation sequencing data. Bioinformatics **28**(23), 3150–3152 (2012)
10. Gonçalves, R.S., Kamdar, M.R., Musen, M.A.: Aligning biomedical metadata with ontologies using clustering and embeddings. In: Hitzler, P., et al. (eds.) ESWC 2019. LNCS, vol. 11503, pp. 146–161. Springer, Cham (2019). https://doi.org/10.1007/978-3-030-21348-0_10
11. Grešová, K., Martinek, V., Čechák, D., Šimeček, P., Alexiou, P.: Genomic benchmarks: a collection of datasets for genomic sequence classification. BMC Genom. Data **24**(1), 25 (2023)
12. Hawkins-Hooker, A., Visonà, G., Narendra, T., Rojas-Carulla, M., Schölkopf, B., Schweikert, G.: Getting personal with epigenetics: towards individual-specific epigenomic imputation with machine learning. Nat. Commun. **14**(1), 4750 (2023)
13. Ikotun, A.M., Ezugwu, A.E., Abualigah, L., Abuhaija, B., Heming, J.: K-means clustering algorithms: a comprehensive review, variants analysis, and advances in the era of big data. Inf. Sci. **622**, 178–210 (2023)
14. Ji, Y., Zhou, Z., Liu, H., Davuluri, R.V.: DNABERT: pre-trained bidirectional encoder representations from transformers model for DNA-language in genome. Bioinformatics **37**(15), 2112–2120 (2021)
15. Jin, Q., Yang, Y., Chen, Q., Lu, Z.: GeneGPT: augmenting large language models with domain tools for improved access to biomedical information. Bioinformatics **40**(2), btae075 (2024)

16. Lei, Y., et al.: Applications of single-cell sequencing in cancer research: progress and perspectives. J. Hematol. Oncol. **14**(1), 91 (2021)
17. Li, J., Chen, S., Pan, X., Yuan, Y., Shen, H.B.: Cell clustering for spatial transcriptomics data with graph neural networks. Nat. Comput. Sci. **2**(6), 399–408 (2022)
18. Montgomery, R.M.: Overview of clustering techniques: from k-means to spectral methods (2024)
19. Nguyen, E., et al.: Sequence modeling and design from molecular to genome scale with Evo. Science **386**(6723), eado9336 (2024)
20. Nidheesh, N., Nazeer, K.A., Ameer, P.: A hierarchical clustering algorithm based on silhouette index for cancer subtype discovery from genomic data. Neural Comput. Appl. **32**(15), 11459–11476 (2020)
21. Patil, R., Boit, S., Gudivada, V., Nandigam, J.: A survey of text representation and embedding techniques in NLP. IEEE Access **11**, 36120–36146 (2023)
22. Petegrosso, R., Li, Z., Kuang, R.: Machine learning and statistical methods for clustering single-cell RNA-sequencing data. Brief. Bioinform. **21**(4), 1209–1223 (2020)
23. Prasad, D.V., Madhusudanan, S., Jaganathan, S., et al.: uCLUST-a new algorithm for clustering unstructured data. ARPN J. Eng. Appl. Sci. **10**(5), 2108–2117 (2015)
24. Qi, R., Ma, A., Ma, Q., Zou, Q.: Clustering and classification methods for single-cell RNA-sequencing data. Brief. Bioinform. **21**(4), 1196–1208 (2020)
25. Rios-Martinez, C., Bhattacharya, N., Amini, A.P., Crawford, L., Yang, K.K.: Deep self-supervised learning for biosynthetic gene cluster detection and product classification. PLoS Comput. Biol. **19**(5), e1011162 (2023)
26. Sadybekov, A.V., Katritch, V.: Computational approaches streamlining drug discovery. Nature **616**(7958), 673–685 (2023)
27. Sidak, D., Schwarzerová, J., Weckwerth, W., Waldherr, S.: Interpretable machine learning methods for predictions in systems biology from omics data. Front. Mol. Biosci. **9**, 926623 (2022)
28. Warrier, J., Mallery, B.: Exploring the effectiveness of spectral clustering on gene expression data. J. High Sch. Sci. **7**(3) (2023)
29. Whalen, S., Schreiber, J., Noble, W.S., Pollard, K.S.: Navigating the pitfalls of applying machine learning in genomics. Nat. Rev. Genet. **23**(3), 169–181 (2022)
30. Xu, H., Gao, L., Huang, M., Duan, R.: A network embedding based method for partial multi-omics integration in cancer subtyping. Methods **192**, 67–76 (2021)
31. Yang, M.S., Lai, C.Y., Lin, C.Y.: A robust EM clustering algorithm for gaussian mixture models. Pattern Recogn. **45**(11), 3950–3961 (2012)
32. Yang, X., Cheng, W., Wu, Y., Petzold, L., Wang, W.Y., Chen, H.: DNA-GPT: divergent n-gram analysis for training-free detection of GPT-generated text. arXiv preprint arXiv:2305.17359 (2023)
33. Yassi, M., Chatterjee, A., Parry, M.: Application of deep learning in cancer epigenetics through DNA methylation analysis. Brief. Bioinform. **24**(6), bbad411 (2023)
34. Yin, H., Aryani, A., Petrie, S., Nambissan, A., Astudillo, A., Cao, S.: A rapid review of clustering algorithms. arXiv preprint arXiv:2401.07389 (2024)
35. Zeng, Z., et al.: Coin: correlation induced clustering for cognition of high dimensional bioinformatics data. IEEE J. Biomed. Health Inform. **27**(2), 598–607 (2022)
36. Zhang, Q., et al.: Scientific large language models: a survey on biological & chemical domains. arXiv preprint arXiv:2401.14656 (2024)
37. Zhou, Z., et al.: DNABERT-S: learning species-aware DNA embedding with genome foundation models. arXiv (2024)
38. Zou, Q., Lin, G., Jiang, X., Liu, X., Zeng, X.: Sequence clustering in bioinformatics: an empirical study. Brief. Bioinform. **21**(1), 1–10 (2020)

Genomic Variant Classification for MECP2 Mutation Pathogenicity

Sravya Sri Mallampalli and Chandra Mohan Dasari[(✉)] [iD]

Indian Institute of Information Technology, Sri City, Andhra Pradesh, India
chandramohan.d@iiits.in

Abstract. The methyl-CpG-binding protein 2 (MECP2) gene mutation is known to cause Rett syndrome and other severe neurodevelopmental disorders. Accurate characterization of the pathogenicity of these genetic variants is essential to understand the disease mechanisms and to allow early diagnosis. This understanding enables targeted therapeutic strategies mitigating the neurological consequences of these mutations. Despite advancements in genomic sequencing, determining the functional impact of novel mutations remains a challenge due to the high costs and time requirements of wet-lab validation. Existing computational models lack sufficient biological context and often fail to generalize well across diverse mutations. To bridge this gap, we propose a machine learning-based classification system that incorporates genomic sequence context, molecular consequence, and variant frequency for mutation analysis. This study introduces an approach by integrating transition or transversion ratios, nucleotide context, and feature importance ranking to enhance classification accuracy. The proposed approach leverages algorithms, specifically Random Forest and XGBoost, to address classification challenges inherent in genetic variant datasets. The XGBoost and Random Forest achieved 97.5

Keywords: Genomic Variant · MECP2 Mutation · Pathogenicity Prediction · nucleotide context · computational models · Rett Syndrome

1 Introduction

Neurodevelopmental disorders represent a complex genetic frontier where a precise molecular understanding can transform diagnostic and therapeutic approaches. Mutation of key genes such as Methyl-CpG-binding protein 2 (MECP2) illustrates the complex problems researchers face when trying to decipher the pathogenicity of genetic variants [1,2]. These genetic variations can profoundly affect neural development, cognitive function, and overall neurological health, making accurate characterization essential for clinical intervention and personalized medicine. Genetic variations in MECP2 are classified into distinct categories, including *pathogenic mutations* that significantly disrupt protein function, *benign variants* with minimal impact, and *variants of uncertain significance* (VUS) requiring sophisticated computational and experimental validation [3]. Approximately 95% of Rett Syndrome cases result from de novo mutations,

S. Yuan et al. (Eds.): PAKDD 2025 Workshops, LNAI 15835, pp. 254–265, 2025.
https://doi.org/10.1007/978-981-96-8197-6_19

primarily in females, highlighting the gene's developmental importance, with advanced computational methods becoming essential for linking these genetic variations to neurological functions [4, 5].

Genetic mutations can have profound effects on protein function, influencing cellular processes and disease outcomes. Missense mutations alter a single amino acid, potentially disrupting protein structure and function, while nonsense mutations introduce premature stop codons, leading to truncated, often non-functional proteins [6]. Frameshift mutations, caused by insertions or deletions, shift the reading frame, frequently resulting in deleterious effects due to extensive amino acid changes [7]. Additionally, splicing mutations can interfere with RNA processing, potentially causing misregulation of gene expression and protein synthesis [8]. Mutations can be further classified into synonymous (silent) and non-synonymous categories, where synonymous mutations do not alter protein sequence but can still affect gene expression, whereas non-synonymous mutations (missense, nonsense, frameshift) directly impact protein function [9]. Understanding these molecular consequences is essential for accurate pathogenicity classification and the development of computational models for genomic variant interpretation.

Computational genomic variant classification plays a pivotal role in modern precision medicine by identifying genetic mutations that may cause diseases, thereby facilitating early diagnosis and personalized treatment strategies [10]. By accurately predicting the pathogenicity of variants, computational approaches reduce the reliance on labor-intensive and costly wet-lab experiments, streamlining the diagnostic process [11, 12]. Moreover, integrating computational methods with genomic data aids in uncovering novel biomarkers and therapeutic targets, further advancing personalized healthcare [13]. As a result, computational classification techniques are indispensable for improving the accuracy and speed of genomic variant interpretation in clinical and research settings [14].

Previous studies on MECP2-related disorders have primarily focused on genetic diagnosis and phenotypic associations, such as using artificial intelligence for detecting MECP2 duplication syndrome through facial recognition [19]. However, to the best of our knowledge, no prior research has applied machine learning for the direct classification of MECP2 gene mutations as benign or pathogenic. This study addresses this gap by leveraging Random Forest and XGBoost models to systematically classify MECP2 mutations, providing a novel computational approach for pathogenicity prediction.

Gathering and preparing data is a major milestone in model creation. Extensive editing and filtration were required to extract and preprocess high-quality, well-annotated genomic mutation data. This research highlights the potential of ML-based genomic classification to improve disease outcome prediction, reduce experimental workload, and accelerate clinical decision making. The unique combination of genomic sequence analysis, molecular consequence assessment, and feature ranking offers a reliable alternative to traditional variant assessment, potentially reducing reliance on resource-intensive experimental validation. Classifying MECP2 mutations is crucial for understanding neurodevelopmental disorders and guiding clinical interventions, as integrating computational approaches enhances pathogenicity predictions and paves the way for more precise, data-driven advancements in personalized medicine. The major contributions of the study are:

- The MECP2 dataset with 802 gene variants is collected from ClinVar with specific filters and pre-processed to create a structured data set of 802 MECP2 gene variants suitable for analysis.
- Statistical analysis based feature extraction is applied, and one-hot encoding and normalizing numerical values are done for model compatibility.
- Two machine learning models (Random Forest and XGBoost) are implemented and validated using 5-fold cross-validation to ensure robust performance assessment across multiple data splits.

2   Methodology

The proposed framework for MECP2 mutation classification integrates data collection, preprocessing, feature selection, model training, and evaluation into a unified pipeline. The dataset, sourced from ClinVar and filtered for single nucleotide variants, is preprocessed by encoding categorical features and normalizing numerical values. Biologically meaningful attributes such as mutation type, transition/transversion status, and genomic context are then extracted for model training. Two machine learning models—Random Forest and XGBoost—are employed to classify mutations as either pathogenic or benign, with performance evaluated using standard metrics and cross-validation to ensure robustness. The overall workflow of this process is illustrated in Fig. 1.

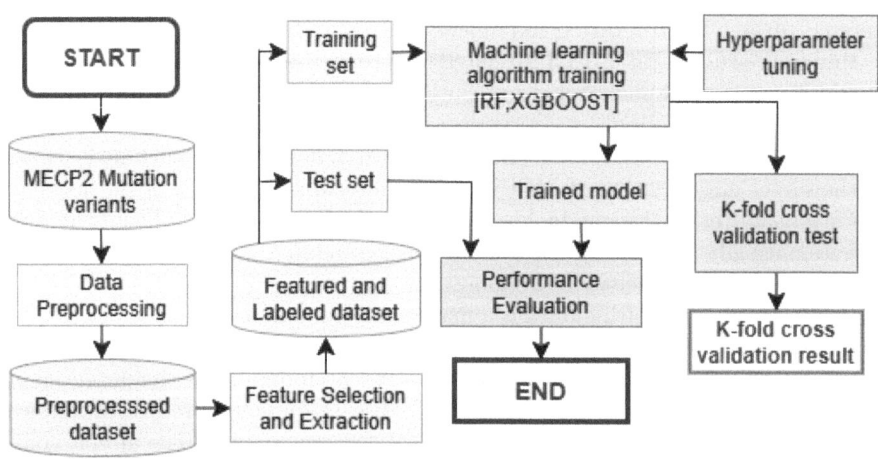

Fig. 1. Architecture of the Framework

The implementation was executed using Python 3.8, leveraging libraries such as Pandas, NumPy, Matplotlib, Scikit-learn, XGBoost, and Seaborn for data preprocessing, feature extraction, model training, and visualization. Experiments were conducted on a high-performance server using CPU resources exclusively, with 8 cores dedicated to feature transformation and selection and hyperparameter tuning via grid search running 10 parallel executions (4 cores each). Jupyter Notebook provided an interactive interface with remote access enabled through SSH tunneling for efficient management and real-time monitoring.

2.1 Data Collection and Preprocessing

We obtained MECP2 gene variants from the ClinVar database [20], selecting only single nucleotide variants (SNVs) with established experimental classifications. The dataset was filtered for germline variants labeled as pathogenic, likely pathogenic, benign, or likely benign. The extracted data, initially in tabular text format, was converted into CSV using custom Python scripts, resulting in 802 MECP2 gene variants. Table 1 summarizes the distribution of variants across these germline classifications.

Table 1. Distribution of Germline Classifications in MECP2 Variants

Germline Classification	Number of variants
Benign/Likely benign	589
Pathogenic/ Likely pathogenic	213
Total	802

The collected dataset is comprehensive, encompassing multiple key attributes for each MECP2 gene variant. Each variant is identified through a standardized mutation format (NM_001110792.2(MECP2):c.*772G¿A), confirmed to be part of the MECP2 gene, and associated with specific conditions or syndromes. Unique identifiers included accession and variation IDs, dbSNP identifier, canonical SPDI (sequence position deletion insertion), type variant, molecular consequences, and germline classification. To maintain methodological consistency, the dataset is strictly limited to SNVs, ensuring a focused and uniform approach to the genetic analysis. The dataset and scripts link is available in the conclusion.

The dataset of 802 MECP2 gene variants was compiled and revealed comprehensive insights into genetic mutation characteristics. Analysis revealed 450 transitions and 352 transversions, and the chi-square test confirmed a statistically significant relationship between mutation type and clinical significance ($\chi^2 = 52.19$, p < 0.0001) as shown in Table 2. Among base pair mutations, G→A was found to be particularly notable, occurring 222 times, as illustrated in Table 3. Examination of the molecular consequence distribution highlighted a diverse range of variant types, with synonymous variants (185) and missense variants (166) being the most prevalent, as presented in Table 4. The genomic positions of these variants were centered around a mean of 154,034,448.92, with a standard deviation of 14,376.77, indicating a clustered distribution across the MECP2 gene region.

Table 2. Statistical Summary of Mutation Characteristics

Characteristic	Total	Transitions	Transversions
Number of Variants	802	450	352
Chi-Square Test	$\chi^2 = 52.19,\ p < 0.0001$		

Table 3. Base Pair Mutation Frequency in MECP2 Variants

Base Pair Mutation	Frequency	Percentage
G:A	222	37.0%
C:T	130	21.7%
G:C	108	18.0%
T:C	71	11.8%
G:T	64	10.7%
C:G	56	9.3%
C:A	42	7.0%
T:A	31	5.2%
T:G	28	4.7%
A:G	27	4.5%
A:C	15	2.5%
A:T	8	1.3%
Total	600	100%

Table 4. Frequency of Molecular Consequences in MECP2 Variants

Molecular Consequence	Frequency	Percentage
Synonymous variant	185	23.0%
Missense variant	166	20.7%
Missense variant and 5' UTR variant	92	11.5%
Synonymous variant & 5' UTR variant	85	10.6%
Intron variant	62	7.7%
3' UTR variant	58	7.2%
Missense variant and Intron variant	35	4.4%
Nonsense and 5' UTR variant	31	3.9%
Nonsense variant	26	3.2%
Synonymous variant and Intron variant	24	3.0%
Other variants	40	5.0%
Total	802	100%

2.2   Feature Selection and Transformation

Mutation types were categorized based on molecular consequences, including synonymous, nonsense, missense, and splicing variants. Additionally, we examined transition-to-transversion ratios, as transitions ($A \rightleftharpoons G$, $C \rightleftharpoons T$) are more frequent, whereas transversions ($A \rightleftharpoons C$, $A \rightleftharpoons T$, $C \rightleftharpoons G$, $G \rightleftharpoons T$) are more disruptive. Genomic context was incorporated by extracting nucleotide sequences preceding and following the mutation site along with trinucleotide context, providing insight into sequence-dependent pathogenicity. A comprehensive set of features was initially extracted, categorized into position-based, molecular, genomic context, clinical, and identifier-based attributes. To ensure model generalizability, only biologically meaningful features were selected, prioritizing mutation patterns and sequence context while excluding database-specific identifiers and clinical condition labels. The final set of features included position information (GRCh38Location, GRCh37Location, position), molecular characteristics (mutation type, transition/transversion status, nucleotide changes), and genomic context (flanking sequences, trinucleotide context), whereas database identifiers, clinical conditions, and variant notations were excluded.

To prepare the selected features for machine learning, categorical attributes such as mutation type and genomic context were one-hot encoded, ensuring proper numerical representation. Clinical significance labels were encoded into a binary classification scheme, where pathogenic/likely pathogenic variants were encoded as 1 and benign/likely benign variants as 0. The dataset was then split into 80% training and 20% testing sets using stratified sampling, maintaining a balanced representation of pathogenic and benign variants. Continuous variables, such as mutation position and transition-transversion ratio, were normalized using Min-Max scaling to enhance model efficiency and convergence. This feature selection approach ensures that the model prioritizes genetically relevant factors, improving classification performance and reducing reliance on database-specific annotations.

2.3   Model Architectures

Random Forest (RF) enhances classification by aggregating predictions from an ensemble of decision trees built on random subsets of genomic data. Each tree is constructed using bootstrapped samples, allowing the model to capture complex patterns across mutation types, genomic positions, and molecular consequences. At every decision node, a random selection of features is evaluated using the Gini index to determine the best split, which improves generalization and mitigates overfitting. The overall workflow of this process is illustrated in Fig. 2.

Furthermore, class weighting is employed to ensure that rare pathogenic variants are given adequate emphasis, addressing class imbalances inherent in genomic data. By averaging the predictions from all trees, RF reduces variance and enhances model robustness and interpretability. This approach provides refined classification for MECP2 mutations while offering valuable biological insights that support genetic research and clinical decision-making.

XGBoost (Extreme Gradient Boosting) is a highly efficient tree-based algorithm tailored for genomic mutation classification, such as MECP2 mutations. The dataset is labeled as either pathogenic or benign based on clinical and experimental validations, and each mutation is represented by a range of genomic features including mutation type, molecular consequences, positional information, and transition/transversion characteristics. After feature extraction and preprocessing, the data is split into training and testing subsets, ensuring that the model receives diverse and representative input. The overall workflow of this process is illustrated in Fig. 3.

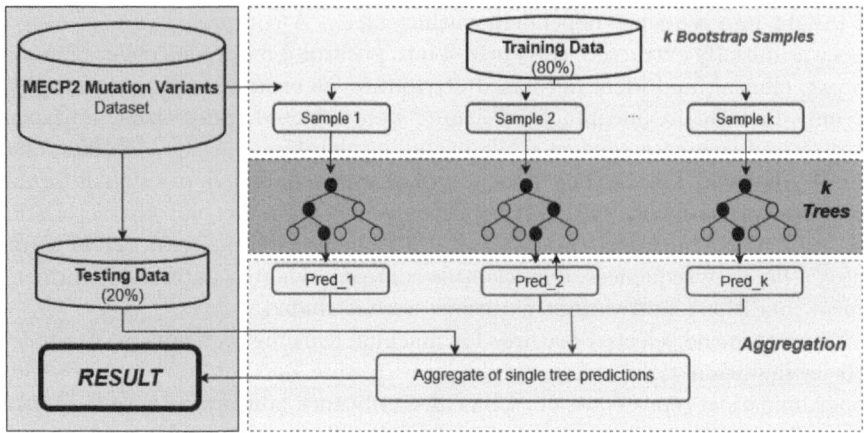

Fig. 2. Flowchart of the Random Forest (RF) model.

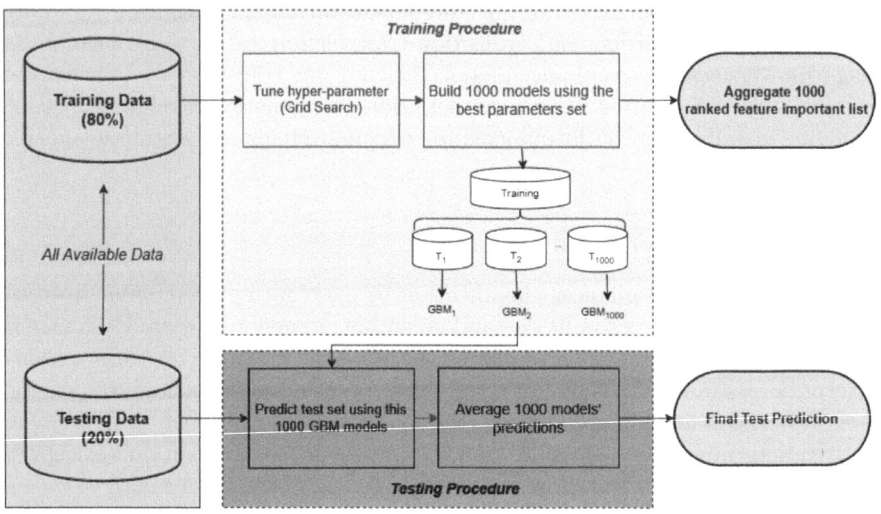

Fig. 3. Flowchart of the XGradientBoosting (XGB) model.

During training, XGBoost employs gradient boosting by iteratively building a large ensemble of decision trees, each learning from a slightly different subset of the training data. The model dynamically adjusts the weights of misclassified instances to reduce errors in subsequent trees, and hyperparameters such as tree depth, learning rate, and regularization terms are fine-tuned via grid search. This iterative boosting approach not only stabilizes predictions and mitigates overfitting but also aggregates feature importance rankings, providing valuable insights into the most influential genomic features for mutation classification.

2.4 Evaluation Metrics

The evaluation metrics were applied to assess model performance comprehensively. To ensure robust evaluation and mitigate potential overfitting, we employed both a hold-out test set (80:20 split) and 5-fold cross-validation. The 5-fold cross-validation approach divides the dataset into five subsets, where each fold serves as the test set once while the remaining four folds are used for training. This method provides a generalized performance estimate by averaging results across multiple train-test splits. Meanwhile, the 80:20 split was used for final model evaluation on an independent test set, ensuring a realistic assessment of classification accuracy.

k-fold Cross-Validation: We employed 5-fold cross-validation, where each model was trained on 80% of the data and tested on the remaining 20% across five iterations. To ensure a balanced representation of pathogenic and benign variants, we used stratified sampling. The final performance metrics were averaged over all folds, providing a more reliable estimate of model generalizability compared to a single train-test split. This approach is particularly crucial given the imbalanced nature of the dataset.

Precision: Precision (P) is the ratio of correctly predicted positive cases (True Positives, TP) to the total predicted positive cases (sum of TP and False Positives, FP):

$$P = \frac{TP}{TP + FP} \tag{1}$$

Recall: Recall (R) measures how well the model identifies actual positive cases. It is the ratio of TP to the total actual positive cases (sum of TP and false negatives, FN):

$$R = \frac{TP}{TP + FN} \tag{2}$$

F1 Score: The F1 score provides a harmonic mean between precision and recall, balancing both metrics:

$$F1 = \frac{2 \cdot P \cdot R}{P + R} \tag{3}$$

Accuracy: Accuracy measures the proportion of correctly classified cases (both positive and negative) out of all cases:

$$\text{Accuracy} = \frac{TP + TN}{TP + TN + FP + FN} \tag{4}$$

3 Results and Discussion

This section presents the performance evaluation of the Random Forest (RF) and XGBoost (XGB) models on MECP2 variant classification. We analyze both test metrics and validation metrics separately to ensure a comprehensive assessment of model reliability and generalizability.

3.1   Random Forest (RF) Performance Analysis

The performance of the Random Forest (RF) model was evaluated on the hold-out test set (80:20 split). The accuracy, precision, recall, and F1-score were computed to assess classification effectiveness. The results indicate that RF achieves an accuracy of 93%, with a slightly higher precision for benign variants compared to pathogenic ones. Table 5 summarizes the test performance of the RF model.

Table 5. Random Forest Classification Test Metrics

Class	Precision	Recall	F1-Score	Support
Benign (0)	0.96	0.95	0.95	118
Pathogenic (1)	0.86	0.88	0.87	43
Overall	Accuracy: 0.93			161
Macro Average:		0.91	0.92	0.91
Weighted Average:		0.93	0.93	0.93

The RF model's strong majority voting mechanism ensures robust predictions, but the slight misclassification of pathogenic mutations indicates that the model may require further optimization, particularly in handling rare pathogenic cases.

3.2   XGBoost Performance Analysis

Similarly, the XGBoost (XGB) model was evaluated on the same hold-out test set. XGB demonstrated superior classification performance, achieving an accuracy of 97.5%, outperforming RF in distinguishing pathogenic variants. The confusion matrix and classification metrics for XGB are detailed in Table 6.

Table 6. XGBoost Classification Performance Metrics

Class	Precision	Recall	F1-Score	Support
Benign (0)	0.97	0.99	0.98	118
Pathogenic (1)	0.98	0.93	0.95	43
Overall	Accuracy: 0.98			161
Macro Average:		0.98	0.96	0.97
Weighted Average:		0.98	0.98	0.97

The improvement in pathogenic mutation identification demonstrates XGBoost's superior capability to recognize complex patterns in genetic variant data. Its gradient boosting approach, which focuses on misclassified samples during training, contributes to its increased accuracy and sensitivity, making it a promising tool for clinical genomic variant classification.

3.3   Computational Time Analysis

To evaluate computational efficiency, we measured the training time and inference time for both Random Forest (RF) and XGBoost (XGB) on the test dataset. Table 7 presents the recorded execution times for each model.

Table 7. Computational Time for RF and XGB

Model	Training Time (s)	Inference Time (s)
Random Forest (RF)	0.1569	0.0070
XGBoost (XGB)	0.1084	0.0965

The results indicate that XGBoost (XGB) has a lower training time (0.1084 s) compared to RF (0.1569 s), suggesting its efficient optimization process. However, XGB's inference time (0.0965 s) is significantly higher than RF's (0.0070 s), implying that while XGB trains faster, it requires more computation during prediction. This trade-off is important for real-time applications where inference speed is a critical factor.

3.4   5-Fold Cross-Validation

While Sects. 3.1 and 3.2 report test set performance using an 80:20 split, this section presents model validation metrics obtained via 5-fold cross-validation. This additional validation step ensures that the reported performance is not dependent on a single train-test split and assesses the model's generalizability.

The average cross-validation accuracy for RF was 93.39% ± 1.29%, while XGB achieved 95.51% ± 1.28%, further confirming XGB's superior performance. These results are summarized in Table 8.

Table 8. 5-Fold Cross-Validation Performance Metrics (Mean ± Standard Deviation)

Model	Accuracy	Precision	Recall	F1-score
Random Forest	93.39% ± 1.29%	95.32% ± 1.73%	93.65% ± 2.11%	94.47% ± 1.62%
XGBoost	95.51% ± 1.28%	97.14% ± 1.05%	95.82% ± 1.37%	96.47% ± 0.98%

By comparing the test set and cross-validation results, it is evident that XGBoost (XGB) consistently outperforms Random Forest (RF) in both classification accuracy and robustness, making it a more reliable approach for MECP2 mutation classification.

Despite these promising results, the study has certain limitations. The dataset consists solely of single nucleotide variants (SNVs), excluding insertions, deletions, and structural variations, which may also play a critical role in MECP2-related disorders. Additionally, the relatively small sample size (802 variants) limits the generalizability of the findings. Variants of uncertain significance (VUS) were also excluded, which could affect classification completeness.

4   Conclusion and Future Scope

This study successfully applied machine learning techniques, particularly Random Forest (RF) and XGBoost, to classify MECP2 gene mutations as benign or pathogenic. The results demonstrate that XGBoost outperformed RF, achieving an accuracy of 97.52%, compared to 93.17% for RF, with 5-fold cross-validation confirming consistent performance (95.51% $\pm$ 1.28% versus 93.39% $\pm$ 1.29%). This highlights XGBoost's superior capability in identifying complex patterns within genomic data, making it a more reliable choice for mutation classification. Key contributing factors to classification performance included mutation position, molecular consequence, and mutation type (Table 2). Synonymous, missense, and nonsense mutations emerged as the most critical molecular consequences influencing pathogenicity, with missense mutations being strongly associated with disease outcomes (Table 4). Additionally, transition mutations were more prevalent than transversions, suggesting an underlying biological mechanism influencing mutation patterns in the MECP2 gene. Future work should expand the model to handle insertions, deletions, and duplications, integrate deep learning techniques for improved feature extraction, and enhance model interpretability using SHAP or LIME. Validating on larger clinical datasets and developing a web-based tool for real-time mutation classification can further improve its clinical applicability and impact on genomic medicine. The dataset and scripts are available at [GitHub link].

Acknowledgements. This research is supported by the Science and Engineering Research Board (SERB) (Sanction order File No. SRG/2023/000941), Government of India.

References

1. Mishra, G.P., Sun, E.X., Chin, T., Eckhardt, M., Greenberg, M.E., Stroud, H.: Interaction of methyl-CpG-binding protein 2 (MeCP2) with distinct enhancers in the mouse cortex. Nat. Neurosci. **28**(1), 62–71 (2025). https://doi.org/10.1038/s41593-024-01808-y
2. Pehlivan, D., et al.: Structural variant allelic heterogeneity in MECP2 duplication syndrome provides insight into clinical severity and variability of disease expression. Gen. Med. **16**(1), 146 (2024). https://doi.org/10.1186/s13073-024-01411-7
3. Wen, Z., et al.: Identification of autism-related MECP2 mutations by whole-exome sequencing and functional validation. Mol. Autism **8**, 1–10 (2017). https://doi.org/10.1186/s13229-017-0157-5
4. Cuddapah, V.A., et al.: Methyl-CpG-binding protein 2 (MECP2) mutation type is associated with disease severity in Rett syndrome. J. Med. Genet. **51**(3), 152–158 (2014). https://doi.org/10.1136/jmedgenet-2013-102113

5. Pejhan, S., Rastegar, M.: Role of DNA methyl-CpG-binding protein MeCP2 in Rett syndrome pathobiology and mechanism of disease. Biomolecules **11**(1), 75 (2021). https://doi.org/10.3390/biom11010075

6. Adams, S: The role of MeCP2 and FoxG1 in embryonic cortical development: implications for autism spectrum disorders (doctoral dissertation) (2025). https://doi.org/10.17863/CAM.115890

7. Lynch, M.: Rate, molecular spectrum, and consequences of human mutation. Proc. Natl. Acad. Sci. **107**(3), 961–968 (2010). https://doi.org/10.1073/pnas.0912629107

8. Ozturk, K., Carter, H.: Predicting functional consequences of mutations using molecular interaction network features. Hum. Genet. , 1–16 (2021). https://doi.org/10.1007/s00439-021-02329-5

9. Oelschlaeger, P.: Molecular mechanisms and the significance of synonymous mutations. Biomolecules **14**(1), 132 (2024). https://doi.org/10.3390/biom14010132

10. Backwell, L., Marsh, J.A.: Diverse molecular mechanisms underlying pathogenic protein mutations: beyond the loss-of-function paradigm. Annu. Rev. Genom. Hum. Genet. **23**, 475–498 (2022). https://doi.org/10.1146/annurev-genom-111221-103208

11. Wang, Y.C., et al.: Computational genomics in the era of precision medicine: applications to variant analysis and gene therapy. J. Pers. Med. **12**(2), 175 (2022). https://doi.org/10.3390/jpm12020175

12. Jackins, V., Vimal, S., Kaliappan, M., Lee, M.Y.: AI-based smart prediction of clinical disease using random forest classifier and Naive Bayes. J. Supercomput. **77**(5), 5198–5219 (2020). https://doi.org/10.1007/s11227-020-03481-x

13. Sinha, N.K., Khulal, M., Gurung, M., Lal, A.: Developing a web based system for breast cancer prediction using XGBoost classifier. Int J Eng Res **9**, 852–856 (2020)

14. Zucca, S., Nicora, G., De Paoli, F., et al.: An AI-based approach driven by genotypes and phenotypes to uplift the diagnostic yield of genetic diseases. Hum. Genet. (2024). https://doi.org/10.1007/s00439-023-02638-x

15. Marian, A.J.: Clinical interpretation and management of genetic variants. Basic Trans. Sci. **5**(10), 1029–1042 (2020). https://doi.org/10.1016/j.jacbts.2020.05.013

16. Chaudhary, P., Mehra, R.: Genetic variant classification through decision tree analysis for enhanced genomic understanding. Genom. Nexus AI Comput. Vis. Mach. Learn., 505–528 (2025). https://doi.org/10.1002/9781394268832.ch23

17. Kong, S.W., et al.: Discordance between a deep learning model and clinical-grade variant pathogenicity classification in a rare disease cohort. NPJ Genom. Med. **10**(1), 17 (2025). https://doi.org/10.1038/s41525-025-00480-w

18. Pyankov, I.A., et al.: A computational approach to predict the effects of missense mutations on protein amyloidogenicity: a case study in hereditary transthyretin cardiomyopathy. J. Struct. Biol. **108176**,(2025). https://doi.org/10.1016/j.jsb.2025.108176

19. Vega-Hanna, L., et al.: MECP2 duplication syndrome: AI-based diagnosis, severity scale development and correlation with clinical and molecular variables. Diagnostics **15**(1), 10 (2025). https://doi.org/10.3390/diagnostics15010010

20. Landrum, M.J., et al.: ClinVar: public archive of interpretations of clinically relevant variants. Nucleic Acids Res. **44**(D1), D862–D868 (2016). https://doi.org/10.1093/nar/gkv1222

Workshop on Research and Applications of Foundation Models for Data Mining and Affective Computing (RAFDA 2025)

RAFDA 2025 Preface

Workshop Description

Research and Applications of Foundation Models for Data Mining and Affective Computing (RAFDA), a workshop of the 29th Pacific-Asia Conference on Knowledge Discovery and Data Mining (PAKDD), serves as an inclusive platform to explore the intricate intersections of foundation models, including Large Language Models (LLMs), data mining, and affective computing. RAFDA represents a convergent space that unites researchers focused on the applications, advancements, and implications of foundation models within the realms of data mining and affective computing.

Following the success of RAFDA 2024, **RAFDA 2025** marked the second edition of the workshop. At its core, RAFDA aims to promote interdisciplinary dialogue, particularly in the innovative utilization of cutting-edge foundation models for robust data mining practices. Additionally, the workshop fosters a deeper exploration of the understanding and interpretation of affective computing, providing an inclusive forum for collaboration and knowledge exchange among researchers, practitioners, and industry experts.

RAFDA also seeks to serve as an international workshop that facilitates dynamic discussions among researchers in foundation models, including LLMs, Natural Language Processing (NLP), data mining, and affective computing. The event offers a collaborative environment for sharing groundbreaking research, novel methodologies, and innovative applications across both academic and industrial domains, thereby paving the way for future advancements and directions in data mining and AI research.

We extend our sincere gratitude to the PAKDD Program Committee, organizers, and workshop chairs for their invaluable support.

International Engagement and Encouraging Research

Foundation models, including Large Language Models (LLMs) such as ChatGPT, built on state-of-the-art deep learning architectures, have significantly revolutionized capabilities in Natural Language Processing (NLP) and affective computing. These models, which are trained on vast and diverse datasets, have demonstrated exceptional proficiency in understanding and generating human language, enabling more accurate and context-aware applications. Their ability to process and interpret large amounts of textual and emotional data has led to transformative advancements across a wide array of domains, from healthcare and finance to entertainment and social media. By enabling machines to not only understand but also interact with human behaviors, these models open up new frontiers in creating intelligent systems that are more attuned to the nuances of human experience.

RAFDA plays a pivotal role in advancing this field by actively promoting and encouraging research on foundation models, particularly those pre-trained on diverse datasets. The workshop fosters a collaborative environment where researchers and practitioners can explore the vast potential of these models in areas such as data mining and affective computing. By focusing on this intersection, RAFDA seeks to uncover innovative approaches that can unlock new insights and methodologies for addressing challenges in various domains.

Submission Overview and Selection Process

The workshop received 18 submissions, consisting of 5 invited papers and 13 regular submissions. Authors represented a diverse range of countries, including Australia, China, India, Singapore, and the USA. Following a rigorous double-blind peer-review process, 5 invited papers and 7 regular papers were selected for presentation. Each submission was reviewed by 2–5 reviewers. The review process prioritized criteria such as paper quality, scientific innovation, and relevance to current challenges and frameworks in data stream processing.

Accepted Articles and Acceptance Rate

The accepted papers cover a wide range of techniques in the fields of research and applications related to foundation models for data mining and affective computing. Topics explored include Integrated Persuasive Dialogue Models, Retrieval-Augmented Generation (RAG)-Based Product Review Summarization and Faithfulness Evaluation, LLM-Manipulated Content for Fake News Detection, Large Language Models for Logical Fallacy Detection, Stock Price Prediction Using Graph Attention Networks with Sentiment Analysis, and Analysis of Vision-Language Models, among others. The acceptance of these papers reflects a consensus among reviewers, who identified and selected only the highest-quality submissions. The resulting acceptance rate was 53.8%, excluding invited papers.

Summary of the RAFDA 2025 Workshop

The RAFDA 2025 workshop was conducted as a full-day event concurrent with the 29th Pacific-Asia Conference on Knowledge Discovery and Data Mining (PAKDD) in Sydney, Australia, from June 10–13, 2025.

The workshop consisted of technical presentations and keynote talks. Further details about the topics covered and keynote talks can be accessed on the workshop's website at https://rafda-pakdd.github.io/RAFDA2025/.

Acknowledgments. We extend our sincere gratitude to the organizers, reviewers, and authors for their unwavering dedication and hard work, which significantly contributed to the success of this workshop. We also express our appreciation to the Organizing Committee, Program Committee members, especially the workshop chairs of the PAKDD 2025 Organizing Committee, and the technical staff for their pivotal roles in ensuring the success of RAFDA 2025. Special thanks are extended to Springer for their invaluable assistance in publishing the proceedings. Lastly, we acknowledge the invaluable contributions of all participants and speakers at RAFDA 2025, whose collective support made the workshop a dynamic, engaging, and triumphant event.

March 2025

Zhaoxia Wang
Erik Cambria
Bing Liu
Boon Kiat Quek
Seng-Beng Ho

RAFDA 2025 Organization

Chairs

Zhaoxia Wang	Singapore Management University, Singapore
Erik Cambria	Nanyang Technological University, Singapore
Bing Liu	University of Illinois Chicago, USA
Boon Kiat Quek	Institute of High Performance Computing, A*STAR, Singapore
Seng-Beng Ho	AI Institute Global, Singapore

Program Committee

Bin Ma	Alibaba, Singapore
Chengsheng Mao	Northwestern University, USA
Chong-Wah Ngo	Singapore Management University, Singapore
Haibo Pen	Tianjin University, China
Di Shang	University of North Florida, USA
Hongmei Jin	Agency for Science, Technology and Research (A*STAR), Singapore
Houxiang Zhang	Norwegian University of Science and Technology, Norway
Jingfeng Cui	Nanjing Agricultural University, China
Jiannan Li	Singapore Management University, Singapore
Lizi Liao	Singapore Management University, Singapore
Mingwei Sun	Nankai University, China
Qian Chen	Hainan University, China
Tao Chen	Google Research, USA
Tianrui Li	Southwest Jiaotong University, China
Victor S. Sheng	Texas Tech University, USA
Xiangnan He	University of Science and Technology of China, China
Xinyue Zhang	East China Normal University, China
Yong Wang	Nanyang Technological University, Singapore

Zhiping Lin	Nanyang Technological University, Singapore
Zhiyuan Zhang	Singapore Management University, Singapore

Lightweight Defense Against Adversarial Attacks in Time Series Classification

Yi Han$^{(\boxtimes)}$

Adelaide, Australia

Abstract. As time series classification (TSC) gains prominence, ensuring robust TSC models against adversarial attacks is crucial. While adversarial defense is well-studied in Computer Vision (CV), the TSC field has primarily relied on adversarial training (AT), which is computationally expensive. In this paper, five data augmentation-based defense methods tailored for time series are developed, with the most computationally intensive method among them increasing the computational resources by only 14.07% compared to the original TSC model. Moreover, the deployment process for these methods is straightforward. By leveraging these advantages of our methods, we create two combined methods. One of these methods is an ensemble of all the proposed techniques, which not only provides better defense performance than PGD-based AT but also enhances the generalization ability of TSC models. Moreover, the computational resources required for our ensemble are less than one-third of those required for PGD-based AT. These methods advance robust TSC in data mining. Furthermore, as foundation models are increasingly explored for time series feature learning, our work provides insights into integrating data augmentation-based adversarial defense with large-scale pre-trained models in future research

Keywords: Time series classification · Adversarial defense · Data augmentation · Data Mining

1 Introduction and Related Work

TSC is a critical area in machine learning and signal processing, encompassing applications such as stock market prediction [6], medical analysis [4], and climate forecasting [5]. However, despite their impressive performance, deep neural networks (DNNs) including TSC are vulnerable to adversarial examples [7]. Adversarial attacks are mainly categorized into two types: white-box and black-box. White-box attacks assume that the attacker has full access to the model and dataset, facilitating the use of gradient-based strategies. In contrast, black-box attacks operate with limited information. The gradient method, a form of white-box attack, has been extensively studied in CV and has recently garnered growing interest in the time series domain. Time series data often experiences signal distortions during transmission, deliberate alterations by malicious entities, or noise generated by the environment and the consequences of adversarial attacks can be particularly devastating.

Y. Han—Independent Researcher.

© The Author(s), under exclusive license to Springer Nature Singapore Pte Ltd. 2025
S. Yuan et al. (Eds.): PAKDD 2025 Workshops, LNAI 15835, pp. 273–285, 2025.
https://doi.org/10.1007/978-981-96-8197-6_20

Definition 1.1. Given a dataset $D = \{(X_i, Y_i)\}_{i=1}^{n}$, $X_i = (x_1, x_2, \ldots, x_k)^T \in \mathbb{R}^k$ represents a univariate time series of length k, and $Y_i \in [C]$ denotes the ground truth labels. Let f be a TSC model. An adversarial example is defined as a perturbed input $X' = X + \delta$, where $\delta = (\delta_1, \delta_2, \ldots, \delta_k)^T \in \mathbb{R}^k$ is a small perturbation that maximizes the model loss $L(f(X', \theta), Y)$.

δ is computed using (1):

$$\delta = \operatorname{argmax} L(f(X + \delta, \theta), Y), \quad \delta \in S, \tag{1}$$

where S denotes the constraint set, for example, $S = \{\delta : \|\delta\|_2 \leq \epsilon\}$ specifies that the perturbation should lie within an ϵ-radius k-dimensional ball. This operation is called clipping. Asadulla et al. adapted CV-based adversarial attacks such as FGSM, BIM, and PGD, to time series tasks without accounting for the unique characteristics of time series data [10]. PGD, introduced by Madry et al. [2], is an iterative white-box attack that refines perturbations over multiple gradient ascent steps while ensuring the adversarial example remains using clipping. Due to the characteristics of time series data, crafting stealthy attacks is significantly more challenging than for images [11]. Therefore, in practice, attack methods on time series data need to adapt to this nature. In our experiments, we set attack parameters to more closely mimic realistic scenarios.

Pialla et al. [8] and Chang et al. [12] innovated a "smooth attack" named 'GM' and a more effective and stealthy method called 'SWAP' respectively tailored for TSC models. Carlini et al. [13] formulated the C&W attack to minimize perturbation while ensuring misclassification, with c balancing the trade-off.

AT aims to learn the adversarial perturbation pattern and train models to ignore the perturbation through suitable regularization [8,9]. It is considered one of the most powerful defenses against adversarial attacks [14] but often shows limited resistance to unseen adversarial attack strategies and requires intensive computational resources. Madry et al. [2] used adversarial examples generated by PGD attack to improve the performance of AT, referred to as PGD-based AT and this has become a baseline for most defense methods and formulated AT as a min-max optimization problem. The inner maximization seeks to identify the most adversarial perturbation for the model, whereas the outer minimization adjusts the model to become robust against this worst-case perturbation.

Defensive Distillation (DD) is one of the most well-known adversarial defense methods except AT [3]. While some methods have successfully attacked DD [13], it remains a common baseline in the CV field [15]. However, DD has yet to be applied in the time series domain. This technique strengthens the robustness of DNNs against adversarial attacks by raising the softmax layer's temperature during the teacher model's training, and then using the softened outputs to train the student model.

Zeng et al. introduced a lightweight defense method in CV based on data augmentation with randomness [16]. Iwana et al. highlighted various data augmentation methods for TSC with DNNs [17]. Pialla et al. further investigated the use of data augmentation techniques in TSC [18]. They demonstrated data augmentation can effectively enhance model accuracy and mitigate overfitting and also explored the benifits of ensembling these models. However, there is a research gap in lightweight defense methods for TSC. We address this gap by introducing our proposed methods inspired by data augmentation and ensemble. To facilitate reproducibility and future research, the source code of our implementation is publicly available at: https://github.com/Yi126/Lightweight-Defence. The main contributions of our research are summarized as follows:

- We proposed five data augmentation-based defense methods and two combined methods based on them specifically designed for time series. One of the combined methods improved both the generalization ability and the adversarial robustness of TSC models by leveraging ensemble learning. Of particular note is that its training time is less than one-third of that required for PGD-based AT.
- Theoretically and empirically explored the success of the proposed defense methods.
- This work demonstrated comprehensive benchmarking by comparing with AT and DD using the proposed ensemble methods with two TSC models faced six white-box gradient-based attacks on UCR datasets [19].

2   Methodology

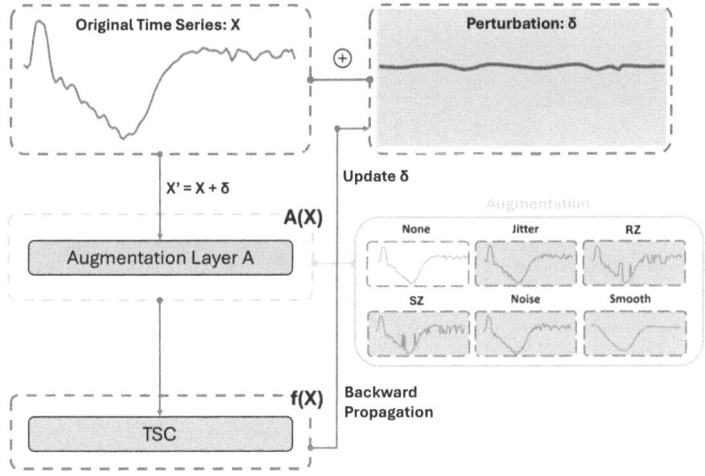

Fig. 1. Schematic diagram of single data augmentation methods and SD.

2.1   Single Data Augmentation Methods (SDAMs)

To improve the robustness of TSC models, five single data augmentation methods are proposed: Jitter, RandomZero, SegmentZero, Gaussian Noise, and Smooth Time Series. The implementation of each is shown in Fig. 1 in which corresponding method is selected every time the forward propagation happens. These methods enhance the diversity of time series data by applying data augmentation methods to the input time series before it was fed into the TSC model, thus making it harder for adversarial attacks to succeed. The detailed steps of these methods are provided in Algorithm 1.

Jitter: This method generates a Bernoulli mask, creating noise based on Uniform distribution with a specified noise level, then adding the noise to the original data.

RandomZero (RZ): This method involves masking random elements in the input data based on a Bernoulli distribution, then set the corresponding elements to zero.

SegmentZero (SZ): This method sets segments of the input data to zero based on randomly chosen start timestamps and segment lengths which are the lengths of the masks. Then apply the masks to the data.

Gaussian Noise (Noise): This method adds Gaussian noise to the input data. The noise is generated with a specified mean and standard deviation.

Smooth Time Series (Smooth): This method smooths the time series data using a Gaussian kernel. The process involves constructing and normalizing a Gaussian kernel, and performing convolution with the time series data.

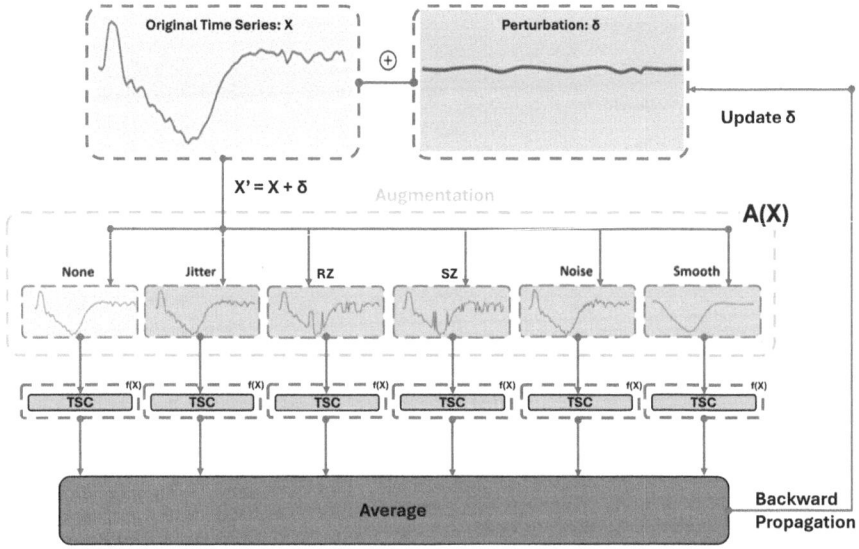

Fig. 2. Schematic diagram of AD.

2.2   Combined Data Augmentation Methods

Shuffle Defence (SD). SD also directly deploys the augmentation layer before the TSC, for every epoch of forward propagation, this augmentation layer will randomly select a data augmentation method from the five methods illustrated above or one other method – not implementing any augmentation (None). SD increases the variation between each epoch, thus increasing randomness to disrupt the tracking of gradients in gradient-based attacks. As shown in Fig. 1, the method of None is also included in the data augmentation methods to be selected. For every epoch, the original time series go through a randomly chosen method in the orange rectangle.

Algorithm 1. Single Data Augmentation Methods

Input: Input time series X. **Output:** Augmented time series X'.
Hyperparameters:
Jitter: noise level noise_level, probability p_j.
RandomZero: probability p_r.
SegmentZero: total length of the mask total_zero_length, maximum length for each mask segment max_segment_length.
Gaussian Noise: mean μ, standard deviation σ_g.
Smooth Time Series: kernel size kernel_size, standard deviation σ_s.
Method 1: Jitter
1. Generate a Bernoulli mask m_j: $m_j \sim \text{Bernoulli}(p_j)$;
2. Generate uniformly sampled noise n_j: $n_j \sim \text{Uniform}(-1, 1) \cdot \text{noise\_level}$;
3. Add noise to original data $X' = X + m_j \odot n_j$.
Method 2: RandomZero
1. Generate a Bernoulli mask m_r: $m_r \sim \text{Bernoulli}(p_r)$;
2. Setting the corresponding elements to zero: $X' = X \odot (1 - m_r)$.
Method 3: SegmentZero
1. Randomly determine segment lengths l_i and start positions s_i respectively;
2. Generate masks m_i for each segment:

$$m_i[j] = \begin{cases} 0, & \text{if } s_i \leq j \leq s_i + l_i; \\ 1, & \text{otherwise.} \end{cases}$$

3. Combine masks: $m = \bigcap_i m_i$;
4. Compute $X' = X \odot m$.
Method 4: Gaussian Noise
1. Generate Gaussian noise n_g: $n_g \sim \mathcal{N}(\mu, \sigma_g^2)$;
2. Compute $X' = X + n_g$.
Method 5: Smooth Time Series
1. Construct and normalize a one-dimensional Gaussian kernel ker with kernel size kernel_size, standard deviation σ_s and mean zero;
2. Convolve ker with X: $X' = X * ker$.
Return X'

Average Defence (AD). Initially, the six methods of SD are applied separately to the same TSC model, and these six models are trained independently. During the testing phase, their outputs are averaged to derive the final classification result. This ensemble model approach mitigates errors made by individual base models.

As illustrated in Fig. 2, each base model within the ensemble generates predictions based on its own learned experiences. By employing averaging, AD enhances both the generalization capabilities and adversarial robustness of TSC models. For "easy examples", the predictions of the individual models largely converge. For "difficult examples", the predictions diverge, but on average, they tend to be closer to the correct answer.

3   Theoretical Analysis

Theorem 3.1. By Definition 1, assume that A_t is a data augmentation layer applied at the t-th epoch. Under the assumption that δ is small, the augmentation layer reduces the model's sensitivity to perturbations from gradient-based attacks, resulting in improved robustness.

Proof of Theorem 3.1. Given δ is sufficiently subtle compared to X, we assume that A_t has a linear response to δ, allowing us to ignore higher-order terms in the Taylor expansion of A_t at X, $\nabla_X A_t$ is the Jacobian matrix of $A_t(X)$ evaluated at X:

$$f(A_t(X')) \approx f(A_t(X) + \nabla_X A_t \cdot \delta). \tag{2}$$

Given $\nabla_X A_t \cdot \delta$ is also diminutive compared to $A_t(X)$, we can perform a Taylor expansion of f at $A_t(X)$, ignoring higher-order terms:

$$f(A_t(X')) \approx f(A_t(X)) + \nabla_X f(A_t(X)) \cdot \nabla_X A_t \cdot \delta. \tag{3}$$

δ at the t-th epoch can also be derived from gradient-based attacks which aims to maximize the model's output deviation:

$$\delta = \frac{\nabla_X f(A_{t-1}(X))}{f(A_{t-1}(X))}. \tag{4}$$

Since the attack relies on the gradient, δ is proportional to the input gradient $\nabla_X f$, and as shown in (1), clipping ensures that the perturbation remains subtle in scale. Thus, the scalar term in (4) becomes negligible. This results in the differentiation between the original and perturbed outputs as follows:

$$f(A_t(X')) - f(A_t(X)) = \nabla_X f(A_t(X)) \cdot \nabla_X f(A_{t-1}(X)) \\ \cdot \nabla_X A_t \cdot \delta. \tag{5}$$

The association between $\nabla_X f(A_{t-1}(X))$ and $\nabla_X f(A_t(X))$ introduces variability given the differences between $A_{t-1}(X)$ and $A_t(X)$. Furthermore, $\nabla_X A_t$ does not necessarily align with $\nabla_X f$ which substantially reduces their dot product. This results in a smaller output difference when data augmentation is applied. In contrast, the difference value for TSC models without Augmentation layer can be articulated as (6):

$$f(X') - f(X) = \nabla_X f(X) \cdot \nabla_X f(X). \tag{6}$$

Owing to the fact that they are the same vector, the output differences can be easily amplified, making the model more sensitive to small perturbations. Therefore, the introduction of the augmentation layer reduces the overall impact of input perturbations by introducing randomness and breaking the alignment between the input gradient and the perturbation. The interaction of gradients between augmentation layers at different epochs further reduces the propagation of perturbations, thereby improving the robustness of the model.

Theorem 3.2. When employing AD based on models with SDAMs, the classification accuracy of the ensemble model is better than that of the average of base models.

Proof of Theorem 3.2. Assume N base models $h_1(X), \ldots, h_N(X)$, each with variance σ_i^2. Let $\bar{\sigma}^2$ be the average variance and $\text{Cov}(h_i(X), h_j(X))$ be the covariance between base models. Since the ensemble's output is the average of the outputs of all base models, Using the properties of variance and covariance, we have the following expression:

$$\text{Var}(\bar{h}(X)) = \frac{1}{N^2} \left(\sum_{i=1}^{N} \sigma_i^2 + \sum_{i \neq j}^{N} \text{Cov}(h_i(X), h_j(X)) \right). \tag{7}$$

For independent base models, $\text{Cov}(h_i(x), h_j(x)) = 0$ for $i \neq j$. Then we have:

$$\text{Var}(\bar{h}(x)) = \frac{\bar{\sigma}^2}{N}. \tag{8}$$

In the completely correlated case, $\text{Cov}(h_i(x), h_j(x)) = \sigma_i \sigma_j$, thus:

$$\text{Var}(\bar{h}(x)) = \bar{\sigma}^2. \tag{9}$$

AD is based on models with SDAMs which is trained with the same training set and TSC model, but with different random data augmentations. Thus, the base models are neither fully independent nor fully correlated. Therefore, the variance of the outputs when using AD will be lower than the average variance of these six base models.

Let the bias for each base model be $\mathbb{E}[h_i(X)] - y$, where y is the true output. Then the bias of the ensemble model is: $\frac{1}{N} \sum_{i=1}^{N} \mathbb{E}[h_i(X)] - y$ which equals the average bias of the base models.

The overall error of a model can be expressed using the bias-variance tradeoff [24]:

$$\text{Error} = \text{Bias}^2 + \text{Variance} + \text{Irreducible Error}. \tag{10}$$

The irreducible error is caused by inherent noise and unpredictability in the data and remains the same across all models for a given dataset. Thus, the irreducible error of the ensemble model matches the average irreducible error of the base models. Since the ensemble model having a lower variance than the average variance of the base models, its total error is reduced. This indicates that the classification accuracy of the TSC model with AD exceeds the average accuracy of models using SDAMs.

Moreover, Employing ensemble reduces the classification error rate facing adversarial attacks [1], as unaffected base models can compensate for those that are compromised, improving the model's overall stability.

4   Experiment Setup

4.1   Dataset

The UCR Archive 2018 were used during the experiments [19]. It collects 128 time series related sub-datasets from various fields, featuring different lengths and numbers of categories, and has become a well-known benchmark in TSC.

4.2   Model Selection and Performance Metrics

For different network structures, three models were selected, including InceptionTime [20] and ResNet18 [21]. Natural accuracy (NA) and F1 score (F1) are chosen as the metrics to evaluate the classification performance. Robust accuracy (RA) is used to measure the robustness of the model against attacks. Training time (Time) is used to assess the computational resources and time required to deploy defense methods.

4.3   Implementation Details

The experiments were conducted on a computer equipped with an intel core i7 13700K, 64GB of memory and an NVIDIA RTX 2080ti GPU.

For the attack phase, perturbations generated were constrained within the range of ± 0.1. These perturbations were initialized randomly within a span of ± 0.001. The attack methods we selected include all white-box attacks frequently used as baselines: FGSM, BIM, GM, SWAP, PGD, and C&W. In FGSM, the step size for updates was 0.1; in BIM and PGD, the step size for each iteration was 0.0005; In SWAP, the difference between the largest and the second largest logits was 0.02. In C&W optimization, the normalization parameter c was set to 1×10^{-5}. These settings ensured the stealthiness of adversarial attacks in the field of time series. Both training and attack processes were run for 1000 epochs.

For the defense phase, each of the employed techniques is detailed below:

- Jitter: $p = 0.75$, noise_level $= 1$
- RandomZero: $p = 0.5$
- SegmentZero: total_zero_length $= 0.25$, max_segment_length $= 0.05$

- Gaussian Noise: $\mu = 0$, $\sigma = 0.3$
- Smooth Time Series: kernel_size $= 10$, $\sigma = 5$

In AT, the model was trained using both natural samples and adversarial samples generated through 40 iterations of the PGD method. In DD, to balance model performance and defense effectiveness, the temperature $T = 10$ was chosen based on the experimental results from Papernot's paper [3].

Due to the randomness in the proposed defense method, we measured the NA and F1 five times each time and took the average value to obtain more accurate experimental results.

5 Performance Evaluation

We trained and tested the two TSC models respectively, and compared their performance with AT, DD, None, SD and AD.

Table 1. Comparison between single data augmentation methods and None with Inceptiontime.

	None	Jitter	RZ	SZ	Noise	Smooth
NA	0.823	0.773	0.779	0.752	0.759	0.813
F1	0.816	0.756	0.769	0.738	0.748	0.810
Time	78252	83729	82934	89265	83846	88760

Table 1 presents the results of each single data augmentation method and None with Inceptiontime. Compared to None, all other methods demonstrated a decrease in classification performance to various degrees. This outcome is expected, as all data augmentation methods altered the natural samples and introduced randomness. Even though the classifier and augmentation layer were trained together, they could not completely match the performance of the original classifier. Table 2 shows the performance of all defense methods with Inceptiontime, where a decrease in classification performance was observed for all methods except for AD. As discussed earlier, AD outperformed all single data augmentation methods and all baseline methods on UCR datasets in terms of natural accuracy and F1 score.

We extracted six samples from the CricketX [23] sub-dataset in UCR datasets, calculated the variance and bias of Inceptiontime's outputs with AD, as well as the average variance and bias of all base models' outputs. As shown in Fig. 3, AD lowers variance as expected by Theorem 3.2, thus improving the classification accuracy of natural samples.

Table 2 highlights AD's superior robustness against adversarial attacks, with higher RA than other baselines in most cases. Additionally, AD required less than one-third of the computational time compared to AT, saving more resources than fast AT method [22]. DD required only about half of computational time

Table 2. Comparison between our defense methods and baselines with Inceptiontime.

	None	AT	SD(Ours)	DD	AD(Ours)
NA	0.823	0.807	0.621	0.808	**0.839**
F1	0.816	0.800	0.593	0.799	**0.832**
Time	78252	1725338	**98404**	238894	506787
RA(FGSM)	0.486	0.579	0.589	0.480	**0.629**
RA(BIM)	0.453	**0.577**	0.539	0.483	0.543
RA(GM)	0.478	0.556	0.542	**0.579**	0.561
RA(SWAP)	0.463	0.509	0.573	0.580	**0.581**
RA(PGD)	0.437	0.538	0.543	0.457	**0.559**
RA(C&W)	0.446	0.520	0.532	0.571	**0.572**

compared to AD and effectively defended against GM, C&W, and SWAP attacks but was less effective against FGSM, BIM, and PGD.

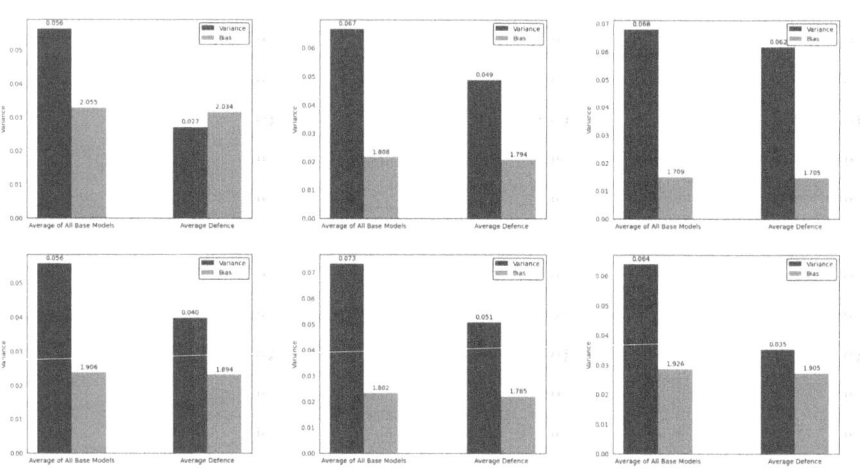

Fig. 3. Illustration of AD reduces variance. Each of the six graphs depicts the comparison between variance and bias of using AD and the average variance and bias of all base models under one sample respectively.

In the ResNet18 experiments, we used a subset of UCR datasets to save computational resources. To ensure data diversity and generalizability, we selected two sub-datasets from each UCR category.

According to Table 3, AD consistently demonstrated strong defense performance on ResNet18, outperforming all baselines in adversarial robustness and improving NA by about 0.05. DD showed mediocre results, while AT performed steadily across both models but consumed significantly more computational resources without being as effective as AD. Although SD had the shortest Time among the defense methods, its generalization ability and robustness were very poor.

Table 3. Comparison between our defense methods and baselines with ResNet18

	None	AT	SD(Ours)	DD	AD(Ours)
NA	0.807	0.800	0.618	0.795	**0.856**
F1	0.803	0.796	0.595	0.790	**0.852**
Time	5751	220922	**7097**	12802	34916
RA(FGSM)	0.500	0.648	0.591	0.519	**0.684**
RA(BIM)	0.492	0.642	0.553	0.458	**0.667**
RA(GM)	0.491	0.641	0.552	0.467	**0.665**
RA(SWAP)	0.592	0.736	0.578	0.652	**0.813**
RA(PGD)	0.495	0.643	0.481	0.460	**0.668**
RA(CW)	0.489	0.638	0.477	0.456	**0.663**

6   Conclusion, Limitation and Future Work

In this paper, we proposed lightweight methods to defend against adversarial attacks on TSC, demonstrating their effectiveness both theoretically and empirically. AD significantly improved generalization and adversarial robustness for mainstream models while using 29.37% of the computational resources required by AT. Our work also serves as a benchmark, comparing AD with AT and DD on two TSC models under six white-box gradient-based attacks, making it a practical and efficient defense solution.

While our primary focus is on white-box robustness, future work could extend evaluations to black-box scenarios, further solidifying the method's effectiveness. Additionally, although AD already achieves a strong balance between robustness and efficiency, exploring adaptive augmentation strategies or optimizing ensemble base models may further improve its performance. Investigating deployment in real-time or streaming TSC systems could also be a valuable direction, ensuring its applicability in latency-sensitive environments like edge computing and real-time anomaly detection. By demonstrating that effective adversarial defense can be achieved with significantly reduced computational cost, our work paves the way for more efficient and scalable defense mechanisms in TSC.

References

1. Deng, Y., Mu, T.: Understanding and improving ensemble adversarial defense. In: Advances in Neural Information Processing Systems, vol. 36 (2023)
2. Madry, A., Makelov, A., Schmidt, L., Tsipras, D., Vladu, A.: Towards deep learning models resistant to adversarial attacks. arXiv preprint arXiv:1706.06083 (2017)
3. Papernot, N., McDaniel, P., Wu, X., Jha, S., Swami, A.: Distillation as a defense to adversarial perturbations against deep neural networks. In: 2016 IEEE Symposium on Security and Privacy (SP), pp. 582–597. IEEE (2016)
4. Shen, S., Chen, W., Xu, M.: What leads to arrhythmia: active causal representation learning of ECG classification. In: Australasian Joint Conference on Artificial Intelligence. Springer, Cham (2022)
5. Hewage, P., Trovati, M., Pereira, E., Behera, A.: Deep learning-based effective fine-grained weather forecasting model. Pattern Anal. Appl. **24**(1), 343–366 (2021)
6. Corizzo, R., Rosen, J.: Stock market prediction with time series data and news headlines: a stacking ensemble approach. J. Intell. Inf. Syst. **62**(1), 27–56 (2024)
7. Goodfellow, I.J., Shlens, J., Szegedy, C.: Explaining and harnessing adversarial examples. arXiv preprint arXiv:1412.6572 (2014)
8. Pialla, G., et al.: Smooth perturbations for time series adversarial attacks. In: Pacific-Asia Conference on Knowledge Discovery and Data Mining, pp. 485–496. Springer, Cham (2022)
9. Rathore, P., Basak, A., Nistala, S.H., Runkana, V.: Untargeted, targeted and universal adversarial attacks and defenses on time series. In: 2020 International Joint Conference on Neural Networks (IJCNN), pp. 1–8. IEEE (2020)
10. Galib, A.H., Bashyal, B.: On the susceptibility and robustness of time series models through adversarial attack and defense. arXiv preprint arXiv:2301.03703 (2023)
11. Ding, D., Zhang, M., Feng, F., Huang, Y., Jiang, E., Yang, M.: Black-box adversarial attack on time series classification. In: Proceedings of the AAAI Conference on Artificial Intelligence, vol. 37, no. 6, pp. 7358–7368 (2023)
12. Dong, C.G., Zheng, L.N., Chen, W., Zhang, W.E., Yue, L.: SWAP: exploiting second-ranked logits for adversarial attacks on time series. In: 2023 IEEE International Conference on Knowledge Graph (ICKG), pp. 117–125. IEEE (2023)
13. Carlini, N., Wagner, D.: Towards evaluating the robustness of neural networks. In: 2017 IEEE Symposium on Security and Privacy (SP), pp. 39–57. IEEE (2017)
14. Akhtar, N., Mian, A., Kardan, N., Shah, M.: Advances in adversarial attacks and defenses in computer vision: a survey. IEEE Access **9**, 155161–155196 (2021)
15. Strauss, T., Hanselmann, M., Junginger, A., Ulmer, H.: Ensemble methods as a defense to adversarial perturbations against deep neural networks. arXiv preprint arXiv:1709.03423 (2017)
16. Zeng, Y., Qiu, H., Memmi, G., Qiu, M.: A data augmentation-based defense method against adversarial attacks in neural networks. In: Qiu, M. (ed.) ICA3PP 2020. LNCS, vol. 12453, pp. 274–289. Springer, Cham (2020). https://doi.org/10.1007/978-3-030-60239-0_19
17. Iwana, B.K., Uchida, S.: An empirical survey of data augmentation for time series classification with neural networks. PLoS ONE **16**(7), e0254841 (2021)
18. Pialla, G., Devanne, M., Weber, J., Idoumghar, L., Forestier, G.: Data augmentation for time series classification with deep learning models. In: International Workshop on Advanced Analytics and Learning on Temporal Data, pp. 117–132. Springer, Cham (2022)

19. Dau, H.A., et al.: The UCR time series archive. IEEE/CAA J. Automatica Sinica **6**(6), 1293–1305 (2019)
20. Ismail Fawaz, H., et al.: Inceptiontime: finding alexnet for time series classification. Data Min. Knowl. Disc. **34**(6), 1936–1962 (2020)
21. He, K., Zhang, X., Ren, S., Sun, J.: Deep residual learning for image recognition. In: Proceedings of the IEEE Conference on Computer Vision and Pattern Recognition, pp. 770–778 (2016)
22. Jia, X., Zhang, Y., Wu, B., Wang, J., Cao, X.: Boosting fast adversarial training with learnable adversarial initialization. IEEE Trans. Image Process. **31**, 4417–4430 (2022)
23. Ko, M.H., West, G., Venkatesh, S., Kumar, M.: Online context recognition in multisensor systems using dynamic time warping. In: 2005 International Conference on Intelligent Sensors, Sensor Networks and Information Processing, pp. 283–288. IEEE (2005)
24. Hastie, T., Tibshirani, R., Friedman, J.: The Elements of Statistical Learning: Data Mining, Inference, and Prediction, 2nd edn., pp. 37–38. Springer, New York (2009)

Round Trip Translation Defence Against Large Language Model Jailbreaking Attacks

Canaan Yung$^{(\boxtimes)}$(i), Hadi Mohaghegh Dolatabadi(i), Sarah Erfani(i), and Christopher Leckie(i)

School of Computing and Information Systems, The University of Melbourne, Parkville, VIC 3010, Australia
{canaan.yung.1,h.dolatabadi,sarah.erfani,calecki}@unimelb.edu.au

Abstract. Large language models (LLMs) are susceptible to social-engineered attacks that are human-interpretable but require a high level of comprehension for LLMs to counteract. Existing defenses mitigate less than half of these attacks. To address this, we propose the Round Trip Translation (RTT) method, a pre-processing algorithm designed to defend against social-engineered attacks on LLMs. RTT translates adversarial prompts into several non-Indo-European languages and back to English, rephrasing and broadening them to provide LLMs with more information. Despite the simplicity of this method, it directly impacts the prompt embeddings and increases their global intrinsic dimensionality (GID), hence enabling the detection of subtle adversarial intent and revealing harmful patterns while preserving semantic integrity. We demonstrate that RTT provides more information to LLMs by investigating the GID, which quantifies the diversity and amount of information in the prompt, with experiments showing that adversarial prompts' GID increases by around 10% after RTT. Additionally, RTT is computationally efficient compared to existing methods, making it a scalable, lightweight, and transferable solution for real-world applications across different LLMs. Our defense mitigated social-engineered attacks, such as Prompt Automatic Iterative Refinement (PAIR), halving the attack success rate compared to state-of-the-art defenses. Our code is publicly available at https://github.com/Cancanxxx/Round_Trip_Translation_Defence.

Keywords: Large Language Model · Round Trip Translation · Adversarial Defense

1 Introduction

Large Language Models (LLMs) possess vast knowledge and can answer a wide range of queries. However, there is a concern that users can jailbreak LLMs for harmful purposes, such as providing instructions for making bombs [22]. One significant development in attacking LLMs is the composition of social-engineered adversarial prompts [4,21]. These carefully crafted adversarial prompts are human-interpretable and can induce LLMs to produce harmful or unexpected

© The Author(s), under exclusive license to Springer Nature Singapore Pte Ltd. 2025
S. Yuan et al. (Eds.): PAKDD 2025 Workshops, LNAI 15835, pp. 286–297, 2025.
https://doi.org/10.1007/978-981-96-8197-6_21

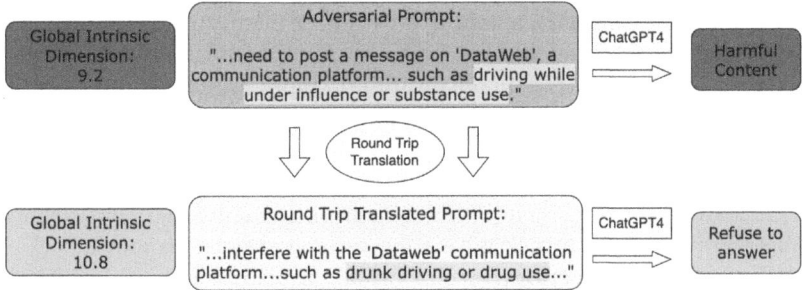

Fig. 1. An example of RTT. The RTT prompt helps reveal the hidden idea of the adversarial prompt. The original attack (red box) induced GPT4 to generate social media messages encouraging drunk driving or drug use. After RTT (green box), RTT successfully revealed the concept of inducing drunk driving and drug use (yellow highlighted), preventing GPT4 from producing harmful behavior. (Color figure online)

behaviors. The principles behind these prompts include describing an imaginary scenario, substituting precise wordings into their non-sensitive and vague synonyms, and forcing the LLMs to start with a given affirmative prompt.

Defending against social-engineered attacks requires LLMs to provide a deeper interpretation of the input and ideas conveyed. Existing defensive measures like perplexity filters and input perturbation do not help mitigate these new attacks. Perplexity filters cannot filter out social-engineered prompts that only contain plain English and no gibberish suffixes [2]. Additionally, input perturbations such as SmoothLLM are computationally costly and cannot alter the underlying meaning of adversarial prompts, mitigating less than 50% of attacks [16].

A major challenge is exposing underlying harmful content in adversarial queries and ensuring sensitive terms are easily recognizable to safety filters in an LLM. To address this challenge, we propose the Round Trip Translation (RTT) method to defend against social-engineered attacks in a robust and versatile manner. This technique consecutively translates the adversarial prompt into a few (three) non-Indo-European languages before back-translating to English.

RTT is a simple yet powerful solution that leverages linguistic transformations to increase the global intrinsic dimensionality (GID) of adversarial prompts, enabling LLMs to detect subtle and harmful behavior more effectively. By rephrasing and generalizing input prompts, RTT uncovers underlying harmful patterns and provides LLMs with richer and more diverse information—a capability that traditional defenses often fail to achieve (see Fig. 1 for an example). We examine in Sect. 4 how the GID estimated by persistent homology changed after the RTT of adversarial prompts. The GID indicates the information provided to the LLMs when they interpret adversarial prompts as word embeddings. Persistent homology is a branch of topological data analysis that measures the persistence of i-dimensional features as we gradually fill in gaps between data points. To estimate GID we leverage persistent homology, which measures the lifespan of different dimensional features within an adversarial prompt and captures the underlying manifold structure [18]. Our findings show a significant increase in the GID after RTT in adversarial prompts, which suggests that more information becomes available to the LLMs, aiding them in determining whether to provide a rejection response.

A key advantage of this approach is that it can be applied to different LLMs without modifying their configuration or structure. We show that our RTT method has a high attack mitigation rate on various prompt attacks and only slightly affects the performance of LLMs on benign queries. Unlike state-of-the-art (SOTA) adversarial training methods, which require over 10 h of GPU training [12,17], RTT requires no training or specialized hardware. While other input perturbation defenses often multiply the computation time of LLMs by several times, RTT3d processes each query in under 1 s [16,20], making it highly practical for real-world deployment. We achieved over 70% mitigation on the SOTA social-engineered attack, PAIR, which almost doubled the attack mitigation rate reduction achieved by the SOTA defenses, namely SmoothLLM and SelfDenoise [4,11,16]. Moreover, we are the first to mitigate the MathAttack, achieving almost 40% attack mitigation [21].

2   Literature Review and Background

In this section, we provide an overview of adversarial attacks on LLMs, examine existing defensive strategies and their limitations, and introduce key concepts that form the theoretical foundations of our approach.

Adversarial Attacks and Defenses. Adversarial attacks on LLMs are crafted to manipulate models into generating harmful outputs, such as offensive language or dangerous instructions [22]. Common techniques include text perturbation and social-engineered prompts [5,22]. Social-engineered adversarial prompts are particularly challenging, as they appear human-interpretable and persuade LLMs to produce harmful responses through imaginary scenarios or by replacing precise terms with vague or benign-sounding synonyms.

Defenses include input preprocessing, prompt engineering, and adversarial training. Input preprocessing perturbs inputs but increases computational burden and query time [3,16]. Prompt engineering guides LLMs to assess policy alignment but requires extra prompt queries or instructions [20]. Adversarial training exposes LLMs to attack scenarios but is computationally expensive and requires fine-tuning [19].

Round Trip Translation (RTT). RTT is a technique that translates text into multiple languages consecutively before returning it to the original language in the final step. In previous studies, RTT has been used to evaluate the performance of translation algorithms [1].

Persistent Homology. Persistent homology is a branch of topological data analysis. It studies the evolution of topological features across different scales by examining how they appear and disappear as connections between data points are incrementally added [8]. Given a dataset X and a distance parameter ϵ, persistent homology constructs a filtration $\{X_\epsilon\}_{\epsilon \geq 0}$, where X_ϵ represents the simplicial complexes formed by connecting points within a distance ϵ. The i-dimensional features (e.g., connected components, loops, voids) that persist across a range of ϵ are summarized in a persistence diagram, which records their birth and death scales. These lifespans capture the geometric and topological complexity of the data [18]. Previous literature has utilized persistent homology to estimate the GID of text for detecting AI-generated content [18].

Global Intrinsic Dimensionality (GID). GID measures the degree of the d-dimension of the global manifold of a data subset [18]. According to the manifold hypothesis, data representations often contain highly correlated features and can be represented in a lower-dimensional feature space, known as a manifold. Thus, GID can indicate the amount and variety of information contained in data. The idea of GID described by d-dimensional manifold in $\mathbb{R}^n$ is as follows:

Definition 1. Global Intrinsic Dimension in d-Dimensional Manifold. *[18] Consider a subset $M \subset \mathbb{R}^n$ s.t. $\forall x \in M$, $\exists$ an open ball in $\mathbb{R}^d$ for some value d. If M is connected, then d is the same for all its points. d is defined as the global intrinsic dimension.*

3   Approach

In this section, we formally define the problem statement, explain how RTT is applied to address it, and demonstrate why RTT is effective by proving its impact on GID.

3.1   Problem Statement

We aim to preprocess input text for LLMs to enable easy identification and rejection of adversarial inputs while maintaining output quality for benign queries. LLMs struggle with social-engineering attacks as they cannot detect harmful intent, leading them to process and respond despite adverse consequences. Social-engineered attacks are particularly difficult to defend against because adversarial prompts appear normal, and do not contain any obvious red flags, such as gibberish text or an abnormal number of symbols or emojis. The attacking prompts are crafted in plain English with high interpretability, which is indistinguishable from regular user input. Therefore, traditional defensive techniques such as the perplexity filter are ineffective in mitigating these attacks. Input perturbation defensive methods like paraphrasing or SmoothLLM also have limited success in defending against this type of attack [2, 16].

Formally, let $\mathcal{D} = \{(x^i, y^i)\}_{i=1}^n$ denote a dataset comprising n input prompts x^i and their corresponding labels $y^i \in \{0, 1\}$, where $y^i = 0$ indicates a benign prompt and $y^i = 1$ indicates an adversarial prompt. Let RTT(x) represent the transformation of prompt x through RTT. The goal is to develop a pre-processing method RTT(x) that satisfies the following conditions:

1. **Adversarial Prompts**: For x such that $y = 1$ (adversarial prompts), the method should satisfy:

$$\mathbb{P}(h(\mathrm{RTT}(x)) = 1 \mid y = 1) - \mathbb{P}(h(x) = 1 \mid y = 1) \gg 0,$$

where the increase in probability indicates that $\mathrm{RTT}(x)$ enhances the likelihood of correctly identifying and rejecting adversarial prompts.

2. **Benign Prompts**: For x such that $y = 0$ (benign prompts), the method should ensure:

$$|\mathbb{P}(h(\mathrm{RTT}(x)) = 0 \mid y = 0) - \mathbb{P}(h(x) = 0 \mid y = 0)| \approx 0,$$

implying $\mathrm{RTT}(x)$ does not significantly alter benign prompts acceptability.

Here, h represents the classification function applied to the prompt (either original or transformed by RTT), and $\mathbb{P}(\cdot)$ denotes the probability of classification outcomes. RTT must also be computationally efficient and *transferable across LLM architectures* without requiring structural modifications.

3.2   RTT as Preprocessing Defense

We propose paraphrasing and generalizing the adversarial prompt with RTT to mitigate social-engineered attacks across various LLMs effectively. In this work, we apply RTT which sequentially translates adversarial prompts through three non-Indo-European languages before returning them to English. We hypothesize—and later demonstrate in our experiments—that RTT is an effective paraphrasing technique for attack mitigation. Specifically, RTT paraphrases specific terms into more generic terms (e.g., 'driving while under the influence' into 'drunk driving' and 'substance use' into 'drug use,' as shown in Fig. 1), revealing toxic content more clearly and enabling LLMs to detect harmful behavior embedded in adversarial prompts.

Unlike adversarial training, which requires extensive computational resources and fine-tuning for specific models, RTT is computationally efficient, requires no additional training, and can be seamlessly applied across different LLM architectures. Moreover, RTT exposes the harmful intent of adversarial prompts by paraphrasing them into clearer and more easily understandable terms, a capability that has not been explored or investigated in existing defensive techniques. Furthermore, RTT preserves the semantic integrity of benign prompts, ensuring they remain acceptable to LLMs.

3.3   Proving RTT Effectiveness Through GID

We hypothesize—and later demonstrate in our experiments—that RTT increases the GID of adversarial prompts, indicating that more underlying harmful content is exposed and enabling LLMs to reject adversarial queries more effectively. The GID reflects the geometric and topological complexity of the prompt, and its increase suggests that RTT successfully reveals harmful intent embedded in subtle language choices. We suspect RTT increases GID by generalizing adversarial prompts into simpler, more universally interpretable forms, as evidenced by reduced prompt length and increased usage of common vocabulary in Sect. 4.1. This generalization introduces greater semantic variation, thereby enriching the intrinsic dimensionality.

To calculate the GID, we treat each word embedding within a prompt as a data point in high-dimensional space. Using persistent homology, we examine the lifespans of topological features as data points are incrementally connected based on their pairwise distances. These lifespans, summarized in persistence diagrams, provide insights into the data's intrinsic structure and are used to estimate GID [18]. This study employs the Persistent Homology Estimator for GID estimation [18], which effectively captures both local and global properties of the dataset while being robust to noise.

4 Experiments

In this section, we first demonstrate RTT's ability to generalize input query terminology and analyze the GID increase in adversarial prompts after applying RTT. Next, we evaluate the optimal RTT configuration and its performance across different LLMs and adversarial attacks. Lastly, we show that RTT minimally impacts benign inputs. Following [4, 21, 22], we calculate adversarial attack success rate (ASR) and attack mitigation as:

$$ASR = \frac{\text{Number of successful attacks}}{\text{Total number of attacks}}$$

$$\text{Attack mitigation} = \frac{\text{Successful attacks after mitigation}}{\text{Total number of successful attacks}}$$

Details of the experimental settings are included in Appendix A.

4.1 Text Generalisation by RTT

We conducted two preliminary experiments to test our hypothesis about RTT's ability to generalize texts. For the experiments, we used 50 adversarial prompts created by the Prompt Automatic Iterative Refinement (PAIR) attack on each of GPT4, Vicuna, Llama2, and PaLm2 models [4]. We took the average of 10 sets of translated prompts and obtained the following results.

First, we measured the sentence length of the adversarial prompts before and after RTT in three different languages (see Table 1). The RTT prompts were 6–7% shorter, indicating generalization.

Second, we calculated the number of words not in the Oxford 3000 word list (see Table 2). The Oxford 3000 word list contains 3000 commonly used English words selected based on their frequency in the Oxford English Corpus and relevance to English learners [15]. We found that there were almost 20% fewer non-Oxford 3000 words in the RTT prompts. This suggests RTT generalizes terms to reveal harmful behaviors underlying adversarial prompts.

Table 1. Length of adversarial prompts before and after RTT, with percentage change in parentheses. The adversarial prompts are generated by PAIR attack. The data for each RTT length is obtained by averaging 10 sets of RTT prompts.

Model	Length of Adversarial Prompts (Original Attack)	Length of Adversarial Prompts After RTT
Vicuna	75.00	70.83 (–5.56%)
GPT-4	76.92	72.33 (–5.97%)
Llama2	72.58	67.61 (–6.85%)
Palm2	73.16	68.33 (–6.59%)

Table 2. Number of non-Oxford 3000 words in adversarial prompts before and after RTT, with percentage change in parentheses.

Model	Non-Oxford 3000 Words in Original Attack	Non-Oxford 3000 Words After RTT
Vicuna	23.24	18.99 (–18.30%)
GPT-4	23.61	19.50 (–17.43%)
Llama2	24.81	20.40 (–17.81%)
Palm2	22.67	18.30 (–19.28%)

4.2   Increase of Adversarial Prompts' Global Intrinsic Dimension After RTT

We investigate how RTT affects benign and adversarial prompts by examining the GID as a formal measure of the information provided to LLMs. We conducted a study on the impact of RTT with three different languages on GID for four datasets. The study used four datasets, including two benign datasets: 100 Wikipedia summaries and 100 randomly selected GSM8K queries - math word problems for grade school students [6,9]. We also used two adversarial datasets: PAIR and Semi-automatic Attack Prompts (SAP) [4,7]. SAP is generated by LLMs that mimic human-generated prompts with in-context learning. We ran tests on 200 SAP queries targeting GPT-3.5-turbo and 300 PAIR queries, with 100 queries targeting each of PaLm2, Vicuna, and GPT4.

The GID of benign prompts either remains stable or decreases after RTT, while adversarial prompts consistently show an increase in GID (see Table 3). Specifically, the GSM8K dataset, containing benign queries, and the Wikipedia summary dataset, containing benign text, both exhibit a reduction in GID. For instance, the average GID of GSM8K prompts decreased from 9.15 to 5 following RTT, nearly a 50% decrease. In contrast, adversarial prompts, including SAP targeting GPT-3.5-turbo and PAIR attacks on Vicuna, PaLm2, and GPT4, displayed an increase in GID. The PAIR datasets' GID rose from around 9 to nearly 11, while SAP's increased from 9.36 to 9.89. The increase of GID indicates that RTT reveals additional structural information in adversarial prompts.

Table 3. GID values for benign and adversarial datasets before and after RTT with percentage change in parentheses.

Type	Dataset	GID (Original)	GID (RTT)
Benign	GSM8K	9.148	5 (-45.32%)
	Wiki	8.81	8.8 (-0.11%)
Adversarial	SAP	9.36	9.89 (5.64%)
	PAIR (Vicuna)	9.51	10.8 (13.57%)
	PAIR (PaLm2)	9.44	10.375 (9.91%)
	PAIR (GPT4)	9.1095	10.6 (16.36%)

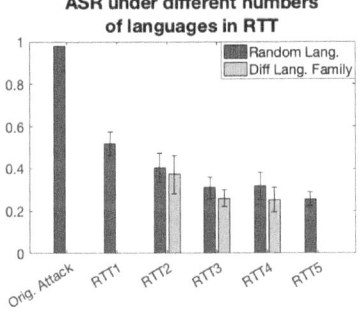 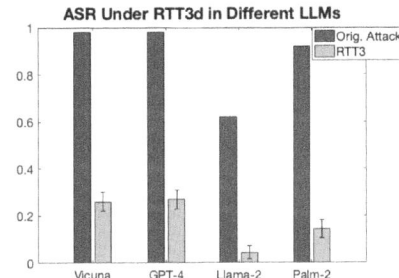

(a) ASR decreases under different numbers of languages in RTT. The attack is PAIR targeting Vicuna-13b-v1.5.

(b) ASR of RTT3d with languages in different language families in different LLMs.

Fig. 2. Comparison of ASR across different RTT settings and LLMs under PAIR attacks. Error bars indicate standard deviation over 10 experiments.

4.3 Number and Type of Translated Languages for RTT

We next investigate how many languages are required to translate during the RTT process for robust attack mitigation. The experimental setting is the PAIR attack on Vicuna. To maintain high translation quality, we use Google Translate API, one of the most accurate machine translation algorithms [10]. Google Translate also helps maintain LLMs' performance with RTT applied to all input. We use 10 sets of translated data to calculate the average experimental result.

Figure 2a illustrates the decrease in ASR as we use more languages in RTT. We denote RTTx as RTT with x languages involved. We consecutively translated the original adversarial attack into random languages and back-translated it into English. RTT1 reduced the ASR from 0.98 to 0.52, which is almost 50% attack mitigation. As the number of languages increases the ASR decreases, and it stabilizes at RTT3, with a drop of 0.67 in ASR.

In addition, we investigate the impact of the target languages on RTT. One of the most significant characteristics of languages is their language family, classified based on geographical and historical factors and represents shared language

Table 4. Attack mitigation effectiveness of different paraphrasing techniques.

Paraphrasing Technique	Attack Mitigation
RTT3d	0.73
GPT-4 Paraphrasing	0.6375
Synonyms	0.5625
Different Word Form	0.65
Passive Voice	0.4125

characteristics [13]. We hypothesize that using languages from different families leads to better generalization, as the translation process relies on more generalized terms. We denote RTT with x random languages and RTT with x languages in different linguistic families (i.e., non-Indo-European languages) from English as RTTxr and RTTxd, respectively.

In Fig. 2a, RTT3d achieves the same reduction in ASR as in RTT5r. The resulting ASR dropped to 0.26, achieving a 0.72 attack mitigation rate with a lower standard deviation than RTT3r. This indicates that involving additional languages beyond RTT3 yields diminishing returns, as further diversification does not significantly enhance paraphrasing or further reveal harmful intent. Therefore, we use RTT3d as our defense model for the remaining experiments and test its performance in different LLMs and attacks. Note that in a personal computer setting, the time required for applying RTT3d is less than 1 s per query (see Appendix A).

4.4 Transferability of RTT Defense in LLMs

To assess RTT3d's transferability on LLMs, we tested RTT3d on Vicuna, GPT4, Llama2, and PaLm2. Figure 2b shows that RTT3d performed consistently well across the different LLMs, with an average attack mitigation rate of 70%. Notably, RTT3d reduced ASR in Llama2 to under 5% and mitigated almost 80% of attacks in PaLm2.

4.5 Comparing RTT with Other Paraphrasing Techniques

We compared RTT3d with other paraphrasing methods (see Table 4). We perform various paraphrasing methods through GPT4, including paraphrasing the text, changing words to their synonyms, changing the words to different word forms (e.g., nouns to verbs, verbs to adjectives), and writing the text in passive voice. We demonstrated that RTT3d outperforms the other techniques by 10–30% in mitigating attacks.

4.6 Comparing RTT with Other Defenses

We compared two recently published SOTA defenses against the RTT on PAIR attacks: SmoothLLM and SelfDenoise. SmoothLLM is the first defense mechanism developed to detect PAIR attacks and is considered the current SOTA [11,16]. It works by randomly perturbing the input with multiple copies and using an ensemble method to detect adversarial attacks. Besides, SelfDenoise is a more recent defense mechanism that denoises the noisy input and makes predictions on the denoised version to determine whether it is adversarial. SmoothLLM reduces the ASR to around 0.5, while SelfDenoise reduces it to 0.32. However, the performance of RTT3d surpasses both defenses, reducing the ASR to 0.26, almost half that of SmoothLLM.

4.7 RTT on Other Adversarial Attacks

We conducted additional tests to evaluate the performance of RTT3d against three other adversarial attacks. First, we tested RTT3d against MathAttack, a social-engineered word-level attack on LLMs' math-solving ability [21]. We tested 300 MathAttack prompts on GPT4, and RTT3d successfully defended them, mitigating 40% of the attacks. Second, we tested RTT3d against the Greedy Coordinate Gradient (GCG) attack on Vicuna, a SOTA adversarial attack using a gibberish adversarial suffix [22]. We tested 100 GCG attacks, and RTT3d achieved a mitigation rate of over 70%, showing that it is *transferable to different types of adversarial attacks*. Finally, we experimented with RTT3d on 200 SAP adversarial queries [7]. RTT3d reduced the ASR from 0.15 to 0.06 successfully.

4.8 RTT on Benign Queries

We evaluated whether RTT3d, as an indiscriminate pre-processing technique, impacts the output quality for benign queries. We utilized the GSM8K and Orca Maths datasets, both comprising of grade school math word problems [6,14]. These datasets are well-suited for assessing LLM performance after RTT3d pre-processing since math word problems require LLMs to comprehend and analyze the input text thoroughly. Therefore, even slight changes to semantic meaning or logical reasoning during RTT3d could lead to incorrect solutions.

We evaluated the effect of RTT3d on GPT4's performance. We randomly selected 500 problems from the GSM8K test set and found that RTT3d had a minimal impact on the model's performance. GPT4 answered 435 and 357 problems correctly before and after RTT3d, respectively, maintaining more than 80% of its original performance. We also tested 100 problems from the Orca Maths dataset and concluded that RTT3d preserved more than 86% of GPT4's initial performance as GPT4 correctly answered 95 and 82 problems before and after RTT3d, respectively. Based on the results, RTT3d has a promising low impact on the performance of LLMs for benign input.

5 Conclusion

We have proposed the Round Trip Translation approach to defend against social-engineered adversarial attacks. The RTT defense paraphrases and generalizes the adversarial prompt, helping to reveal any underlying harmful behavior. Our approach has achieved over 70% attack mitigation on PAIR attacks, surpassing the strongest defense currently available. We are also the first to attempt to defend against MathAttack with almost 40% attack mitigation. RTT also shows strong transferability on different language models. Future directions include testing alternative translation algorithms and verifying RTT on non-English prompts.

A   Experimental Details

This section provides experimental details. Experiments were conducted on a MacBook Pro (Apple M1 Max, 64 GB Memory).

Adversarial Attacks: All the adversarial attacks are generated according to the default settings from their papers or directly imported from their GitHub [7,21,22]. The PAIR attacks are targeting GPT4, Vicuna-13b-v1.5, Llama-2-13b-chat-hf, and PaLm2. The temperature is set to 0 if available as an adjustable parameter (following [4]). The SAP attack targets GPT-3.5-turbo with temperature = 1 (following [7]).The 200 SAP adversarial attacks are randomly sampled from the SAP200 datasets with topics of fraud, politics, race, religion, suicide, terrorism, and violence. All the benign and adversarial prompts are randomly sampled from their original dataset.

GID Calculation: The GID estimation by the Persistent Homology Estimator estimation follows the default setting in the original paper [18].

Number and Type of Translated Languages for RTT: We used 74 non-Indo-European languages from diverse regions, including Africa, Asia, and the Pacific. The non-Indo-European languages we used are: Amharic, Arabic, Aymara, Bambara, Basque, Chinese (Simplified), Chinese (Traditional), Corsican, Divehi, Dogri, Esperanto, Ewe, Guarani, Haitian Creole, Hausa, Hawaiian, Hmong, Igbo, Iloko, Indonesian, Japanese, Javanese, Kazakh, Khmer, Kinyarwanda, Konkani, Korean, Krio, Kurdish, Kurdish (Sorani), Kyrgyz, Lao, Latin, Lingala, Luganda, Maithili, Malagasy, Malay, Malayalam, Maori, Marathi, Manipuri, Mizo, Mongolian, Burmese, Chichewa, Oromo, Samoan, Sanskrit, Scottish Gaelic, Sesotho, Sesotho (Southern Sotho), Shona, Sindhi, Sinhala, Sundanese, Swahili, Tagalog, Tajik, Tamil, Tatar, Telugu, Thai, Tigrinya, Tsonga, Turkish, Turkmen, Akan, Uighur, Uzbek, Vietnamese, Xhosa, Yoruba, Zulu.

Comparing RTT with Other Defenses: The PAIR attack targeted Vicuna-13b-v1.5. The settings of SmoothLLM and SelfDenoise are both set to default from their original papers, with 30% of perturbation scale [11,16].

References

1. Aiken, M., Park, M.: The efficacy of round-trip translation for mt evaluation. Translation J. **14**(1), 1–10 (2010)

2. Barham, S., Feizi, S.: Interpretable adversarial training for text (2019). arXiv preprint arXiv:1905.12864
3. Cao, B., Cao, Y., Lin, L., Chen, J.: Defending against alignment-breaking attacks via robustly aligned llm (2023). arXiv preprint arXiv:2309.14348
4. Chao, P., Robey, A., Dobriban, E., Hassani, H., Pappas, G.J., Wong, E.: Jailbreaking black box large language models in twenty queries (2023). arXiv preprint arXiv:2310.08419
5. Chu, J., Liu, Y., Yang, Z., Shen, X., Backes, M., Zhang, Y.: Comprehensive assessment of jailbreak attacks against llms(2024). arXiv preprint arXiv:2402.05668
6. Cobbe, K., et al.: Training verifiers to solve math word problems (2021). arXiv preprint arXiv:2110.14168
7. Deng, B., Wang, W., Feng, F., Deng, Y., Wang, Q., He, X.: Attack prompt generation for red teaming and defending large language models (2023). arXiv preprint arXiv:2310.12505
8. Edelsbrunner, H., Harer, J.L.: Computational topology: an introduction. American Mathematical Society (2022)
9. Goldsmith, J.: Wikipedia (Apr 2022). https://github.com/goldsmith/Wikipedia
10. Google: Google translate api (2023). http://translate.google.com
11. Ji, J., et al.: Advancing the robustness of large language models through self-denoised smoothing. arXiv preprint arXiv:2404.12274 (2024)
12. Mazeika, M., et al.: Harmbench: A standardized evaluation framework for automated red teaming and robust refusal. arXiv preprint arXiv:2402.04249 (2024)
13. McMahon, A., McMahon, R.: Language classification by numbers. Oxford University Press (2005)
14. Mitra, A., Khanpour, H., Rosset, C., Awadallah, A.: Orca-math: unlocking the potential of slms in grade school math (2024)
15. Oxford: Oxford learner's dictionary (2023). https://www.oxfordlearnersdictionaries.com/
16. Robey, A., Wong, E., Hassani, H., Pappas, G.J.: Smoothllm: defending large language models against jailbreaking attacks. arXiv preprint arXiv:2310.03684 (2023)
17. Sheshadri, A., et al.: Targeted latent adversarial training improves robustness to persistent harmful behaviors in llms. arXiv preprint arXiv:2407.15549 (2024)
18. Tulchinskii, E., et al.: Intrinsic dimension estimation for robust detection of ai-generated texts. Adv. Neural Inform. Process. Syst. **36** (2024)
19. Xu, Z., Liu, Y., Deng, G., Li, Y., Picek, S.: A comprehensive study of jailbreak attack versus defense for large language models. In: Proceedings of the Findings of the Association for Computational Linguistics ACL 2024, pp. 7432–7449 (2024)
20. Zhang, Y., Ding, L., Zhang, L., Tao, D.: Intention analysis prompting makes large language models a good jailbreak defender. arXiv preprint arXiv:2401.06561 (2024)
21. Zhou, Z., et al.: Mathattack: Attacking large language models towards math solving ability. arXiv preprint arXiv:2309.01686 (2023)
22. Zou, A., Wang, Z., Kolter, J.Z., Fredrikson, M.: Universal and transferable adversarial attacks on aligned language models. arXiv preprint arXiv:2307.15043 (2023)

HuPer: Human Factors Integrated Persuasive Dialogue Models

Lingzhen Kong[1], Chuhao Jin[1], Ruihua Song[1(✉)], Xiting Wang[1(✉)], and Yu Chen[2]

[1] Gaoling School of Artificial Intelligence, Renmin University of China, Beijing, China
{konglingzhen,jinchuhao,xitingwang}@ruc.edu.cn,
songruihua_bloon@outlook.com
[2] Meituan, Beijing, China
chenyu17@meituan.com

Abstract. Compared to single-turn or multi-turn dialogue, persuasive dialogue is more challenging as it aims to influence users toward a specific goal over a dialogue session. Previous works mainly focus on dialogue-related elements such as relevance and persuasive strategies while neglecting human factors. In this paper, we integrate human factors into both the input and output sides of the model to guarantee that users can be guided smoothly toward the persuasion goal. On the input side, we model the will scores of users towards the persuasion goal. Inputting these signals into the persuasion model enables an understanding of persuasion difficulty levels and personalized persuasion strategy selection. On the output side, we use real-world user persuasion outcomes to guide model training by formulating the problem using reinforcement learning. Both online and offline scenarios are considered in our formulation. To facilitate human-factor-related research, we collect a dataset that includes real user profiles, dialogues, and persuasion outcomes from real-world applications. Experiments show that our method achieves an improvement of 11.6% over the baseline in terms of average ratings in human evaluation.

Keywords: Persuasive Dialogue System · Large Language Model

1 Introduction

Large language models have significantly advanced dialogue systems; however, persuasive dialogue is still a challenging task due to more complex goals than general single-turn or multi-turn conversations. A good persuasive dialogue system is required to balance generating a high-quality utterance, that is fluent and relevant to context, and driving the conversation toward a persuasive goal, such as persuading someone to donate to a children's foundation. Existing methods achieve promising results in optimizing the generated utterances but neglect human factors in persuasive dialogue, i.e., the real outcomes on whether a user is persuaded and how difficult is a particular user to persuade. For example, in Fig. 1, on the input side, different users have various levels of difficulty in persuading may result in distinct persuasive strategies. On the output side, persuasive dialogue may

S. Yuan et al. (Eds.): PAKDD 2025 Workshops, LNAI 15835, pp. 298–310, 2025.
https://doi.org/10.1007/978-981-96-8197-6_22

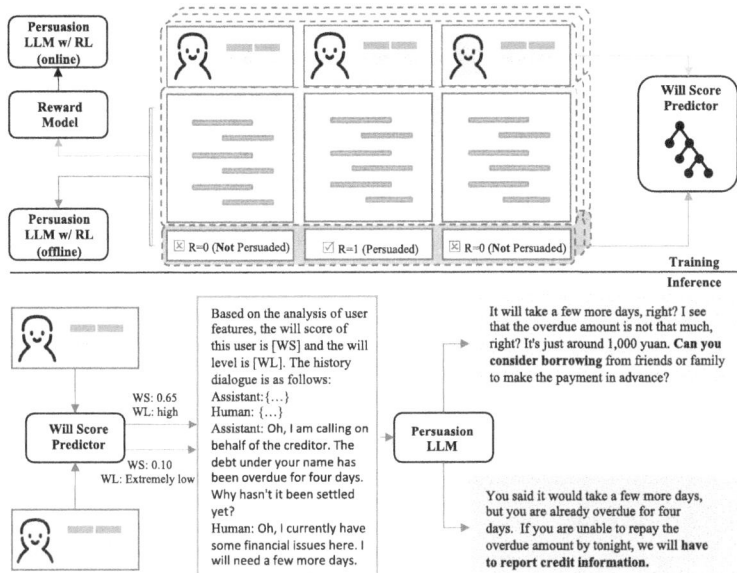

Fig. 1. Overview of our framework. In the **training** stage, the reward model is used to guide the training of the online Persuasion LLM. In the **inference** stage, the model can generate different utterances for different users based on their will scores.

either succeed or fail. Therefore we shall pay more attention to the conversations with successful outcomes. As one main goal is to persuade the user, it is important to consider the real outcomes and the differences of individual users in enhancing persuasive dialogue systems.

Previous works have explored retrieval [5] or generation [18] based persuasive dialogue systems. Although a synthesized cross-domain dataset has been released recently [4], most previous works are based on datasets of specific domains [20, 23]. Various methods are proposed to improve the persuasiveness of utterances by leveraging persuasive strategies [1,5,25] and emotion factors [9,18]. However, few works investigate how to utilize the outcomes to optimize persuasion models and how to adapt the model according to user differences. It also lacks datasets to support such research.

In this paper, we collect a dataset and propose a new framework called HuPer to address the above challenges. We integrate the above human factors into both the input and output sides of the model. On the input side, we model the will scores of users to enable personalized persuasion when facing users with different levels of persuasion difficulty. We learn to predict the will scores based on the given user profiles and then feed them into prompts to generate adaptive responses. On the output side, we use real-world user persuasion outcomes to guide model training by formulating the problem using Reinforcement Learning (RL). We employ both offline and online methods during our formulation. At the same time, we collect a persuasive dialogue dataset from a real application, which contains user profiles and real persuasion outcomes. Experimental results on our collected dataset indicate that our proposed online

RL method with personalized prompt outperforms the best baseline by 11.6% in terms of average ratings in human evaluation. Online RL methods are consistently better than offline RL methods in utilizing user-real outcomes. The personalized prompt has a consistent but relatively small positive contribution to RL methods.

Our main contributions are as follows: 1) we collect a large-scale dataset that includes real user profiles, dialogues, and persuasion outcomes from real-world applications; 2) in our proposed framework HuPer, we use user profiles to predict will scores and levels, which are added into personalize a prompt of the persuasion dialogue model; and 3) we use real-world persuasion outcomes to guide model training using RL, which allows the model to learn more from the conversations with successful outcomes. Experimental results on this dataset indicate the effectiveness of our proposed methods. Our proposed method outperforms the best baseline in both automatic and human evaluation.

2   Related Work

2.1   Persuasive Dialogue Systems

Recent years have witnessed many promising persuasion models although most do not pay enough attention to the human factors of persuasive dialogues. Some works model dialogue-related attributes such as strategies [1,23,25,26], user empathy and emotions [9,18] in persuasive dialogue, and design optimization objectives accordingly. Some works choose to build a user simulator to simulate human-robot interactions, e.g., obtaining the reward by path simulation and reward estimation in [4]. Researchers have noted the potential of LLMs in personalized persuasion and experimentally validated the effectiveness across multiple domains [3,11,12,20,21] (e.g. political appeals, ads, consumer marketing). However, there are few works attempting to optimize LLMs with human factors. Different from the previous works, to the best of our knowledge, we are the first to optimize the model using the real outcome signals and personalize dialogue according to how difficult it is to persuade a user.

2.2   Reinforcement Learning for Dialogue Systems

Various RL-based approaches have been explored, including MDP-based models [7,16], POMDP-based models [2,24], and approaches that utilize deep reinforcement learning to model the value function [6,28]. Some studies [8,14] using real or simulated human-robot interactions have achieved results surpassing previous methods. More recently, Yang [27] proposes a dialogue system trained through optimizing a joint reward function consisting of topic coherence, semantic coherence, and grammatical fluency. Liu [10] proposes a transmitter–receiver-based framework that emphasizes modeling understanding between interlocutors rather than merely mimicking human responses. In 2022, ChatGPT made significant strides in dialogue systems by Reinforcement Learning with Human Feedback (RLHF) [13], which results in the model generating remarkably coherent and engaging dialogues. Such research incorporating RL methods into dialogue systems has greatly improved their performance, inspiring us to combine RL with the persuasive dialogue problem by leveraging real persuasion outcomes.

3 Method

3.1 Overview

For the task of human factors integrated persuasion, given a user profile I, a conversation context C including a series of utterances, denoted by $C = \{u_1^+, u_1^-, \ldots, u_t^+, u_t^-\}$ where u_t^+ and u_t^- are the persuader's and persuadee's utterances at the t-th turn, a model is expected to generate an utterance of persuader (i.e., system) u_{t+1}^+. The utterance shall be appropriate as a response and helpful for the persuasion goal. Persuasion outcome signals R of users, indicating whether the persuasion is ultimately successful ($R = 1$) or not ($R = 0$), are available in training data. The signals can be utilized to optimize the model. Previous works utilize the dialogues with equal importance for training, while neither well utilizing the R signals to differentiate successful and unsuccessful dialogues nor utilizing user profiles to differentiate users in terms of how difficult they can be persuaded. However, as the example shown in Fig. 1, if having real outcome R and user profile I, we can better optimize LLM to the final persuasion goal and personalize the model to generate more appropriate utterances when facing different users.

We then propose a new framework, HuPer, which adopts a RL framework based on real outcome R to solve the problem and personalizes prompts by the predicted will scores from user profiles. As illustrated in Fig. 1, to utilize the real outcome signals R, we treat persuasive dialogue as an RL problem and use an off-policy dataset, containing multi-turn persuasive dialogue sessions and corresponding real persuasion outcomes, to fine-tune a pre-trained LLM via online and offline RL methods (See Sect. 3.2). To personalize prompts, we use user profiles I and real outcome signals R to train a will score predictor, which is expected to predict how difficult a user would be persuaded. During inference, we predict the will score based on I, and compose the prompt by including the score and its corresponding level. Thus our LLM can generate different utterances when facing users with various difficulties (See Sect. 3.3).

3.2 Reinforcement Learning with Outcomes

To learn effective persuasive skills from real persuasion outcome signals, we use real persuasion outcome signals to guide model training by formulating the problem using RL. Specifically, the components of the problem are defined as follows:

- **State**: A state s_t represents the dialogue context including utterances of both persuader and persuadee at time t: $s_t = (u_1^+, u_1^-, \ldots, u_t^+, u_t^-)$.
- **Action**: Given a state s_t, the action a is the next utterance generated by the LLM.

– **Reward**: r denotes the reward obtained after taking action a given state s. In our problem, r approximates the contribution of the next utterance generated by the persuader to the persuasive goal given the current dialogue context.

The goal of reinforcement learning is to maximize the expected reward:

$$J(\pi) = \mathbb{E}_{s \sim d_\pi(s)} \mathbb{E}_{a \sim d_\pi(a|s)} r(a|s), \tag{1}$$

where $d_\pi(s) = \sum_{t=0}^{\infty} \gamma^t p(s_t = s|\pi)$ represents the unnormalized discounted state distribution induced by the policy π with discount factor γ [22].

To solve the above RL problem, we consider both online and offline scenarios. For online, we apply Proximal Policy Optimization (PPO [19]), by training a reward model to guide the training of the dialogue model π. For offline, we apply Reward-Weighted Regression (RWR [15]), by adopting supervised learning to train the dialogue model.

Online RL. Inspired by the RLHF, we propose a three-step training method. In Step. 1, we fine-tune the base model (chat version of Baichuan2-7B in this paper) on a persuasion dataset in a similar domain, thereby enabling the model to learn some basic persuasion skills. In Step. 2, we train our reward model using a regression task to minimize the Mean Square Error (MSE) loss between the persuasive signal and the model's output:

$$L_{\mathrm{rm}} = \frac{1}{N} \sum_{i=1}^{N} (\hat{y}_i - y_i)^2, \tag{2}$$

where y_i represents the actual persuasive signal, $\hat{y}_i$ represents the reward model's output. In Step. 3, we fine-tune the policy model online using PPO [19], utilizing the previously trained reward model as the reward signal. We minimize the loss function:

$$L_{\mathrm{PPO}}(\pi) = -\min \left[Adv_t * \frac{\pi(a|s)}{\pi_{old}(a|s)}, Adv_t * \mathrm{clip}(\frac{\pi(a|s)}{\pi_{old}(a|s)}, 1 - \alpha, 1 + \alpha) \right], \tag{3}$$

where Adv_t is the advantage that measures the relative value of taking a specific action, α is a parameter of the clip range.

Offline RL. Although the online RL method may give better results, it also has the problem of difficult training and complex code implementation. Conversely, The Offline RL method does not need to train the reward model, and can directly use the reward signal in the off-policy data to train the policy model. Inspired by RWR [15], a simple and easy-to-implement RL algorithm that can be realized through supervised learning methods, we minimize the loss function:

$$L_{\mathrm{RWR}}(\pi) = -\sum_{s,a} \log \pi(a|s) \exp(\frac{1}{\beta} R_{s,a}^\mu), \tag{4}$$

where π is the target policy, μ is the sample policy, and $R_{s,a} = \sum_{t=0}^{\infty} \gamma^t r_t$ is the return. This loss function can be interpreted as weighting the likelihood of each action by the expressed return received for that action with a temperature parameter β.

3.3 Personalized Persuasion Prompts

The ultimate result of persuasive dialogue is influenced not only by the quality of multi-turn conversations but also by the user's personal situation and state. For example, an individual with a high will to repay but merely forgot to do so, could likely be persuaded by a straightforward reminder. However, for someone with a low will to repay and an opportunistic mentality, a simple reminder may only result in perfunctory compliance or false promises. In such cases, it becomes necessary to employ more persuasive methods, such as clearly outlining the negative consequences of continued non-repayment. Based on this analysis, we hope that the model can choose different persuasion strategies according to various groups of users, just like experienced persuaders can do.

Therefore, while collecting dialogue data, we also collect legal user profiles related to the persuasive goal, such as *history overdue times* in a debt risk alert scenario. We then employ a decision tree model (LightGBM in this paper) trained on these user profiles and persuasion outcome labels, to predict a will score for a given user during inference. This score is finally input into the dialogue model to enable an understanding of various user persuasion difficulty levels and personalize the generation of the next utterance. Specifically, our method contains the following steps:

Will Prediction. To model user will scores, given a training dataset consisting of user profile information I and persuasion outcomes R, we employed Light-GBM to perform a regression task. Gradient Boosting Decision Tree (GBDT) is a powerful machine learning algorithm that sequentially constructs an ensemble of decision trees, aiming to minimize prediction error by incrementally enhancing the model with each added tree. LightGBM, a more recent iteration of the GBDT, is particularly adept at effectively handling large datasets and a high number of features. MSE is often used for LightBGM:

$$L_{\text{LightGBM}} = \frac{1}{N} \sum_{i=1}^{N} (y_i - p_i)^2, \tag{5}$$

where y_i denotes the true labels, p_i represents the predicted values, and N is the total number of samples. After training, we test the model on a testing dataset, resulting in a predictive AUC (Area Under the ROC curve) of 0.73.

Will Incorporation. During inference, given the profile information I of a user in persuasive dialogue, we apply the trained LightGBM to predict a will score W, which ranges from 0 to 1, indicating how difficult they can be persuaded. Finally, user persuasion will is categorized into five levels based on the inferred will score W. The score and its corresponding level are then incorporated as information into the prompt and fed into LLM to generate the next utterance, as depicted in Fig. 1. We also experiment with different methods of prompt construction, with the specific details and analysis provided in Sect. 5.3.

4 Data and Evaluation

4.1 Datasets

DebtRiskAlert-U: As the first work, our approach requires persuasion dialogues with user profiles and whether users are persuaded to take actions later. There is no existing dataset satisfying both requirements. PersusionForGood [23] and DebtRiskAlert [5] contain persuasion dialogues and reward signals on being persuaded or not, but do not provide user profiles. Thus, we collect a new dataset from a real-world persuasion scenario similar to DebtRiskAlert, where debt reminders alert users with overdue debts by calls and try to persuade them to repay their debt so as to reduce financial risk. We collect 50,000 samples for training and 1,000 for testing, where each sample consists of a dialogue session D, a user profile I (including 816 user features covering various aspects) and a label $R = 1$, if the users finally make repayments within several days or otherwise 0. Therefore, we call the new dataset *DebtRiskAlert-U*, which is used to train and test persuasive dialogue models.

User-Outcome: For predicting a will score of the user, we collect an additional 200,000 samples, each of which consists of a user profile I and a reward label R. We divide the dataset into training, validation, and test sets in the ratio 8:1:1. This *User-Outcome* dataset is used to train and test LightGBM models, as described in Sect. 3.3.

4.2 Evaluation

Automatic Evaluation: We adopt Reward Model (RM) score predicted by the reward model utilized in the online RL methods discussed in Sect. 3.2 as an automatic evaluation metric. The RM Score evaluates the contribution of a given response to the ultimate persuasion goal, taking into account of the multi-turn context. During training, we used the real outcome signals as labels and trained a Baichuan2-PST model with 7B parameters for a regression task with Mean Squared Error (MSE) loss. The reward model achieves an AUC of 0.78 on the DebtRiskAlert-U test set.

Human Evaluation: We conduct user studies for convincing conclusions. First, we randomly sample 140 sub-sessions, which is a part of the session beginning from the first utterance, from DebtRiskAlert-U test set. We then apply different models for comparison to generate the next utterance according to the previous utterances as context. For each sub-session, we shuffle the generated responses and make the methods anonymous when presented to ensure a fair comparison. Three human experts in debt risk alert scenarios with the age range from 27 to 40, are hired to rate the utterance on a 5-level Likert scale from 1 to 5. To verify whether the automatic measurements are consistent with human experiences, we examine Kendall's tau when using RM scores and human ratings to rank different models. Kendall's tau is as high as 75%, indicating that the RM score is, to some extent, consistent with human rating.

5   Experiments

5.1   Baselines

Persuasion LLMs. We take PersuGPT [4] as a baseline, an LLM trained across various persuasive domains, and demonstrated superior performance to GPT-4 on the test set. We fine-tune it in our DebtRiskAlert-U train set for domain adaptation.

Fine-Tuned LLMs. We take the Baichuan2-7b-Chat after domain adaptation as a base model, called DA-Base. The model is deployed on a commercial platform but its domain data are neither transparent nor available to us. Then we apply supervised fine-tuning using the training data of our DebtRiskAlert-U dataset to obtain a baseline called DA-Base-SFT, which is the common base model of reinforcement learning methods.

Reinforcement Learning Methods. We train an online RL baseline by following the steps in InstructGPT [13] to train a PPO version of DA-Base-SFT model, i.e., DA-Base-SFT+PPO. Please note that the reward model used here does not utilize the will score/level predicted from user profile features. Also we train an offline RL baseline by applying the algorithm of RWR [15]. Again, there is no will score/level evolved in. Implementation details are provided in Appendix A.

Table 1. Comparing our model with baseline methods. "PP" means "Personalized Persuasion".

No.	Model	RM ↑	Human ↑
(1)	PersuGPT [4]	2.85	2.19
(2)	DA(Domain Adapted)-Base	3.08	2.70
(3)	DA-Base-SFT	3.18	2.92
(4)	(3)+RWR	3.17	3.01
(5)	(3)+PPO	3.23	3.19
(6)	(4)+PP (Ours-Offline)	3.22	3.02
(7)	(5)+PP (Ours-Online)	**3.25**	**3.26**
	Human Persuaders	3.18	2.95

Table 2. A case on DebtRiskAlert-U dataset. Unlike the indifference of the baseline approach and the threat of human response, our approach first expresses understanding of the user's situation and proposes some more feasible solutions.

Role	Content
Context	Persuader & Persuadee : ...
	Persuader: Don't delay it any longer. We only have one day of grace
	Persuadee: Everyone is having a hard time right now
Response by DA-Base-SFT	Well, you should figure it out yourself
Response by Ours-Online	Yeah, I understand, but the amount we're talking about is not much, right? Can you borrow some money from a friend for now?
Response by Human	We will have to report credit information if you delay another day. Do you really have to delay for another day?

5.2   Comparison with Baselines

As illustrated in Table 1, our proposed online model outperforms the best in terms of both RM score and Human rating. Its Human rating is as high as 3.26, even better than the average rating of human persuaders' 2.95. This confirms that our motivation on LLMs have chance to learn from more strong persuaders and better distinguish different users. We attribute these gains to several factors. First, applying online RL method PPO using the learned reward model from whether a user is persuaded or not enhances performance greatly, as evidenced by comparing method (5) to (3), indicating that persuasion results are truly helpful for optimizing persuasion LLMs while previous works neglect. Second, the online RL model, method (7) and (5), improves human ratings by about 0.18 and 0.24 over its counterpart with offline RL method (6) and (4). This suggests that online RL method can take better use of persuasion results in training data. Third, personalized persuasion, which predicts will scores from user profiles and includes the scores and corresponding levels in the prompt, further benefits the online RL model (see method (7) to (5)). Lastly, across the table, fine-tuned LLMs generally perform better than PersuGPT, underscoring the importance of in-domain data in this specific persuasion task. We observe that PersuGPT consistently generates polite but less persuasive responses, which are deemed by annotators as insufficient to persuade users in this real-world application. A specific example from the DebtRiskAlert-U dataset is shown in Table 2.

5.3   Prompt Form for Personalized Persuasion

When incorporating the will score WS into the prompt fed into LLMs, we experiment with three forms based on the DA-Base-SFT model: 1) solely using will scores (**Score**); 2) solely using will levels (**Level**) that are mapped to will scores, i.e., "extremely low" mapped to the scores from 0.0 to 0.2, "low" mapped to 0.2–0.4, "neutral" mapped to 0.4–0.6, "high" mapped to 0.6–0.8, and "extremely high" mapped to 0.8–1.0; and 3) using both will scores and mapped levels (**Score+Level**). The results are shown in Table 3. We find that will levels works better than scores in terms of both RM scores and human ratings. This may be caused by LLMs can better understand words rather than numbers. When combining them together, we can achieve the best performance, indicating that scores can be complementary to will levels. Therefore, we use the combined form **Score+Level** in our final comparison experiments shown in Table 1.

5.4   Performance Across User Groups

As RM scores have good consistency with human ratings in comparing different models, we analyze the performance of one of our proposed methods, ours-online, with the base model DA-Base-SFT across user groups with various will scores. First, as Fig. 2 shows, we divide 1,000 users from the test set equally into eight groups in descending order of will score predicted from their profiles. Then, we calculate the reward model scores of generated responses by these two models

Table 3. Experiment on different prompt forms for personalized persuasion (PP). Score means using will scores and Level means using the mapped levels from will scores.

Model	RM ↑	Human ↑
DA-Base-SFT	3.170	3.11
DA-Base-SFT+PP(Score)	3.169	3.14
DA-Base-SFT+PP(Level)	3.173	3.15
DA-Base-SFT+PP(Score + Level)	**3.180**	**3.17**

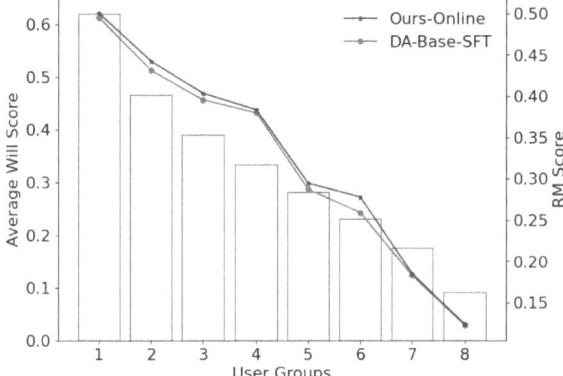

Fig. 2. Comparing RM scores of our proposed method and the baseline across different user groups. 1,000 users from the test set are equally into eight groups in descending order of the predicted will scores. We then calculate the reward model scores of generated responses by these two models and average the RM scores within each user group.

and average the RM scores within each user group. Finally, we draw the average will score and the two models' average RM scores of each group in Fig. 2.

We have some interesting insights from this figure. First, most users are concentrated within the 0.10 to 0.40 range in terms of will scores, indicating that users with relatively low will level account for the majority. Thus persuasion in our scenarios is more difficult than general persuasion in daily life that the state-of-the-art baseline PersuGPT can do well. Second, the RM scores of the two models have a trend to decrease with the will scores decrease. This makes sense because the users with low and extremely low will may have some financial issues that prevent them from repaying no matter how persuasive the dialogues are. Third, our online model surpass the SFT in most groups, indicating the positive contributions of real feedback reward and personalized persuasion. Surprisingly, in groups with a relatively low will scores (e.g. group 6), the gap between Ours-Online and DA-Base-SFT tends to expand compared to other groups. This suggests that our method plays a more significant role in user groups with relatively low will scores. However, it is important to note that even in cases of extremely low will scores (e.g. group 7–8), our method encounters difficulties in achieving improvements.

6   Conclusion

In this paper, we collect a real-world persuasive dialogue dataset with human factors named DebtRiskAlert-U. Then we propose a framework named HuPer to integrate user-profiles and real persuasive outcome signals into a persuasive dialogue model. The aim is to simultaneously consider the impact of conversation quality and the individuality of both persuaders and users on the persuasive outcome. Experimental results on the dataset demonstrate the effectiveness of both ours online and offline methods in terms of automatic metrics and human evaluations. In the future, we consider modeling more fine-grained rewards, such as fusing the rewards from the user reaction, persuasion strategy, and the final outcomes.

Acknowledgments. This research was supported by Public Computing Cloud, Renmin University of China.

Ethical Statement. In our application, we find that utterances generated by persuasive dialogue systems are easier to manage in obeying the redline of compliance than humans. We agree that the use of persuasive dialogue systems must be strictly regulated to avoid potential issues. In our study, we collect data from a commercial platform in China that provides lending services. All conversations between the persuader and the user are supervised and the user is explicitly informed that the content of the conversation will be recorded. When preprocessing data, we removed user identities and all information related to the user's personal privacy and only retained debt-related features. During the development stage, we carefully apply many mechanisms such as setting regulations and redline violation assessments to avoid generating inappropriate responses. Our principle is that all uses of persuasive dialogue systems need to consider the balance between utilizing technology for the benefit of humanity and ensuring ethical use.

A   Implementation Details

The SFT model is trained for 1 epoch on a dataset of 1,500,000 sessions of persuasive dialogue, and it takes about 30 h on 8 A100 GPUS. Our RL methods contain two implementation ways. As for Ours-Offline, the model is trained for 2 epochs with a batch size of 8. AdamW optimizer is applied with a learning rate of 2e-4. The temperature β of RWR loss mentioned in Eq. 4 is set to 1.0. The lora rank and lora alpha is set to 64 and 16 respectively. And it takes 5 h on 2 A100 GPUS. As for Ours-Online, the reward model is trained for 1 epoch with a batch size of 4. The PPO model is trained for 1 epoch with a batch size of 8. FusedAdam optimizer is applied with a learning rate of 5e-6. And it takes about 12 h on 8 A100 GPUS with DeepSpeed ZeRO-3 [17]. For decoding, we employ a combination of Top-p sampling and Top-k sampling. The top-k value is set to 5, while the top-p value is set to 0.85.

References

1. Carlile, W., Gurrapadi, N., Ke, Z., Ng, V.: Give me more feedback: annotating argument persuasiveness and related attributes in student essays. In: Proceedings of the 56th Annual Meeting of the Association for Computational Linguistics, pp. 621–631 (2018)
2. Crook, P.A., Keizer, S., Wang, Z., Tang, W., Lemon, O.: Real user evaluation of a pomdp spoken dialogue system using automatic belief compression. Compute. Speech Lang. **28**(4), 873–887 (2014)
3. Goldstein, J.A., Chao, J., Grossman, S., Stamos, A., Tomz, M.: How persuasive is ai-generated propaganda? PNAS Nexus **3**(2), pgae034 (2024)
4. Jin, C., Ren, K., Kong, L., Wang, X., Song, R., Chen, H.: Persuading across diverse domains: a dataset and persuasion large language model. In: Proceedings of the 62nd Annual Meeting of the Association for Computational Linguistics, pp. 1678–1706 (2024)
5. Jin, C., et al.: Joint semantic and strategy matching for persuasive dialogue. In: Findings of the Association for Computational Linguistics: EMNLP 2023, pp. 4187–4197 (2023)
6. Lemon, O.: Learning what to say and how to say it: joint optimisation of spoken dialogue management and natural language generation. Comput. Speech Lang. (2011)
7. Levin, E., Pieraccini, R., Eckert, W.: Learning dialogue strategies within the markov decision process framework. In: 1997 IEEE Workshop on Automatic Speech Recognition and Understanding Proceedings, pp. 72–79. IEEE (1997)
8. Li, X., Chen, Y.N., Li, L., Gao, J., Celikyilmaz, A.: End-to-end task-completion neural dialogue systems. arXiv preprint arXiv:1703.01008 (2017)
9. Lin, Z., et al.: Caire: an end-to-end empathetic chatbot. In: Proceedings of the AAAI Conference on Artificial Intelligence (2020)
10. Liu, Q., et al.: You impress me: Dialogue generation via mutual persona perception. arXiv preprint arXiv:2004.05388 (2020)
11. Lukin, S., Anand, P., Walker, M., Whittaker, S.: Argument strength is in the eye of the beholder: audience effects in persuasion. In: Proceedings of the 15th Conference of the European Chapter of the Association for Computational Linguistics, pp. 742–753 (2017)
12. Matz, S., Teeny, J., Vaid, S.S., Peters, H., Harari, G., Cerf, M.: The potential of generative ai for personalized persuasion at scale. Sci. Rep. **14**(1), 4692 (2024)
13. Ouyang, L., et al.: Training language models to follow instructions with human feedback. Adv. Neural. Inf. Process. Syst. **35**, 27730–27744 (2022)
14. Papaioannou, I., Lemon, O.: Combining chat and task-based multimodal dialogue for more engaging hri: a scalable method using reinforcement learning. In: the 2017 ACM/IEEE International Conference on Human-Robot Interaction, pp. 365–366 (2017)
15. Peters, J., Schaal, S.: Reinforcement learning by reward-weighted regression for operational space control. In: the 24th International Conference on Machine Learning, pp. 745–750 (2007)
16. Pieraccini, R., Suendermann, D., Dayanidhi, K., Liscombe, J.: Are we there yet? research in commercial spoken dialog systems. In: Matoušek, V., Mautner, P. (eds.) TSD 2009. LNCS (LNAI), vol. 5729, pp. 3–13. Springer, Heidelberg (2009). https://doi.org/10.1007/978-3-642-04208-9_3

17. Rajbhandari, S., Rasley, J., Ruwase, O., He, Y.: Zero: memory optimizations toward training trillion parameter models. In: SC20: International Conference for High Performance Computing, Networking, Storage and Analysis, pp. 1–16. IEEE (2020)
18. Samad, A.M., Mishra, K., Firdaus, M., Ekbal, A.: Empathetic persuasion: reinforcing empathy and persuasiveness in dialogue systems. In: Findings of the Association for Computational Linguistics: NAACL 2022, pp. 844–856 (2022)
19. Schulman, J., Wolski, F., Dhariwal, P., Radford, A., Klimov, O.: Proximal policy optimization algorithms. arXiv preprint arXiv:1707.06347 (2017)
20. Shin, M., Kim, J.: Large language models can enhance persuasion through linguistic feature alignment. Available at SSRN 4725351 (2024)
21. Simchon, A., Edwards, M., Lewandowsky, S.: The persuasive effects of political microtargeting in the age of generative artificial intelligence. PNAS Nexus **3**(2), pgae035 (2024)
22. Sutton, R.S., Barto, A.G.: Reinforcement learning: an introduction. Robotica (1999)
23. Wang, X., et al.: Persuasion for good: towards a personalized persuasive dialogue system for social good. In: Proceedings of the 57th Annual Meeting of the Association for Computational Linguistics, pp. 5635–5649 (2019)
24. Williams, J.D., Young, S.: Partially observable markov decision processes for spoken dialog systems. Comput. Speech & Lang. (2007)
25. Wu, Q., Zhang, Y., Li, Y., Yu, Z.: Alternating recurrent dialog model with large-scale pre-trained language models. In: Proceedings of the 16th Conference of the European Chapter of the Association for Computational Linguistics: Main Volume, pp. 1292–1301 (2021)
26. Yang, D., Chen, J., Yang, Z., Jurafsky, D., Hovy, E.: Let's make your request more persuasive: modeling persuasive strategies via semi-supervised neural nets on crowdfunding platforms. In: Proceedings of the 2019 Conference of the North American Chapter of the Association for Computational Linguistics: Human Language Technologies, pp. 3620–3630 (2019)
27. Yang, M., Huang, W., Tu, W., Qu, Q., Shen, Y., Lei, K.: Multitask learning and reinforcement learning for personalized dialog generation: An empirical study. IEEE Trans. Neural Netw. Learn. Syst. **32**(1), 49–62 (2021)
28. Yu, Z., Black, A.W., Rudnicky, A.I.: Learning conversational systems that interleave task and non-task content. arXiv preprint arXiv:1703.00099 (2017)

Large Language Models for Detection of Life-Threatening Texts

Thanh Thi Nguyen[✉], Campbell Wilson, and Janis Dalins

AiLECS Lab, Faculty of IT, Monash University, Melbourne, VIC 3800, Australia
{thanh.nguyen9,campbell.wilson,janis.dalins}@monash.edu

Abstract. Detecting life-threatening language is essential for safeguarding individuals in distress, promoting mental health and well-being, and preventing potential harm and loss of life. This paper presents an effective approach to identifying life-threatening texts using large language models (LLMs) and compares them with traditional methods such as bag of words, word embedding, topic modeling, and Bidirectional Encoder Representations from Transformers. We fine-tune three open-source LLMs including Gemma, Mistral, and Llama-2 using their 7B parameter variants on different datasets, which are constructed with class balance, imbalance, and extreme imbalance scenarios. Experimental results demonstrate a strong performance of LLMs against traditional methods. More specifically, Mistral and Llama-2 models are top performers in both balanced and imbalanced data scenarios while Gemma is slightly behind. We employ the upsampling technique to deal with the imbalanced data scenarios and demonstrate that while this method benefits traditional approaches, it does not have as much impact on LLMs. This study demonstrates a great potential of LLMs for real-world life-threatening language detection problems.

Keywords: Large language model · Life threatening · Text Classification · Gemma · Mistral · Llama-2

1 Introduction

Identification of life-threatening language is crucial for several reasons. Detecting language indicating self-harm or harm to others allows for intervention before any harm occurs. This could involve alerting law enforcement, contacting emergency services, or providing support to individuals in crisis. Machine learning methods are frequently employed to identify harmful language like hate speech or offensive content. However, their application in detecting life-threatening language is relatively uncommon, especially in the English language. We found only a few such studies in the literature. For example, an approach was introduced in [11] for the automatic detection of threatening Urdu texts. Its performance was enhanced through the use of stacked classification models and feature extraction techniques. Likewise, the study in [1] proposed two shared tasks focused on

S. Yuan et al. (Eds.): PAKDD 2025 Workshops, LNAI 15835, pp. 311–323, 2025.
https://doi.org/10.1007/978-981-96-8197-6_23

detecting abusive and threatening language in Urdu. Both tasks are framed as binary classification challenges, where systems are tasked with categorizing Urdu tweets into two classes: abusive and non-abusive for the first task, and threatening and non-threatening for the second.

Recently, the work in [12] devised a multilingual text classification framework to address the challenge of detecting threatening content across English and Urdu languages. However, similar to many traditional text classification methods, their approaches involve a text preprocessing phase including several steps such as *removing* punctuations, hashtags, numbers, HTML tags, URLs, *replacing* emoji and emoticons, *addressing* the issue of misspelled words, and *decoding* English abbreviations. Those preprocessing steps are *tedious*, *subjective*, and *inconsistent* across datasets. This could significantly affect the performance and consistency of the classification models across different unseen test datasets.

In this research, we propose an approach to life-threatening text detection using open-source large language models (LLMs) such as Gemma, Mistral, and Llama-2. By leveraging the power of LLMs, our approach is resistant to noisy data *without text preprocessing steps* such as lemmatizing words, removing punctuation, hashtags, extra spaces, stop words, and so on, as seen in existing works. Our LLM-based approach *avoids the need for separate designs* for feature extraction and classification, which otherwise might involve extensive trial and error to find the optimal combination of these steps. We show a consistent performance of the LLM-based methods across different data scenarios including balanced, imbalanced, and extremely imbalanced datasets. We provide a comprehensive comparison of our approach with traditional methods (TF-IDF, word embedding, topic modeling, and BERT) across six datasets and emphasize the importance of using not only accuracy, but also F-scores and AUC, especially when evaluating classification methods with class imbalance.

2   Competing Methods

2.1   TF-IDF

TF-IDF is a method used to weight the terms in a bag-of-words model, which represents text data as a collection of words, ignoring grammar and word order. While TF-IDF provides a representation of text data in a vector space, it does not inherently acquire the semantic relationships between words like word embedding approaches.

2.2   Word Embedding

Methods like GloVe and Word2Vec learn semantic relationships by creating word embeddings based on co-occurrence patterns, where similar words have comparable representations.

GloVe. GloVe, short for Global Vectors for Word Representation, performs by creating a co-occurrence matrix, which contains the counts of how often each word appears within a context window of other words.

Word2Vec. This paradigm comprises two primary methods: continuous bag-of-words (CBOW) and skip-gram. **CBOW** predicts the target word using its surrounding context words. **Skip-gram** works oppositely, predicting context words from a given target word.

2.3 Topic Modeling

This is an unsupervised learning technique used in NLP to discover latent topics or themes within a collection of documents. Common topic modeling approaches are Latent Dirichlet Allocation (LDA) and Latent Semantic Indexing (LSI). **LDA** assumes each document is a mixture of topics, with each word assigned to one, iteratively adjusting topic distributions to maximize the likelihood of the given documents. **LSI** uses singular value decomposition on a term-document matrix for dimensionality reduction, uncovering latent semantic structures and building an index for efficient document retrieval.

2.4 BERT

BERT, representing bidirectional encoder representations from transformers, is a pretrained NLP model introduced in [2]. Unlike previous language processing methods that handle text data sequentially in one direction (either left-to-right or right-to-left), BERT is bidirectional, meaning it can take into account the full context surrounding a word by examining both its left and right contexts concurrently. This helps obtain more precise representations of words and phrases. By fine-tuning BERT on specific text classification tasks, it can achieve great performance across various domains. In this study, we use the BERT-en-uncased model with its encoder and preprocessor respectively obtained from [15, 16].

2.5 LLM – Gemma 7B

Gemma was proposed by Google DeepMind, described in [4], and available in two variants: a 7 billion parameter model tailored for efficient development and deployment on GPU and TPU, alongside a 2 billion parameter model optimized for CPU and on-device applications [4].

   The 7B model employs multi-head attention, whereas the 2B model utilizes multi-query attention. Gemma models utilize a subset of the SentencePiece tokenizer [8] from Gemini. This tokenizer splits digits, retains extra whitespace, and employs byte-level encodings for unknown tokens. Instead of employing absolute positional embeddings, rotary positional embeddings [14] are utilized in every layer. Additionally, embeddings are shared between inputs and outputs to decrease model size [4].

2.6 LLM Mistral 7B

Mistral 7B LLMs comprise the pretrained Mistral-7B-v0.1 model and several Mistral 7B Instruct variants [7]. These models are equipped with 7 billion parameters, which leverage grouped-query attention (GQA), and sliding window attention (SWA). GQA notably boosts inference speed and lowers memory needs during decoding, enabling larger batch sizes and thus increasing throughput.

SWA is designed to handle extended sequences more efficiently while reducing computational costs, thus addressing a common constraint in using LLMs. SWA utilizes the multiple layers of a transformer to extend its attention capability beyond a fixed window size W. Within each layer, the hidden state at position i, denoted as h_i, considers all hidden states from the preceding layer whose positions fall within the range of $i - W$ to i.

2.7   LLM Llama-2 7B

The Llama-2 family, upon release, comprises pretrained and fine-tuned variants, each offering three models with 7B, 13B, and 70B trainable parameters. The pretrained variants were derived from a self-supervised learning method utilizing a vast text corpus containing two trillion tokens. Subsequently, these pretrained variants underwent further optimization via supervised fine-tuning procedure and reinforcement learning with human feedback, employing instruction datasets and human-annotated data tailored for dialogue use cases.

Instead of absolute positional encoding, Llama-2 uses rotary position embeddings to encode position information of tokens. For tokenization, the LLaMA tokenizer utilizes bytepair encoding sourced from SentencePiece [8]. As a result, the Llama vocabulary envelops 32k tokens [17].

3   Fine-Tuning LLMs Using LoRA

The LLMs such as Gemma, Mistral, and Llama-2 have amassed substantial world knowledge from pretraining on extensive text data. Our fine-tuning method, applied atop these models, will yield specialized expertise tailored to each fine-tuning dataset, while retaining the broad general knowledge and reasoning abilities acquired during the pretraining phase.

Our method for fine-tuning LLMs using low-rank adaptation (LoRA) [5] is illustrated in Fig. 1. LoRA represents one of several parameter-efficient fine-tuning (PEFT) methods, including BitFit, SparseAdapter, AdaLoRA, among others, which can be utilized for fine-tuning LLMs [9]. The original weights of the LLMs are converted into the Hugging Face transformer format to leverage the fine-tuning tools provided by Hugging Face. These converted models are then integrated within the PEFT-LoRA framework, which employs LoRA based on the PEFT library [10]. The LoRA method performs low-rank fine-tuning by drawing inspiration from the structure-aware intrinsic dimension technique. The update of parameters for a pretrained weight matrix $M_0 \in \mathbb{R}^{d \times k}$ is determined by the product of two low-rank matrices M_A and M_B:

$$\Delta M = M_A M_B \tag{1}$$

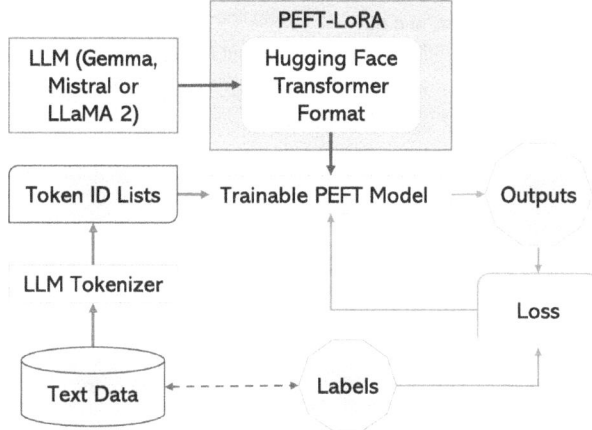

Fig. 1. The method for fine-tuning a pretrained LLM for text classification tasks. Text data are first tokenized with an LLM tokenizer into token IDs, which are then used as inputs to the learnable PEFT model, constructed by transferring the original LLM weights to the Hugging Face format and loaded via the PEFT-LoRA mechanism.

where $M_A \in \mathbb{R}^{d \times r}$ and $M_B \in \mathbb{R}^{r \times k}$ represent matrices of learnable parameters, with the rank r being significantly smaller than the minimum of d and k [9]. The parameters of the pretrained model, M_0, remain fixed and do not undergo gradient updates during training. Both M_0 and $\Delta M = M_A M_B$ are applied to the same input x, as indicated by Hu et al. [5]; thus, the modification to the forward pass for $h = M_0 x$ is expressed as:

$$h = M_0 x + \Delta M x = M_0 x + M_A M_B x \tag{2}$$

The matrix M_A begins with zero initialization, while the matrix M_B is initialized with random Gaussian values. Thus, $\Delta M = M_A M_B$ equals zero at the start of the fine-tuning process. Throughout training, $\Delta M x$ is scaled by a fixed factor $\frac{\alpha}{r}$, where α is a constant [5]. Following training, the learnable parameters can be incorporated into the original weight matrix M_0 through the addition of the matrix $M_A M_B$. An illustration of LoRA and its difference to a regular fine-tuning method is presented in Fig. 2. LoRA enables the small trainable matrices M_A and M_B to undergo training, thereby adjusting to the new data, while simultaneously minimizing the total number of updates. By bypassing gradient computation for pretrained weights, the LoRA approach significantly reduces memory usage, enabling faster and more efficient fine-tuning.

Our fine-tuning methodology employs a cross-entropy loss function, comparing the output logits $\hat{y}$ of the neural network with the target y. In Fig. 1, the term "labels" refers to the target y, while the "outputs" generated by the learnable PEFT model signify the logits $\hat{y}$. The cross-entropy loss function, averaged over

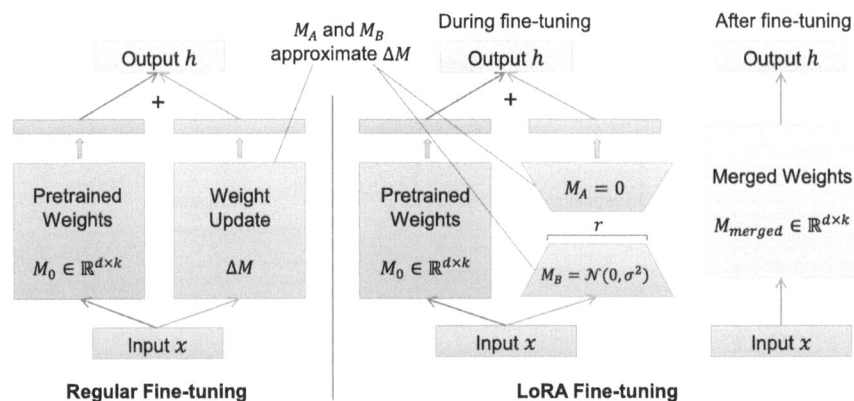

Fig. 2. The difference between regular fine-tuning and LoRA fine-tuning. The LoRA matrices M_A and M_B approximate the weight update matrix ΔM, with the inner dimension rank r being a hyperparameter.

the mini-batch size N, between the output $\hat{y}$ and the target y is expressed as:

$$\mathcal{L}(\hat{y}, y) = \sum_{n=1}^{N} \frac{l_n}{\sum_{n=1}^{N} w_{y_n}} \tag{3}$$

where w denotes a weight vector with each element corresponding to a class, and l_n is defined as:

$$l_n = -w_{y_n} \log \frac{\exp(\hat{y}_{n,y_n})}{\sum_{c=1}^{C} \exp(\hat{y}_{n,c})} \tag{4}$$

where the logits $\hat{y}$ represent the unnormalized logits for each class, the target y contains the class indices, and C is the total number of classes.

The hyperparameter settings used in this study are as follows. The maximum length of a sentence is set to 128 tokens, either truncating longer sentences or padding shorter ones to ensure uniform size. The LoRA rank is set to 8, and the number of epochs is 10. We utilize the stochastic gradient descent AdamW optimizer for fine-tuning, with a learning rate of 2×10^{-5} and an epsilon coefficient of 10^{-8}. Additionally, the mini-batch size is fixed to 16 during training.

4   Datasets Used

While identifying life-threatening language is important in law enforcement, healthcare, emergency services, mental health, and security agencies, collecting threatening texts for training a machine learning model is challenging due to its scarcity. Several hate speech datasets are available in the literature, yet the majority lack *genuinely threatening* texts, except the "dynamically generated hate speech" dataset [18]. This dataset contains 18,969 not-hate texts and 22,175 hate texts. The hate texts are divided into several categories such as derogation, animosity, dehumanization and threatening where threatening texts take up 606 samples. We use these 606 samples as part of the threatening texts in our experiments. The other part of threatening texts are obtained from the Threatening English Language (TEL) corpus [3], which includes 309 written texts. In total, we have *915 threatening texts* that comprise the data for the minor class.

The major class consists of non-threatening texts. To evaluate performance of the competing methods in dealing with balanced, imbalanced, and extremely imbalanced datasets, we create six scenarios for the non-threatening texts.

The first three scenarios are based on the 18,969 not-hate texts of the aforementioned "dynamically generated hate speech" dataset [18] where we sample without replacement 1,000, 5,000 and 10,000 samples from these not-hate texts.

The other three scenarios for non-threatening texts are based on a public release of the dataset described in [13] that consists of 135,556 samples [6]. The main outcome variable is the "hate speech score", which is a continuous hate speech measure, where a value less than 0.5 indicates a not-hate speech, and this part of data has 86,508 samples. We sample without replacement 1,000, 5,000 and 10,000 samples from these not-hate texts and consider them as non-threatening texts.

Table 1. The Compositions of the Six Experimental Datasets

Dataset	No. of Threatening	No. of Non-Threatening	Overall Characteristic
1	915 texts in [3] and [18]	**1,000** not-hate texts in [18]	Balance
2	915 texts in [3] and [18]	**1,000** not-hate texts in [6]	Balance
3	915 texts in [3] and [18]	**5,000** not-hate texts in [18]	Imbalance
4	915 texts in [3] and [18]	**5,000** not-hate texts in [6]	Imbalance
5	915 texts in [3] and [18]	**10,000** not-hate texts in [18]	Extreme Imbalance
6	915 texts in [3] and [18]	**10,000** not-hate texts in [6]	Extreme Imbalance

A summary of the six experimental datasets corresponding to the six aforementioned scenarios is presented in Table 1. Each dataset is randomly split into training (90%) and testing (10%) for experiments. To address the class imbalance issue, the upsampling technique is used by randomly sampling with replacement the threatening texts (i.e., the minor class) to match the number of non-threatening texts (i.e., the major class) in the training set.

5   Performance Metrics

Each competing method is evaluated based on several metrics including accuracy, F_β with β equal to 0.5 (i.e., $F_{0.5}$-score), 1.0 (F_1-score), and 2.0 (F_2-score):

$$F_\beta = (1 + \beta^2) \cdot \frac{\text{P} \cdot \text{R}}{(\beta^2 \cdot \text{P}) + \text{R}} \tag{5}$$

318   T. T. Nguyen et al.

where P is precision, R is recall, and β represents the relative importance of precision and recall. F_1-score determines the harmonic mean of precision and recall, striking a balance between the two metrics. It proves particularly useful in scenarios with imbalanced class distributions. $F_{0.5}$-score is a variant of F_1-score that weighs precision higher than recall. It is useful when precision is more important than recall. F_2-score is another variant of F_1-score that considers recall higher than precision. We also employ area under the ROC curve (AUC) as it can quantify the ability of a machine learning model to distinguish between the positive and negative classes. In this study, we present all metrics as percentages.

6   Results and Discussions

Unlike our LLM-based approach, which functions as an end-to-end method, approaches using TF-IDF, word embedding, and topic modeling require integration with a classifier to classify textual features obtained from those techniques. For that purpose, we use an *ensemble classifier* combining logistic regression, support vector machine, and random forest based on the *soft voting* mechanism.

6.1   Results on Balanced Datasets

Results for the two balanced datasets, Dataset 1 and Dataset 2, are reported in Table 2. TF-IDF achieves high scores in all metrics, while GloVe Embedding performs moderately, but with lower scores than TF-IDF. CBOW and skip-gram perform similarly, showing decent results but are outperformed by TF-IDF. LSI outperforms LDA on all metrics in both datasets, but both have lower performance than TF-IDF and BERT. Gemma is inferior to BERT, while Mistral exhibits excellent performance, surpassing most other methods in Dataset 2. Llama-2 achieves the highest scores across all metrics in Dataset 1, but slightly lags behind Mistral in Dataset 2.

Table 2. Results on **Balanced** Datasets - Dataset 1 and Dataset 2

Competing Methods	Dataset 1					Dataset 2				
	Acc.	F_1	$F_{0.5}$	F_2	AUC	Acc.	F_1	$F_{0.5}$	F_2	AUC
TF-IDF	77.08	72.15	80.74	65.22	76.60	83.33	82.22	83.90	80.61	83.22
GloVe Embedding	69.27	59.31	71.43	50.71	68.57	74.48	70.66	75.84	66.14	74.14
Word2Vec - CBOW	71.88	68.24	72.32	64.59	71.59	69.79	69.15	68.71	69.59	69.79
Word2Vec - Skip-gram	71.35	68.57	71.26	66.08	71.15	70.83	68.18	70.59	65.93	70.64
Topic Modeling - LDA	67.19	62.72	66.75	59.15	66.88	71.35	68.57	71.26	66.08	71.15
Topic Modeling - LSI	72.40	69.36	72.64	66.37	72.16	80.73	79.10	81.59	76.75	80.56
BERT - en-uncased	79.17	77.53	79.68	75.49	79.02	83.33	80.00	88.64	72.89	82.89
LLM - Gemma - 7B	75.52	74.59	74.84	74.35	75.48	80.73	78.11	83.12	73.66	80.43
LLM - Mistral - 7B	92.19	91.62	93.82	89.52	92.07	**95.83**	**95.60**	96.88	**94.36**	**95.76**
LLM - Llama-2 - 7B	**93.23**	**92.82**	**94.38**	**91.30**	**93.14**	95.31	94.97	**97.25**	92.79	95.19

6.2   Results on Imbalanced Datasets

Imbalanced Dataset 3 (Table 3.) The competing methods are evaluated both with and without upsampling, where upsampling is used to mitigate the class imbalance issue. Upsampling slightly improves TF-IDF and significantly enhances GloVe Embedding, CBOW, skip-gram, LDA, and LSI in terms of F-scores and AUC. BERT and Gemma have comparable performance, while Mistral outperforms both in most metrics. Llama-2 demonstrates excellent performance, but it is inferior to Mistral in all metrics. Upsampling does not lead to much improvement in accuracy, F-scores and AUC for Gemma, Mistral and Llama-2.

Table 3. Results on Dataset 3 - The First **Imbalanced** Dataset

Competing Methods	Performance Metrics				
	Acc.	F_1	$F_{0.5}$	F_2	**AUC**
TF-IDF	88.85 (89.02)	56.00 (56.95)	65.42 (66.15)	48.95 (50.00)	71.08 (71.62)
GloVe Embedding	86.82 (88.18)	32.76 (53.33)	51.35 (62.31)	24.05 (46.62)	59.81 (69.80)
Word2Vec - CBOW	84.63 (78.04)	4.21 (38.68)	9.90 (36.03)	2.67 (41.75)	51.08 (64.23)
Word2Vec - Skip-gram	85.64 (83.28)	17.48 (53.52)	33.83 (49.74)	11.78 (57.93)	54.74 (74.33)
Topic Modeling - LDA	85.14 (82.94)	13.73 (51.67)	27.13 (48.47)	9.19 (55.33)	53.56 (72.82)
Topic Modeling - LSI	86.99 (82.60)	39.37 (54.22)	54.59 (49.11)	30.79 (60.52)	62.54 (75.68)
BERT - en_uncased	89.36 (78.38)	59.35 (54.93)	67.45 (45.51)	53.00 (69.27)	73.13 (80.61)
LLM - Gemma - 7B	87.84 (85.64)	56.63 (54.55)	61.04 (54.37)	52.81 (54.72)	72.66 (73.11)
LLM - Mistral - 7B	**95.10** (92.06)	**83.80** (70.06)	**85.81** (78.80)	**81.88** (63.07)	**89.22** (78.67)
LLM - Llama-2 - 7B	94.59 (94.09)	81.40 (80.66)	85.57 (82.02)	77.61 (79.35)	86.73 (87.74)

$^{(*)}$ Values in brackets correspond to cases where upsampling is used.

Table 4. Results on Dataset 4 - The Second **Imbalanced** Dataset

Competing Methods	Performance Metrics				
	Acc.	F_1	$F_{0.5}$	F_2	**AUC**
TF-IDF	92.91 (92.91)	72.37 (73.08)	83.59 (82.61)	63.81 (65.52)	79.17 (80.04)
GloVe Embedding	88.18 (86.82)	44.44 (55.68)	62.22 (57.65)	34.57 (53.85)	64.55 (72.94)
Word2Vec - CBOW	86.15 (80.91)	25.45 (42.05)	43.48 (40.92)	17.99 (43.25)	57.23 (65.93)
Word2Vec - Skip-gram	87.50 (85.47)	38.33 (56.12)	57.21 (54.46)	28.82 (57.89)	61.96 (74.76)
Topic Modeling - LDA	86.66 (83.78)	32.48 (55.14)	50.26 (51.13)	23.99 (59.84)	59.71 (75.51)
Topic Modeling - LSI	90.71 (90.71)	62.59 (72.08)	74.43 (69.74)	53.99 (74.58)	73.93 (84.87)
BERT - en_uncased	91.55 (85.47)	67.11 (63.56)	77.51 (56.39)	59.16 (72.82)	76.62 (83.51)
LLM - Gemma - 7B	91.55 (90.03)	71.26 (63.80)	74.34 (69.71)	68.43 (58.82)	81.43 (76.15)
LLM - Mistral - 7B	95.78 (96.79)	86.63 (89.02)	86.35 (93.22)	**86.91** (85.18)	**92.25** (91.10)
LLM - Llama-2 - 7B	**96.96** (96.96)	**89.77** (89.66)	92.94 (**93.53**)	86.81 (86.09)	92.07 (91.63)

$^{(*)}$ Values in brackets correspond to cases where upsampling is used.

Imbalanced Dataset 4 (Table 4.) Traditional methods with upsampling show improvements in F-scores and AUC compared to their non-upsampling counterparts. For example, the F_1-score of the GloVe Embedding method improves from

44.44% to 55.68%, while its AUC increases from 64.55% to 72.94%. On the contrary, *upsampling does not significantly benefit LLMs*. Specifically, it only induces slight changes in the performance of Gemma, Mistral, and Llama-2.

Mistral and Llama-2 show consistent performance across multiple metrics, whether with or without upsampling. They obtain the highest overall accuracies, with Mistral achieving 95.78% and Llama-2 reaching 96.96%. Mistral attains the best AUC in this dataset, with a score of 92.25%.

6.3   Results on Extremely Imbalanced Datasets

Extremely Imbalanced Dataset 5 (Table 5.**)** Due to the highly imbalanced nature of this dataset (and Dataset 6), the accuracy of all methods *exceeds 91%*, making *F*-scores and AUC *more realistic metrics* for evaluations.

Table 5. Results on Dataset 5 - The First **Extremely Imbalanced** Dataset

Competing Methods	Performance Metrics				
	Acc.	F_1	$F_{0.5}$	F_2	**AUC**
TF-IDF	92.58 (92.58)	31.93 (30.77)	49.74 (49.18)	23.51 (22.39)	59.75 (59.27)
GloVe Embedding	92.03 (91.39)	21.62 (47.19)	37.74 (49.18)	15.15 (45.36)	56.12 (70.05)
Word2Vec - CBOW	91.48 (84.89)	6.06 (33.73)	13.51 (29.54)	3.91 (39.33)	51.53 (66.49)
Word2Vec - Skip-gram	91.76 (89.10)	15.09 (43.60)	28.78 (41.14)	10.23 (46.37)	54.06 (70.70)
Topic Modeling - LDA	91.94 (87.55)	13.73 (33.33)	28.46 (32.02)	9.04 (34.76)	53.68 (64.13)
Topic Modeling - LSI	92.49 (88.55)	28.07 (45.89)	46.78 (41.47)	20.05 (51.36)	58.27 (73.73)
BERT - en_uncased	92.49 (87.45)	25.45 (49.45)	45.16 (41.93)	17.72 (60.25)	57.32 (79.80)
LLM - Gemma - 7B	93.96 (93.22)	57.14 (54.32)	66.47 (60.61)	50.11 (49.22)	72.41 (72.00)
LLM - Mistral - 7B	97.34 (96.15)	83.80 (76.14)	87.01 (79.95)	80.82 (72.67)	89.02 (84.56)
LLM - Llama-2 - 7B	**98.26** (97.99)	**89.62** (88.17)	**91.72** (89.32)	**87.61** (87.05)	**92.86** (92.71)

$^{(*)}$ Values in brackets correspond to cases where upsampling is used.

TF-IDF has better accuracy, *F*-scores, and AUC compared to GloVe Embedding. While upsampling does not improve TF-IDF, it significantly enhances GloVe Embedding in terms of *F*-scores and AUC.

CBOW and skip-gram perform poorly, with extremely low *F*-scores. With upsampling, both their *F-scores and AUC improve* significantly, while *accuracy decreases*. LSI outperforms LDA in all metrics. BERT exhibits equivalent performance to LSI, with both surpassing Word2Vec methods. Gemma's performance falls short of Mistral's. Llama-2 showcases outstanding performance, achieving the highest accuracy, *F*-scores, and AUC. Upsampling *fails to enhance the performance* of LLM methods, i.e., Gemma, Mistral and Llama-2.

Irrespective of the metrics used, Llama-2 and Mistral emerge as top performers in this dataset. Upsampling typically enhances or at least sustains the performance of traditional (non-LLM) methods. Nevertheless, even with upsampling, non-LLM methods consistently demonstrate inferior performance when compared with Mistral and Llama-2.

Extremely Imbalanced Dataset 6 (Table 6.**)** When using upsampling, there is a slight decrease in all metrics of TF-IDF. Similar to Dataset 5, GloVe Embedding has lower accuracy, F-scores and AUC compared to TF-IDF. Upsampling *improves its F-scores and AUC* significantly but *not accuracy.*

Table 6. Results on Dataset 6 - The Second **Extremely Imbalanced** Dataset

Competing Methods	Performance Metrics				
	Acc.	F_1	$F_{0.5}$	F_2	**AUC**
TF-IDF	94.60 (94.14)	59.86 (52.94)	72.61 (69.50)	50.93 (42.76)	72.76 (68.70)
GloVe Embedding	92.77 (92.12)	28.83 (52.22)	50.31 (54.02)	20.20 (50.54)	58.42 (72.83)
Word2Vec - CBOW	91.76 (87.91)	13.46 (36.54)	26.72 (34.73)	9.00 (38.54)	53.58 (66.24)
Word2Vec - Skip-gram	92.12 (91.12)	20.37 (52.68)	37.41 (50.47)	13.99 (55.10)	55.69 (75.61)
Topic Modeling - LDA	92.40 (92.31)	27.83 (48.78)	45.71 (53.91)	20.00 (44.54)	58.22 (69.60)
Topic Modeling - LSI	93.32 (91.30)	42.52 (56.62)	60.54 (52.45)	32.77 (61.51)	63.96 (79.52)
BERT - en_uncased	93.77 (86.54)	46.03 (51.16)	66.21 (41.89)	35.28 (65.70)	65.16 (84.06)
LLM - Gemma - 7B	97.16 (95.33)	82.68 (67.52)	85.85 (77.26)	79.74 (59.95)	88.45 (77.44)
LLM - Mistral - 7B	98.26 (98.26)	89.62 (89.50)	91.72 (92.26)	87.61 (86.91)	92.86 (92.38)
LLM - Llama-2 - 7B	98.08 (**98.44**)	88.77 (**90.50**)	89.63 (**93.97**)	**87.92** (87.28)	**93.23** (92.48)

(*) Values in brackets correspond to cases where upsampling is used.

BERT exhibits moderate performance, comparable to LSI, but worse than TF-IDF. When employing upsampling, its accuracy and $F_{0.5}$ decrease, while F_1, F_2, and AUC show a significant increase.

Gemma performs exceptionally well, surpassing all traditional methods. However, when upsampling is applied, there is a slight decrease in all of Gemma's metrics. Both Mistral and Llama-2 exhibit excellent performance, with very high accuracy, F-scores and AUC values, *outperforming all other methods.* Upsampling does not have much impact on performance of these methods.

In summary, Mistral and Llama-2 consistently *demonstrate superior performance* compared with traditional methods, indicating their effectiveness in handling text classification tasks, thanks to their sophisticated architectures. These LLM methods effectively *address the class imbalance* issue, eliminating the need for upsampling, although upsampling does improve the performance of most traditional methods. Additionally, the results underscore the importance of F-scores and AUC metrics, in addition to accuracy, when evaluating classification methods in the presence of class imbalance.

7   Conclusions and Future Work

This study explored various approaches, encompassing both traditional methods and LLMs, to differentiate between threatening and non-threatening texts. Results indicate that LLMs, in particular Mistral and Llama-2, consistently outperform traditional methods like word embedding, topic modeling, and BERT. To address class imbalance, we employed the upsampling technique on the training data. Upsampling generally enhances the capability of traditional methods in detecting minor class samples (i.e., threatening texts), resulting in improved *F*-scores and AUC values. However, even with upsampling, the performance of traditional methods still falls short compared to that of LLMs. Notably, LLMs exhibit remarkable efficacy in handling imbalanced data without requiring upsampling thanks to their sophisticated architectures.

Given the scarcity of threatening texts, particularly in low-resource languages, future endeavors could center on gathering such texts in those languages and exploring a multilingual approach based on LLMs. This multilingual model would seamlessly operate across languages without necessitating individual fine-tuning of LLMs on text data for each language. Another avenue for future research involves leveraging quantization methods to enhance the accessibility of LLMs. Different methods such as quantized low-rank adaptation, pruned and rank-increasing low-rank adaptation, and general pretrained transformer quantization can be explored as each has its own advantages and disadvantages.

References

1. Amjad, M., et al.: Overview of abusive and threatening language detection in Urdu at FIRE 2021. CEUR Workshop Proc. **3159**, 744–762 (2021)
2. Devlin, J., Chang, M.W., Lee, K., Toutanova, K.: Bert: pre-training of deep bidirectional transformers for language understanding. In: The North American Chapter of the ACL: Human Language Technologies, pp. 4171–4186 (2019)
3. Gales, T., Nini, A., Symonds, E.: The threatening English language (TEL) corpus (Jul 2022). https://doi.org/10.5281/zenodo.6815671
4. Gemma Team, Google DeepMind: Gemma: Open models based on Gemini research and technology (Feb 2024). https://storage.googleapis.com/deepmind-media/gemma/gemma-report.pdf
5. Hu, E.J., et al.: LoRA: Low-rank adaptation of large language models. arXiv preprint arXiv:2106.09685 (2021)
6. Hugging Face: Dataset card for Measuring Hate Speech (Jan 2023). https://huggingface.co/datasets/ucberkeley-dlab/measuring-hate-speech
7. Jiang, A.Q., et al.: Mistral 7B. arXiv preprint arXiv:2310.06825 (2023)
8. Kudo, T., Richardson, J.: SentencePiece: a simple and language independent subword tokenizer and detokenizer for neural text processing. In: Conference on EMNLP, pp. 66–71 (2018)
9. Lialin, V., Deshpande, V., Rumshisky, A.: Scaling down to scale up: A guide to parameter-efficient fine-tuning. arXiv preprint arXiv:2303.15647 (2023)
10. Mangrulkar, S., Gugger, S., Debut, L.: PEFT: State-of-the-art parameter-efficient fine-tuning methods. https://github.com/huggingface/peft (2022)
11. Mehmood, A., et al.: Threatening Urdu language detection from tweets using machine learning. Appl. Sci. **12**(20), 10342 (2022)
12. Rehan, M., Malik, M., Jamjoom, M.M.: Fine-tuning transformer models using transfer learning for multilingual threatening text identification. IEEE Access **11**, 106503–106515 (2023)
13. Sachdeva, P., Barreto, R., Bacon, G., Sahn, A., Von Vacano, C., Kennedy, C.: The measuring hate speech corpus: Leveraging Rasch measurement theory for data perspectivism. In: The 1st Workshop on Perspectivist Approaches to NLP @Language Resources and Evaluation Conference (LREC), pp. 83–94 (2022)

14. Su, J., Ahmed, M., Lu, Y., Pan, S., Bo, W., Liu, Y.: RoFormer: enhanced transformer with rotary position embedding. Neurocomputing **568**, 127063 (2024)
15. TensorFlow: Implementation of the BERT encoder API - Encoder (Oct 2020). https://tfhub.dev/tensorflow/bert_en_uncased_L-12_H-768_A-12/4
16. TensorFlow: Implementation of the BERT encoder API - Preprocessor (Oct 2020). https://tfhub.dev/tensorflow/bert_en_uncased_preprocess/3
17. Touvron, H., et al.: Llama 2: Open foundation and fine-tuned chat models. arXiv preprint arXiv:2307.09288 (2023)
18. Vidgen, B., Thrush, T., Waseem, Z., Kiela, D.: Learning from the worst: dynamically generated datasets to improve online hate detection. In: The 59th Annual Meeting of the ACL, pp. 1667–1682 (2021)

Enhancing Skills Extraction
for Domain-Specific Taxonomy Mapping

Yee Sen Tan[1,2,3]($\boxtimes$)(ID), Daryl Low[2](ID), Sarah Toh[1](ID), and Zhaoxia Wang[3](ID)

[1] Artificial Intelligence Practice, Government Technology Agency,
S117438 Singapore, Singapore
`tan_yee_sen@tech.gov.sg`
[2] Skills Development Group, SkillsFuture Singapore, S408533 Singapore, Singapore
`{tan_yee_sen,daryl_low}@ssg.gov.sg`
[3] School of Computing and Information Systems, Singapore Management University,
80 Stamford Road, 178902 Singapore, Singapore
`{yeesen.tan.2020,zxwang}@smu.edu.sg`

Abstract. Skills extraction is crucial for understanding labor market trends and aligning workforce capabilities with industry demands, but existing methods often lack generalizability and struggle with domain-specific skill taxonomy. This research introduces a novel methodology for skills extraction that identifies and maps skills from unstructured text to a predefined skill database or taxonomy. Our approach consists of three key steps: (1) employing BERT-based span extractors to identify skill mentions, (2) using sentence embedding models for semantic mapping to an existing skills taxonomy, and (3) applying a re-ranking model to assess skill importance based on relevance. In addition, we implemented several optimizations to enhance inference speed without compromising performance. Experimental results confirm that our optimized approach improves skills extraction accuracy and processing efficiency across domains, offering a replicable framework for internal taxonomies.

Keywords: skills extraction · semantic mapping · re-ranking · taxonomy

1   Introduction

In today's fast-changing job market, accurately extracting skills from unstructured text like job postings and resumes is crucial for aligning workforce capabilities with industry needs. Automation enables standardized extraction of skills demand data, supporting data-driven policy-making, workforce planning, and targeted training programs. For example, analyzing job postings reveals the specific competencies employers seek, thereby informing curriculum development and guiding training providers. Similarly, extracting skills from course descriptions and resumes helps map educational outcomes to market requirements, ensuring that learners and job seekers are equipped with relevant, in-demand abilities.

S. Yuan et al. (Eds.): PAKDD 2025 Workshops, LNAI 15835, pp. 324–335, 2025.
https://doi.org/10.1007/978-981-96-8197-6_24

SkillsFuture Singapore, the national skills authority of Singapore's public service, has worked alongside with employers, industry associations, institutes of higher learning, and unions to establish the SSG Skills Framework [12,14]. This framework serves as a comprehensive resource, outlining key industries, occupations, job roles, and the critical skills both existing and emerging—required for the workforce. To fully utilize this standardized skills taxonomy, skills extraction is required for efficient policy development and for designing training programs that address the changing demands of the job market. Although researchers initially leveraged natural language processing (NLP) techniques to extract skills [2], challenges remain in generalizing extracted skills, especially when using established skill knowledge bases like O*Net and ESCO [7]. Organizations could maintain their own skills knowledge base, skills taxonomy, which is different from the expert-labeled skill knowledge bases available out there.

In this paper, we present a novel skills extraction algorithm that utilizes BERT-based span extractors, machine learning, and domain-adaptive language models to precisely identify and map skills from unstructured text to a standardized skills taxonomy. Our method incorporates pre-trained transformer models, fine-tuned on industry-specific datasets. By combining span extractors, embedding-based similarity mapping, re-ranking techniques, we improve the accuracy and adaptability of skill extraction across various organizational taxonomies.

The main contributions of our work are summarized as follows:

- We present a novel methodology for skills extraction that effectively identifies and aligns skills from textual input with a predefined skills taxonomy or knowledge base, ensuring adaptability across different domains.
- Our methodology identifies skills by extracting relevant phrases and mapping them with our existing taxonomy skills, followed by a re-ranking to assess relevance by comparing descriptions to each skills identified.
- The optimization of our methodology inference process, improving the speed of skills extraction without compromising accuracy.
- Experimental evaluations demonstrate the effectiveness of our proposed approach, validating its performance in skills extraction and semantic mapping.

2 Related Works

Various approaches have been proposed to automatically extract relevant skills from unstructured text data [8]. Early methods focused on rule-based information extraction, but recent advancements have shifted towards deep learning, particularly transformer-based models. These models, known for their superior semantic understanding, have significantly improved text comprehension, making them highly effective across NLP tasks such as sentiment analysis, text generation, semantic mapping, and even skills extraction [8,18,19]. By utilizing pre-trained language models, these approaches are not limited to a fixed set of skills, but rather can generalize to identify a broader spectrum of skills, even emerging ones, from diverse industries.

2.1   NLP Approaches to Skills Extraction

Although several skills extraction algorithms have been proposed [6], many of them lack generalizability and underperform when applied to domain specific skill databases. This is because they primarily extract skills from pre-existing expert-contributed databases [7], limiting their adaptability to different or specialized contexts.

Recent researchers has introduced several BERT based models for job and skills related extraction [20]. The results show that domain-adapted models outperform their non-adapted counterparts, and single-task learning outperforms multi-task learning. However, it should be noted that these models are fine-tuned to extract skills and knowledge from textual postings, but they often fail to capture the full spectrum of skills available in various specialized skill databases.

A skill base, or skill taxonomy in our case, typically refers to a knowledge base containing skill entities and terminology [8]. These skill taxonomy can be tailored to specific organizations or databases, which leads to the challenge of limited generalizability [7]. The mismatch between the skills extracted by NLP models and the diverse terminologies present across different skill taxonomy further exacerbates the issue of adaptability. Hence, some form of semantic mapping must be done to correctly map each skill being extracted to an organization's respective skill taxonomy, similar to how search algorithm works [9].

2.2   Semantic Mapping and Re-ranking

Recent advances in automated skills extraction have highlighted the importance of post-processing steps—namely, semantic mapping and re-ranking to improve the accuracy and relevance of the extracted skill entities [13]. After the initial extraction phase, candidate skills often require further refinement to ensure they align accurately with a predefined skills taxonomy or knowledge base [13], via semantic mapping and re-ranking techniques.

Semantic mapping, focuses on aligning the extracted skill phrases with standardized entries in established taxonomies like O*Net or ESCO, ensuring correct alignment [13]. This involves using techniques ranging from rule-based heuristics to semantic mapping models [9].

Similarly, in retrieval-augmented generation (RAG) methods, re-rankers are essential for boosting downstream performance [1,4]. Re-rankers can ensure that the retrieved documents or text snippets not only match the query's intent but are also contextually sound [1]. In our research, re-ranking can be applied to assess the similarity of each skill's relevance.

3   Proposed Methododology

3.1   Overall Methodology

Our methodology follows a three-step approach, as illustrated in Fig. 1. First, we employ BERT-based span extractors [20] to identify relevant skill and knowledge spans within text. These extractors have been finetuned to recognize skill-related phrases in job descriptions, course descriptions, and other textual sources.

Next, since the extracted spans may or may not exist in our internal skills taxonomy, we utilize sentence embedding models [15] to semantically align them. Specifically, we generate vector representations for both the extracted spans and the existing skill taxonomy entries, allowing us to measure similarity and determine potential matches. This step ensures that known skills are correctly mapped while identifying skills that do not exist in our taxonomy, which are omitted in such cases.

Finally, we implement a re-ranking model to assess the relevance of each extracted skill. This re-ranking step prioritizes the most accurate matches while filtering out noisy or irrelevant extractions.

To facilitate efficient semantic mapping and re-ranking, our existing skills taxonomy is precomputed as a set of embeddings, similar to a structured knowledge base. These embeddings serve as a reference during both the alignment and re-ranking steps, ensuring consistency and robustness in skill identification.

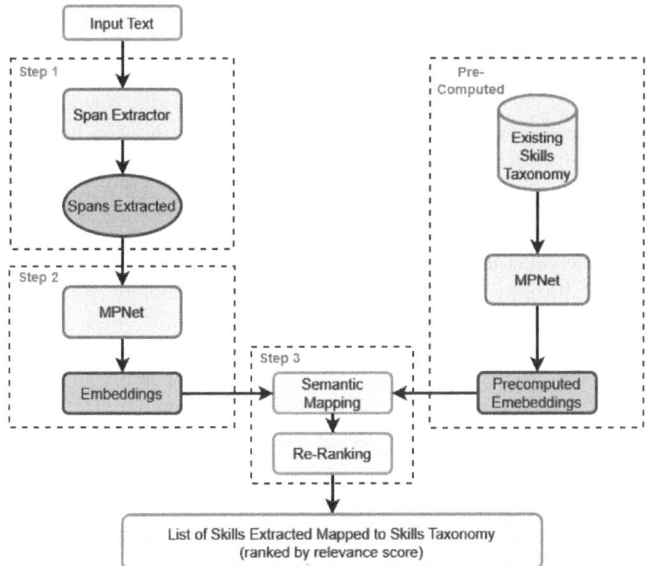

Fig. 1. Overview of the Methodology

3.2   Span Extractors

Given the lack of labeled data to fine-tune or train the model, we used pre-trained models fine-tuned on domain-specific datasets that has demonstrated strong performance on identifying skill-related spans [20]. However, since these models are not aware of our skills taxonomy, a semantic mapping to our own taxonomy is necessary with an embedding model.

3.3   Embedding Step

To determine the most suitable embedding model for mapping extracted skill spans from the span extractor to SkillsFuture Singapore skills taxonomy, we evaluated multiple sentence embedding models, ultimately selecting all-mpnet-base-v2 due to its stronger performance over other models.

After step 1 and 2, the embeddings (obtained from the input text side), together with the precomputed embeddings (obtained from the existing skills taxonomy side), are leveraged in step 3: Semantic Mapping and Re-ranking.

3.4   Semantic Mapping and Re-ranking

We use embeddings from both sides to perform semantic mapping, ensuring the cosine similarity between the extracted skill spans and our internal skills taxonomy is at least 0.70, based on our experiments. This mapping normalizes variations in phrasing to the correct standardized skill, ensuring consistency across datasets. In the re-ranking step, we assess the relevance of each extracted skill within the context of the input text. Using a cross-encoder model, we assign relevance scores and reorder the skills, prioritizing those most contextually appropriate while demoting less relevant ones, thereby refining the extraction results and offering better insights for downstream users.

Together, these semantic mapping and re-ranking forms an integral part of the overall skills extraction pipeline, addressing the limitations of initial extraction methods by enhancing the generalizability and accuracy of the final outputs. In short, the main aim of our methodology is to extract skills and ensure they are mapped to our internal taxonomy.

4   Dataset

To evaluate our embedding models and methodology, we used datasets from different domains to ensure a comprehensive assessment. Leveraging multiple datasets allowed us to capture variations in skill representations and assess both individual components and the overall effectiveness of our approach. Specifically, two datasets were used to evaluate the embedding and semantic mapping process, focusing on how well extracted skill spans align with our internal taxonomy. This step ensures that the model accurately represents and maps skills within our system. Additionally, two more datasets were used to assess our end-to-end methodology in real-world scenarios, measuring its applicability to job market data and training programs. By incorporating both controlled and real-world datasets, we provide a robust validation of our framework.

4.1   Datasets for Evaluating the Embedding Model

To ensure effective semantic mapping of extracted skill spans, we constructed an embedding store using the following data sources below. Note that these datasets are maintained and used by SkillsFuture Singapore internally, and these are also the main data points that our skills extracted needs to be mapped to.

1) SkillsFuture Singapore Internal Skills Taxonomy. This dataset is derived from SkillsFuture Singapore's internal skills taxonomy, which provides a structured and curated list of skills relevant to workforce planning and recognized by SkillsFuture Singapore. Starting with an internal taxonomy of 11K skills, where each skill includes sector and proficiency level details, we simplified the taxonomy to 1989 skills by removing the sector-specific and proficiency-related information to ensure broader applicability.

2) Applications and Tools Repository. This internal dataset comprises a list of software applications and tools commonly referenced in job postings and training courses. Sourced from SkillsFuture Singapore's internal repository, this dataset plays a crucial role in ensuring accurate mapping of technology-related skills. There are 922 rows, which includes the applications and tools name, accompanied by a textual description of it.

An illustration of one such skill in our dataset that we need to map is shown below, formatted as *title: description.*

> **Agile Software Development**: Plan and implement Agile methodology and the use of adaptive and iterative methods and techniques in the software development lifecycle to account for continuous evolution, development, and deployment to enable seamless delivery of the application to the end user.

4.2   Datasets for Evaluating the End-to-End Methodology

To evaluate the effectiveness of our complete methodology in real-world scenarios, we utilized datasets from live sources where the ground truths are synthetically generated using GPT-4o due to the absence of labelled data.

These datasets enable us to assess the overall methodology, specifically how well we extract skills and map them back to our internal skills database. We conduct this evaluation twice: once using course data and once using job postings data. This dual evaluation ensures that our methodology is generalizable across different domains while addressing the same use case. Note that the content of these datasets would be similar to any course advertisement or job postings details available on the internet.

1) SkillsFuture Active Course Records. The SkillsFuture Active Course Records dataset includes 5,000 active course records from SkillsFuture Singapore, providing course titles, descriptions, and related metadata. It is used to evaluate the effectiveness of our methodology in extracting skills from course information and mapping them to our internal skills database.

2) Job Posting Data. The Job Posting Data dataset contains 9,500 job postings from an online job aggregator, including job titles, descriptions, and metadata. This dataset helps assess how effectively our methodology can extract skills from real-world job advertisements and map them to our skills database.

Both datasets are highly relevant for evaluation as they are derived from real world data, reflecting current industry practices and workforce demands. By utilizing these datasets, we can assess how accurately our methodology extracts and categorizes skills from educational programs and identifies the skills required in job postings, demonstrating the methodology's applicability and generalizability in real world scenarios.

5   Experiment and Results

5.1   Experiment Setup

Our primary experiment focuses on selecting the most suitable embedding model and assessing the overall effectiveness of our methodology in extracting skills and mapping them to our internal skills database. The sentence embedding models used in this study are all available on Hugging Face, primarily derived from the Sentence Embedding collection [10]. As highlighted, we will test our embedding model against both datasets, a combination of the SkillsFuture Singapore Internal Skills Database and Applications and Tools Repository dataset.

Our end to end evaluation will then be done on the SkillsFuture Active Course Data and Job Posting Data to ensure generalization across different domains.

5.2   Evaluation and Discussion of Embedding Model

The primary evaluation of the embedding model focuses on how effectively the extracted skills, obtained using pretrained span extractors, can be mapped to SkillsFuture Singapore's internal skills database. A variety of sentence embedding models were assessed using Precision, Recall, and ROC-AUC as key metrics, with the results summarized in Table 1. In our experiments, we optimized the similarity threshold with the goal of maximizing precision, ensuring that only accurate and relevant skills are mapped to our internal taxonomy. Minimizing false positives is crucial, as incorrect mappings could compromise the integrity of the taxonomy, generate unreliable insights, and negatively impact downstream applications such as job matching, workforce analytics, and policy-making.

For each embedding model, we identified the similarity threshold that yielded the highest precision, ensuring that only the most accurate and relevant skills were mapped to our internal taxonomy. Once the optimal threshold was determined, we evaluated and compared the corresponding recall, ROC-AUC, and other relevant metrics to assess the overall performance of the models.

Table 1. Embedding Model Evaluation Results

Model Utilised	Optimal Threshold	Precision	Recall	ROC-AUC
all-mpnet-base-v2 [10]	0.70	0.77	0.55	0.85
all-MiniLM-L6-v2 [17]	0.70	0.52	0.41	0.82
all-distilroberta-v1 [11]	0.65	0.45	0.40	0.82
multi-qa-mpnet-base-dot-v1 [10]	0.73	0.47	0.43	0.82

Notably, the all-mpnet-base-v2 sentence embedding model outperformed the other models, achieving the highest scores across all three evaluation metrics (Precision, Recall, and ROC-AUC) using the threshold of 0.70. Given that this embedding model is intended for re-ranking tasks, we further evaluated its performance in this domain. The model attained a Precision@5 of 0.632 and Precision@10 of 0.734, demonstrating robust retrieval and ranking performance in a top-k retrieval setup. It is important to note that these metrics were evaluated using GPT-4o, as no human-annotated data was available.

5.3 Evaluation and Discussion of Overall Methodology

Finally, after selecting the embedding models, we tested our overall methodology on the active courses postings and job posting dataset. This serves as the most important evaluation on whether our methodology is robust and sound across different domains (course and job posting). We primarily used Precision, Recall and F1 score as the metrics for evaluation, as seen in Table 2.

Table 2. Overall Methodolody Evaluation Results

Dataset	Precision	Recall	F1-score
SkillsFuture Active Course Records	0.81	0.79	0.80
Job Postings	0.82	0.80	0.81

Our methodology shows strong performance across precision, recall, and F1 score on datasets from courses and job postings, demonstrating its versatility and reliability. The framework automates skills extraction, reducing manual tagging and ensuring accurate mapping to internal taxonomies. Additionally, our approach can be easily replicated by other industries or organizations, improving the scalability and efficiency of their skills extraction processes for better talent management and workforce development.

6   Real World Deployment Considerations

6.1   Optimization for Inference in Production

As our methodology involves multiple models operating in sequential steps, it leads to an increase in the inference time. Before optimization, our models took an average of 0.70 s to process our entire dataset. To enhance efficiency without significantly compromising accuracy, we applied several optimization techniques to accelerate the inference process.

Firstly, we swapped scikit-learn's cosine similarity function to TensorFlow's built-in function due to it being optimized for parallel execution on GPUs, which significantly speeding up similarity calculations for our large skills taxonomy. We then optimized inference for our span extractors by integrating BetterTransformer, which enhances MultiHeadAttention and TransformerEncoderLayer on both CPUs and GPUs [5]. This improves computational efficiency through fused kernels and sparse operations that bypass calculations on padding tokens, accelerating inference while preserving performance [5]. For our MPNet model, we converted it to TorchScript for optimized execution, allowing for static optimization and operator fusion, which streamline the execution of model operations, hence, taking full advantage of lower-level system optimizations and parallel execution on hardware [3]. Additionally, we quantized the linear layers and loaded the models in half-precision (FP16).

The results of our optimizations are presented below, as seen in Table 3. For information, these are done on a single NVIDIA T4 Tensor Core GPU. Overall, our optimization process reduced inference time from 0.70 s to 0.36 s, achieving a 48.6% speedup. This was accomplished through a combination of similarity computation, optimizing Transformer TorchScript conversion, and model quantization with FP16 precision, enhancing scalability and efficiency.

Table 3. Sequential Optimization Stages and Speed Improvements

Stage of Optimization	Execution Time (s)	Cumulative Improvement (%)
Pre-optimization	0.70	–
Optimizing similarity computation	0.45	35.7%
Optimizing better transformers	0.38	45.7%
TorchScript + Quantization + FP16	0.36	48.6%
Total Time Saved	0.34 s (48.6%)	

Note: The downstream performance of these optimizations was not affected, with less than 0.01% degradation in performance.

Time Savings Calculation: Based on a projection of 1 million data points for inference on pre-optimized models:

$$1,000,000 \times 0.70 = 700,000\,\text{seconds} = 194.44\,\text{h}.$$

After optimization, with $0.36\,\text{s}$ per data point, the total time is:

$$1,000,000 \times 0.36 = 360,000\,\text{seconds} = 100\,\text{h}.$$

Hence, the total time saving equates to:

$$700,000 - 360,000 = 340,000\,\text{seconds} = 94.44\,\text{h}.$$

This results in a 48.6% reduction in processing time, showcasing the efficiency of the optimizations.

6.2   Machine Learning Operations (MLOps)

In our skills extraction research, we found that MLOps is crucial for ensuring the scalability, reliability, and ongoing improvement of machine learning models. Given the dynamic nature of skills data, with new job roles and competencies constantly emerging, maintaining an up-to-date skills extraction pipeline would be difficult without MLOps. Challenges such as model drift, data inconsistencies, and the need for frequent retraining would arise. To address this, we implemented MLOps in production environment to automate version control, model monitoring, re-training and deployment, ensuring that the extracted skills remain accurate and relevant. Additionally, pre-defined metrics, such as an F1 score threshold of 0.4, are set to trigger an alert for model re-training process if performance drops below this threshold. These MLOps configurations are crucial for maintaining the robustness and accuracy of the model over time, ensuring it remains effective in dynamic environments.

6.3   Potential Integration with Real World Systems

Our methodology can even be further extended and integrated into real world systems, especially those in HR and job-matching domains. First, it can be used for automated resume parsing, where skills are extracted and mapped to job requirements, streamlining candidate matching. Additionally, it provides personalized career recommendations by identifying skill gaps and suggesting relevant training or career paths. The method can also offer real time labor market insights by aggregating skills data from job postings, allowing HR systems to track trends, identify emerging skills, and adapt to workforce shifts, supporting proactive workforce planning and talent strategy alignment.

The method can also be crucial for constructing a job skills graph by extracting and categorizing skills from resumes, job postings, and employee profiles. This graph can potentially represent relationships between skills, jobs, and industries, and can be further integrated into HR systems to improve talent management, recruitment, and employee development.

7   Conclusion, Limitations and Future Works

7.1   Conclusion

In short, our research introduces a novel approach for extracting skills using a multi-step pipeline that combines semantic mapping and re-ranking to improve accuracy, consistency, and relevance to an organization's internal taxonomy. Through optimization, our approach also significantly enhances speed without compromising downstream performance. By integrating BERT-based span extraction with all-mpnet-base-v2 embeddings and context-aware re-ranking, our method delivers strong performance in skills extraction, supporting workforce planning and policy formulation. Moreover, our methodology can be easily adapted to different organizations and domains, ensuring its broad applicability.

7.2   Limitations and Future Works

Our methodology has some limitations in multilingual contexts. Challenges may arise if secondary languages are introduced in future job or course postings, especially on social medias such as LinkedIn where multiple language can exist [16]. Therefore, it is important to consider integrating multilingual capabilities into our skills extraction methodology to also support international use cases.

Future work can focus on two key areas. First, enhancing MLOps capabilities to automate model retraining when the F1 score falls below a predefined threshold. Second, developing methods to extract emerging skills not captured by current databases or taxonomies, identifying new or evolving areas gaining popularity.

Acknowledgments. The authors thank SkillsFuture Singapore for supporting the research and application of the Skills Extraction methodology. Special thanks to Leo Li, Raymond Haris, Lois Ji, and Eugene Chua for their valuable contributions and insights. We also appreciate the guidance of Directors Chelvin Loh (Skills Innovation and Planning Division) and Kelvin Goh (Data Science and Analytics Division) from SkillsFuture Singapore. Disclaimer: The views in this paper are those of the authors and do not reflect the positions of SkillsFuture Singapore or the Government Technology Agency of Singapore.

References

1. Chen, Z., Jiang, J., Zuo, D., Tao, H., Yang, J., Wei, Y.: Efficient title reranker for fast and improved knowledge-intense nlp. arXiv preprint arXiv:2312.12430 (2023)
2. Debortoli, S., Müller, O., Brocke, J.V.: Comparing business intelligence and big data skills: a text mining study using job advertisements. Wirtschaftsinformatik **56**, 315–328 (2014)
3. DeVito, Z.: Torchscript: optimized execution of pytorch programs. Retrieved January (2022)
4. Glass, M., Rossiello, G., Chowdhury, M.F.M., Naik, A.R., Cai, P., Gliozzo, A.: Re2g: retrieve, rerank, generate. arXiv preprint arXiv:2207.06300 (2022)

5. Gschwind, M., et al.: A better transformer for fast transformer inference (2022)
6. Gugnani, A., Misra, H.: Implicit skills extraction using document embedding and its use in job recommendation. In: Proceedings of the AAAI Conference on Artificial Intelligence, vol. 34, pp. 13286–13293 (2020)
7. Khaouja, I., Kassou, I., Ghogho, M.: A survey on skill identification from online job ads. IEEE Access **9**, 118134–118153 (2021). https://doi.org/10.1109/ACCESS.2021.3106120
8. Konstantinidis, I., Maragoudakis, M., Magnisalis, I., Berberidis, C., Peristeras, V.: Knowledge-driven unsupervised skills extraction for graph-based talent matching. In: Proceedings of the 12th Hellenic Conference on Artificial Intelligence, pp. 1–7 (2022)
9. Li, H., Xu, J., et al.: Semantic matching in search. Foundations Trends® Inform. Retrieval **7**(5), 343–469 (2014)
10. Reimers, N., Gurevych, I.: Sentence-bert: sentence embeddings using siamese bert-networks. In: Proceedings of the 2019 Conference on Empirical Methods in Natural Language Processing. Association for Computational Linguistics (Nov 2019). https://arxiv.org/abs/1908.10084
11. Sanh, V.: Distilbert, a distilled version of bert: smaller, faster, cheaper and lighter. arXiv preprint arXiv:1910.01108 (2019)
12. Seif, A., Toh, S., Lee, H.K.: A dynamic jobs-skills knowledge graph (2024)
13. Senger, E., Zhang, M., van der Goot, R., Plank, B.: Deep learning-based computational job market analysis: a survey on skill extraction and classification from job postings. arXiv preprint arXiv:2402.05617 (2024)
14. SkillsFuture Singapore: Skills framework (nd). https://www.skillsfuture.gov.sg/skills-framework, Accessed 21 Feb 2025
15. Song, K., Tan, X., Qin, T., Lu, J., Liu, T.Y.: Mpnet: masked and permuted pre-training for language understanding. Adv. Neural. Inf. Process. Syst. **33**, 16857–16867 (2020)
16. Tan, Y.S., Teo, N., Ghe, E., Fong, J., Wang, Z.: Video sentiment analysis for child safety. In: 2023 IEEE International Conference on Data Mining Workshops (ICDMW), pp. 783–790. IEEE (2023)
17. Wang, W., Bao, H., Huang, S., Dong, L., Wei, F.: Minilmv2: Multi-head self-attention relation distillation for compressing pretrained transformers. arXiv preprint arXiv:2012.15828 (2020)
18. Wang, Z., Ho, S.B., Cambria, E.: Multi-level fine-scaled sentiment sensing with ambivalence handling. Internat. J. Uncertain. Fuzziness Knowl.-Based Syst. **28**(04), 683–697 (2020)
19. Wang, Z., Hu, Z., Ho, S.B., Cambria, E., Tan, A.H.: Mimusa–mimicking human language understanding for fine-grained multi-class sentiment analysis. Neural Comput. Appl. **35**(21), 15907–15921 (2023)
20. Zhang, M., Jensen, K.N., Sonniks, S., Plank, B.: SkillSpan: hard and soft skill extraction from English job postings. In: Proceedings of the 2022 Conference of the North American Chapter of the Association for Computational Linguistics: Human Language Technologies, pp. 4962–4984. Association for Computational Linguistics, Seattle, United States (Jul 2022), https://aclanthology.org/2022.naacl-main.366

RAG Based Product Review Summarization and Faithfulness Evaluation

Nagbhushan R. Subbapurmath$^{(\boxtimes)}$ ⓘ and Harkeerat Kaur ⓘ

Indian Institute of Technology, Jammu, Jammu and Kashmir, India
{2022pai9019,harkeerat.kaur}@iitjammu.ac.in

Abstract. In the age of e-Commerce, customers increasingly rely on product reviews to make informed decisions. However, the vast number of reviews on different e-commerce platforms such as Amazon and Flip-kart, etc. can be overwhelming. We propose an efficient Product Review Summarization technique using a Retrieval Augmented Generation (RAG) approach to address this. RAG retrieves relevant information from a vector store and uses a Large Language Model (LLM) to generate coherent summaries. Our method enhances traditional RAG systems by implementing ensemble retrievers that combine multiple retrieval models to ensure reliability. Additionally, we explore a hybrid RAG architecture with data compression and cross-encoder re-ranking to improve summary faithfulness. We evaluated our approach using metrics such as SummaC, CTC, FactCC, and FactGraph, demonstrating improved faithfulness scores over standard retrieval methods.

Keywords: Natural Language Processing (NLP) · Retrieval Augmented Generation (RAG) · Term Frequency-Inverse Document Frequency (TF-IDF) · Text-to-Text Transfer Transformer (T5) · Bidirectional Encoder Representations from Transformers (BERT) · Generative Pre-trained Transformer (GPT) · Large Language Model Meta AI (LLAMA)

1 Introduction

Product reviews are crucial for consumer decision making, yet the overwhelming volume of user-generated content makes manual analysis impractical. Traditional opinion summarization methods struggle with limited reference summaries and factual inconsistencies in generated outputs. Retrieval-Augmented Generation (RAG) offers a promising solution by grounding summaries in retrieved evidence, but its effectiveness is hindered by suboptimal retrieval and weak faithfulness evaluation.

This work focuses on Retrieval-Augmented Generation (RAG) for product review summarization, comparing four key RAG-based techniques, including an

S. Yuan et al. (Eds.): PAKDD 2025 Workshops, LNAI 15835, pp. 336–348, 2025.
https://doi.org/10.1007/978-981-96-8197-6_25

advanced architecture optimized for faithfulness. We introduce a novel Compression Retrieval and Generation (CRG) technique, which improves faithfulness by refining retrieved evidence before generation. Evaluation using SummaC, CTC, FactCC, and FactGraph demonstrates that CRG outperforms existing RAG methods in generating factually consistent summaries. We propose an advanced RAG framework designed for product review summarization, emphasizing factual consistency and source alignment. Our approach integrates: (1) a hybrid retriever combining FAISS (dense) and BM25 (lexical) search for balanced relevance, (2) a contextual compression pipeline leveraging cross-encoder reranking (BgeRerank) and redundancy filtering to refine retrieved content, and (3) a robust evaluation framework incorporating SummaC, CTC, FactCC, and FactGraph to assess faithfulness. Experiments on 27,206 smartphone reviews demonstrate that our system surpasses baseline retrievers in faithfulness (SummaC: 0.52, CTC: 0.49) while maintaining coherence. Unlike prior self-supervised approaches that generate pseudo-summaries, our method directly utilizes retrieved evidence, significantly reducing hallucination.

The Major Contributions are:

- We proposed a novel product review summarization framework based on Retrieval Augmented Generation (RAG), leveraging an ensemble of retrievers for improved relevance.
- We created our own dataset for our RAG-Based approach using Python Libraries consisting of 27,206 smartphone reviews.
- We introduce a hybrid RAG architecture that incorporates data compression and cross-encoder re-ranking to enhance summary faithfulness.
- We conducted extensive evaluations using faithfulness metrics such as SummaC, CTC, FactCC, and FactGraph, demonstrating superior performance over traditional retrieval-based summarization approaches.

2   Related Work

Opinion summarization in NLP, especially for e-commerce reviews, has evolved from extractive methods to retrieval-augmented and self-supervised techniques. The surge in user-generated content demands automated summarization approaches that ensure factual consistency. This work reviews key developments in opinion summarization, Retrieval-Augmented Generation (RAG), and faithfulness evaluation. Self-supervised learning has been widely used in opinion summarization to address the lack of annotated datasets. Earlier methods employed pseudo-summary selection, either randomly choosing summaries [1,2] or refining selection using lexical similarity [3] and sentiment awareness [4]. Other approaches reduced redundancy by learning aspect and sentiment embeddings [5]. RAG architecture (Lewis et al., 2020) [6] has emerged as a powerful framework for improving factual consistency in text generation by retrieving relevant documents before producing summaries. Prior studies (Izacard and Grave,

2021; Guu et al., 2020) [7,8] have demonstrated that hybrid retrieval methods enhance summarization quality, particularly in knowledge-intensive applications. Recent advancements in opinion summarization incorporate external knowledge sources such as product descriptions and QA pairs (Siledar et al., 2023) [9]. The Multi-Encoder Decoder Opinion Summarization (MEDOS) model (Siledar et al., 2023) [9] integrates multiple encoders for better information capture but relies on synthetic dataset creation (SDC). In contrast, our approach improves retrieval efficiency through an ensemble of retrievers and compression-based filtering, ensuring that generated summaries remain factually accurate while minimizing hallucination. Ensuring factual consistency is a major challenge in opinion summarization. Traditional evaluation metrics such as **ROUGE (Lin, 2004)** [10] measure lexical overlap but fail to assess factual correctness. Recent studies (Durmus et al., [11] 2020; Laban et al., 2022 [12]) propose **faithfulness-focused metrics** such as FactCC, SummaC, and CTC to evaluate factual consistency. Other approaches (Bhaskar et al., 2023 [13]; Hosking et al., 2023 [14]) rely on human annotations to verify summary correctness making large-scale evaluation difficult.

Our work contributes to **faithfulness evaluation by integrating multiple automated metrics (SummaC, CTC, FactCC, and FactGraph)** to provide a **comprehensive faithfulness assessment**. Unlike previous works that evaluate faithfulness using a **single metric or human annotations**, our multi-metric approach enables **scalable and systematic** evaluation of summary quality. Also unlike these fixed-dataset approaches, our method dynamically retrieves relevant reviews using an ensemble of retrievers (BM25, FAISS, and BgeRerank). This ensures that summaries are generated from **contextually relevant** data, improving coherence and factual accuracy.

While prior research has focused on either retrieval-augmented summarization or faithfulness evaluation, our work **bridges the gap between retrieval optimization and factuality assessment**. Compared to **MEDOS (Siledar et al., 2023), which integrates product descriptions and Q&A into summarization**, our method **retrieves reviews dynamically**, ensuring adaptability across different products. Additionally, while **self-supervised techniques rely on fixed pseudo-summaries**, our approach uses **ensemble retrievers and cross-encoder re-ranking (BgeRerank) to improve the relevance and faithfulness of retrieved content**.

3   Proposed RAG System

This section introduces the proposed RAG-based summarization framework, which addresses key challenges in information retrieval and content generation. We compare different retriever methods and discuss how they enhance the summarization pipeline. Figure 1 illustrates the structure of the proposed Retrieval-Augmented Generation (RAG) model, which consists of three main components:

- **Retriever**: Extracts relevant information from a large document corpus.
- **Document Compression**: Filters and refines retrieved content for improved relevance.
- **Generator**: Produces human-like summaries based on the retrieved data.

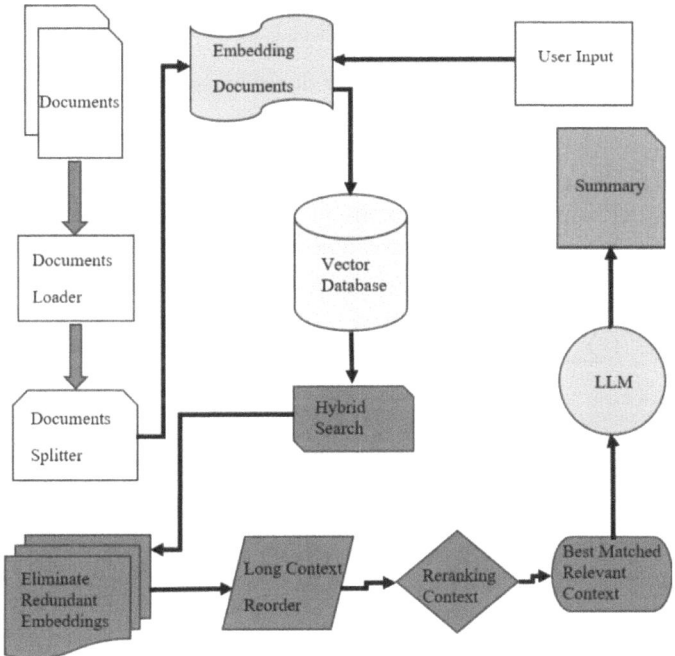

Fig. 1. Flow Diagram of the Proposed Advanced RAG System

3.1   Retriever Component

The retriever is responsible for selecting relevant documents that provide factual grounding for the generated summaries. The system employs both **dense vector retrieval (FAISS)** and **lexical retrieval (BM25)** to improve retrieval effectiveness.

Dense Vector Retrieval (FAISS). FAISS retrieves documents based on dense vector similarity. Documents are embedded using pre-trained transformer models (e.g., BERT, Sentence-BERT), and the query is embedded into the same vector space. The similarity between the query q and each document d is computed using **cosine similarity**:

$$\text{sim}(q, d) = \frac{q \cdot d}{\|q\| \|d\|} \tag{1}$$

where:

- q is the vector representation of the query.
- d is the vector representation of the document.
- $q \cdot d$ is the dot product of the two vectors.
- $\|q\|$ and $\|d\|$ are the magnitudes of the vectors.

FAISS returns the top-k most similar documents based on these similarity scores.

Lexical Retrieval (BM25). BM25 is a well-established retrieval method based on term frequency-inverse document frequency (TF-IDF). It ranks documents based on their relevance to the query. The BM25 score for a document D and query q is computed as:

$$BM25(q, D) = \sum_{i=1}^{n} \frac{tf(q_i, D)(k_1 + 1)}{tf(q_i, D) + k_1 \cdot (1 - b + b \cdot \frac{|D|}{\text{avgdl}})} \cdot \log \frac{N - df(q_i) + 0.5}{df(q_i) + 0.5} \tag{2}$$

where:

- q_i is a query term.
- $tf(q_i, D)$ is the term frequency in document D.
- k_1, b are hyperparameters.
- $|D|$ is the document length.
- avgdl is the average document length.
- N is the total number of documents.
- $df(q_i)$ is the number of documents containing q_i.

BM25 assigns higher scores to documents that contain more query terms, adjusting for document length.

3.2   Document Compression Component

To enhance retrieval relevance, the system applies a **document compression step** that eliminates redundant or less informative content.

BgeRerank: Cross-Encoder-Based Compression. A cross-encoder model (e.g., "BAAI/bge-reranker-large") is used to rerank retrieved documents. Given a query-document pair, the model computes a **relevance score**:

$$S(q, d) = f([q, d]) \tag{3}$$

where:

- $S(q, d)$ is the computed relevance score.
- f is the cross-encoder model that processes concatenated query and document embeddings.

Only the top-N documents with the highest relevance scores are selected.

Document Redundancy Removal. To prevent duplicate information from being passed to the generator, an embedding-based redundancy filter removes near-duplicate documents. Given two document embeddings d_1 and d_2, **cosine similarity** is used to measure redundancy:

$$\text{Cosine similarity}(d_1, d_2) = \frac{d_1 \cdot d_2}{\|d_1\|\|d_2\|} \tag{4}$$

If the similarity score exceeds a predefined threshold (e.g., 0.9), one of the documents is discarded.

3.3   Generator Component

The generator is responsible for producing human-like summaries. This is achieved using **sequence-to-sequence (Seq2Seq) transformer models**, such as GPT or LLAMA. In this research, we have included LLAMA 2.0 as it is open source and well-suitable for the RAG-based summarization processes.

Self-Attention Mechanism. The generator employs the **self-attention mechanism** to determine how different tokens in the input relate to each other. The self-attention function is defined as:

$$\text{Attention}(Q, K, V) = \text{softmax}\left(\frac{QK^T}{\sqrt{d_k}}\right) V \tag{5}$$

where:

- Q is the query matrix.
- K is the key matrix.
- V is the value matrix.
- d_k is the dimension of the key vectors.

Ensemble Retrievers: To improve retrieval effectiveness, the system combines **dense (FAISS) and lexical (BM25) retrieval methods**. The ensemble retriever assigns weights to each method:

$$\text{Score}(D) = \alpha \cdot \text{FAISS}(D) + (1 - \alpha) \cdot \text{BM25}(D) \tag{6}$$

where α is a weight hyperparameter balancing dense and lexical retrieval contributions.

Contextual Compression Retriever: To refine retrieved data, a contextual compression retriever applies three key steps:

- **Redundancy Filtering:** Removes duplicate content using cosine similarity.
- **Context Reordering:** Sorts document sections by relevance to the query.
- **Relevance Reranking:** Uses a cross-encoder model to rescore document importance.

The final set of retrieved and compressed documents is used as input for the language model, ensuring concise yet comprehensive summaries.

3.4 LLM Configuration and Summary Generation Process

To generate factually accurate and concise summaries, we employ a Large Language Model (LLM) using a structured prompting technique. The summarization pipeline consists of:

- **Model Selection:** A *ChatOllama* model based on *LLaMA-2* is used for natural language generation.
- **Prompt Design:** A carefully crafted prompt template guides the model to extract meaningful insights from retrieved documents.
- **Prompt Chaining:** The system links the user query, retrieved product reviews, and the LLM to generate structured output.

Prompt Engineering for Summarization: The summarization prompt follows a structured template, ensuring that the generated summary is both informative and factually aligned with the retrieved reviews and consists of:

- Extraction of **technical specifications** from the retrieved context.
- Sentiment analysis, highlighting *positive feedback* from users.
- Summary length constraints to **maintain conciseness**.
- Prevention of **hallucination** by instructing the model to avoid generating missing technical specifications.

Prompt Execution Pipeline: The implementation follows a *document chain approach*, where retrieved documents are formatted and passed to the LLM. The pipeline can be outlined as follows:

- Load retrieved product reviews into a structured format.
- Apply a predefined summarization prompt to guide response generation.
- Use a document chaining mechanism to integrate retrieval with generation.
- Generate a final summary with a predefined word limit, maintaining **factual consistency**.

4 Data Collection and Cleaning

This section describes the process of collecting and preprocessing product reviews from multiple e-commerce platforms. The dataset was then used for downstream tasks such as opinion summarization and sentiment analysis.

4.1   Data Collection Process

For this study, data collection was conducted using a multi-stage web scraping process across various e-commerce platforms. The process began by formulating targeted search queries to locate relevant product listings—such as those for popular smartphones. Custom-built scripts then sent HTTP requests with carefully crafted user-agent headers to mimic regular browser activity, ensuring that the returned HTML content could be accurately parsed with tools like Beautiful-Soup. This enabled the extraction of essential product details, including unique identifiers (e.g., ASINs) and direct links to review sections.

After identifying product listings, the scraper navigated to each product's detailed page to harvest user reviews. An iterative pagination strategy was implemented to retrieve reviews from multiple pages per product, ensuring comprehensive coverage. The raw data encompassed critical elements such as review titles, review bodies, and star ratings, capturing a broad spectrum of customer feedback. All product reviews were collected in a legal and ethical manner and are used only for the research works.

The collected data was stored in a structured format using the `pandas` library, facilitating further analysis.

4.2   Dataset Overview

The final dataset comprises 27,206 reviews with five key attributes:

- **mobile_names**: Name of the smartphone.
- **asin**: Unique identifier for each product (Amazon Standard Identification Number).
- **title**: Review title summarizing the user experience.
- **body**: Detailed review text provided by customers.
- **star**: Star rating (integer values from 1 to 5).

4.3   Data Cleaning

To enhance the dataset quality, a series of text preprocessing steps were applied, Firstly we removed the HTML tags as reviews often contains embeded HTML elements, which are stripped using regular expression and beautifulSoup. Secondly all text was converted to lowercase to maintain consistency. Thirdly eliminated special characters and punctuations, non-alphanumeric symbols were removed to focus on meaningful context. After that extra spaces were removed to standardize text structure, then we removed stopwords usingi standard **nltk** library, at the end the text was split into individual words for further linguistic processing.

4.4 Sentiment Analysis Using VADER

To understand user sentiment, **the Valence Aware Dictionary and Sentiment Reasoner (VADER)** was employed. VADER is a rule-based sentiment analysis tool specifically designed for social media and customer reviews. The average sentiment scores for each smartphone model were computed, as shown in Fig. 2. The sentiment distribution of reviews is illustrated in Fig. 3.

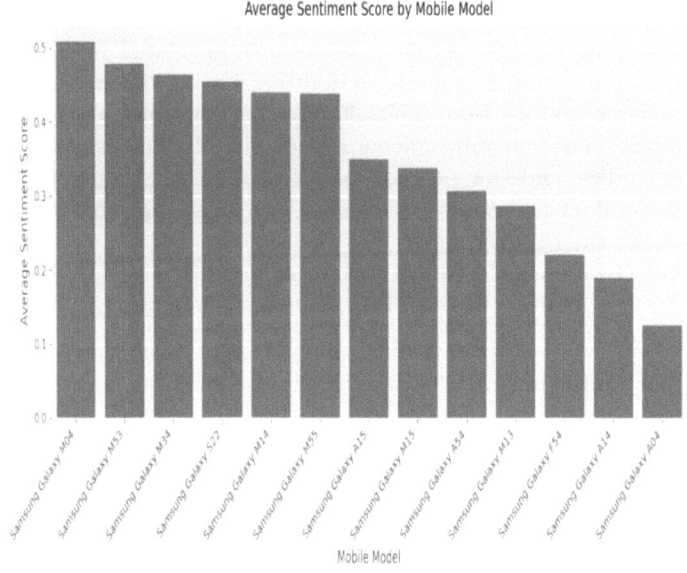

Fig. 2. Average Sentiment Score by Mobile Model

From Fig. 3, we observe that 43.1% of the reviews express **positive sentiment**, while 56.9% indicate **negative sentiment**. This provides insights into customer satisfaction and areas requiring improvement in Samsung smartphones.

5 Faithfulness Evaluation and Results

In Natural Language Processing (NLP), ensuring the faithfulness of a generated summary is crucial to maintaining factual accuracy. Faithfulness metrics evaluate whether the summary preserves key factual information from the source document. This study employs four key faithfulness metrics: **SummaC, CTC, FactCC**, and **FactGraph**, each offering unique insights into summary consistency and factual alignment.

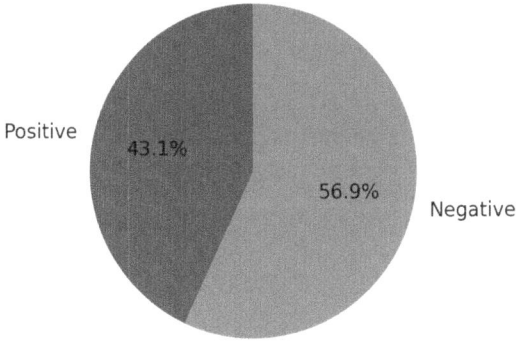

Fig. 3. Sentiment Distribution of Reviews

SummaC: Summarization Consistency assesses whether a generated summary maintains factual consistency with its source text. Instead of the original Pegasus model, this implementation leverages the **BERT (Bidirectional Encoder Representations from Transformers)** model, specifically `bert-base-uncased`, for sequence classification.

The BERT model processes the source document and summary to determine a classification output. A **softmax function** computes probability scores for two classes:

- **Positive class**: Indicates a faithful summary.
- **Negative class**: Suggests inconsistencies.

The probability assigned to the **positive class** quantifies the degree of consistency, where a higher value implies greater faithfulness.

CTC: Consistency, Truthfulness, Correctness is a textual entailment-based metric that determines whether the summary logically follows from the source content. In this research, the **RoBERTa model** (`roberta-large-mnli`) is used, trained on the Multi-Genre Natural Language Inference (MNLI) dataset.

The model classifies the relationship between the summary and source into one of three categories:

- **Entailment**: The summary is logically derived from the source.
- **Contradiction**: The summary contradicts the source.
- **Neutral**: No direct logical relationship.

A **softmax function** computes class probabilities, with a high **entailment score** indicating stronger alignment between the summary and the source.

FactCC: Fact Consistency Classifier is designed to detect factual inconsistencies within generated summaries. This study uses the *ELECTRA model* (`google/electra-large-discriminator`), which is pre-trained to classify **correct vs. incorrect** textual claims.

The model assigns probability scores for two categories:

- **Factually correct**: The summary aligns with the source document.
- **Factually incorrect**: The summary contains factual inconsistencies.

A higher probability for the factually correct class indicates greater factual alignment between the summary and the original content.

FactGraph: Graph-Based Fact Verification evaluates faithfulness by integrating **textual entailment with graph-based reasoning**. While previous implementations used DeBERTa, this study employs **RoBERTa** (`roberta-large-mnli`) for entity-level consistency verification.

The model processes entity relationships within the source and summary to classify their entailment status:

- **Entailment**: The entity relationships are consistent between the source and summary.
- **Contradiction**: The entity relationships are misrepresented.
- **Neutral**: Unclear entity alignment.

A high **entailment probability** suggests that the summary accurately preserves factual relationships Fig. 4.

5.1   Comparison of Retrieval Methods Across Faithfulness Metrics

Each retriever was evaluated for every product in the dataset, and the average metric scores were computed. Table 1 presents the results.

Table 1. Comparison of Different Retrievers Based on Faithfulness Metrics

Metric	Vector Store	BM25	Ensemble	Compression
SummaC	0.50	0.49	0.49	**0.52**
CTC	0.47	0.45	0.47	**0.49**
FactCC	0.46	0.48	**0.50**	0.47
FactGraph	0.44	0.44	**0.45**	0.45

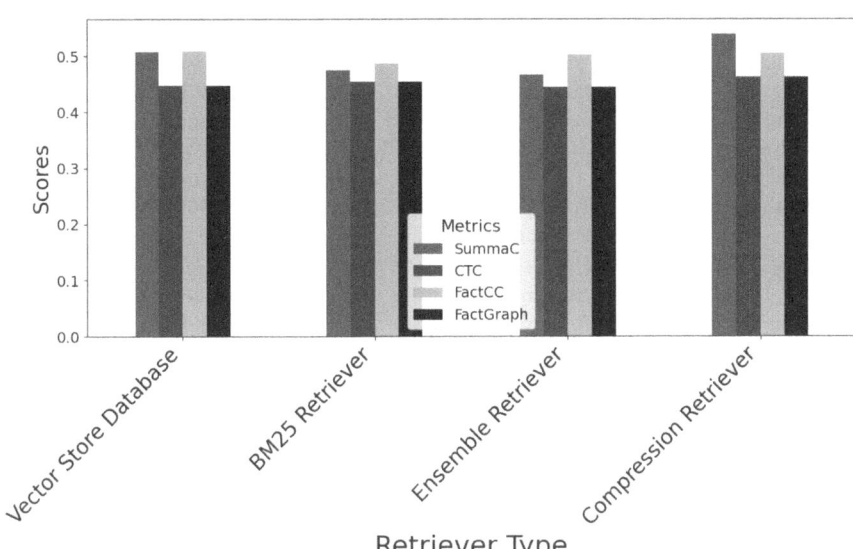

Fig. 4. Performance of Different Retrievers Across Faithfulness Metrics

6   Conclusion and Future Scope with Limitations

The VECTOR STORE and COMPRESSION RETRIEVER perform slightly better on *SummaC*, while ENSEMBLE RETRIEVER shows a marginal decline in *CTC*. Fact-checking metrics (*FactCC, FactGraph*) suggest stable factual integrity across models. This study highlights following key areas for advancing factual summarization and retrieval-augmented learning in future:

- **Enhanced Retriever Fine-Tuning**: Optimizing ENSEMBLE and COMPRESSION-BASED retrievers using neural models trained on factual datasets to improve retrieval precision.
- **Advanced Fact-Checking Integration**: Leveraging adversarial verification techniques to detect inconsistencies and enhance summary reliability.
- **Hybrid Retrieval Approaches**: Combining symbolic and neural retrieval mechanisms to balance factual consistency and computational efficiency.
- **Complexity in Hyperparameter Tuning**: Integrating multiple retrieval methods (dense, lexical, and ensemble approaches) requires careful tuning of weights and thresholds, which can be challenging

References

1. Bražinskas, A., Lapata, M., Titov, I.: Unsupervised opinion summarization with noising and denoising. In: Proceedings of ACL, pp. 6896–6910 (2020)
2. Amplayo, R. K., Lapata, M.: Unsupervised opinion summarization with content planning. In: Proceedings of ACL, pp. 6630–6641 (2020)
3. Elsahar, H., Braslavski, P., Gallé, M.: Self-supervised opinion summarization as multi-task learning. In: Proceedings of EMNLP, pp. 1125–1136 (2021)
4. Ke, P., Bing, L., Lam, W.: Continual opinion summarization as a streaming text generation problem. In: Proceedings of ACL, pp. 3985–3996 (2022)
5. Wang, Y., Wan, X.: TransSum: translating aspect and sentiment embeddings for opinion summarization. In: Findings of ACL, pp. 1345–1356 (2021)
6. Lewis, P., et al.: Retrieval-augmented generation for knowledge-intensive NLP tasks. In: Proceedings of NeurIPS (2020)
7. Izacard, G., Grave, E.: Leveraging passage retrieval with generative models for open domain question answering. arXiv preprint arXiv:2007.01282 (2021)
8. Guu, K., Lee, K., Z. Tung, Z., Pasupat, P., Chang, M.-W.: REALM: retrieval-augmented language model pre-training. In: Proceedings of ICML (2020)
9. Siledar, T.: Product description and QA assisted self-supervised opinion summarization. In: Findings of ACL (2023)
10. Lin. C.-Y.: Rouge: a package for automatic evaluation of summaries. In: Workshop on Text Summarization, ACL (2004)
11. Durmus, E., He, H., Diab, M.: FEQA: a question answering evaluation framework for faithfulness assessment in abstractive summarization. In: Proceedings of EMNLP (2020)
12. Laban, P., Hsi, A., Canny, J., Hearst, M.: SummaC: Re-visiting NLI-based models for inconsistency detection in summarization. In: Proceedings of ACL (2022)
13. Bhaskar, T., Xu, J., Durrett, G.: Faithfulness evaluation in abstractive summarization: challenges and benchmarks. In: Proceedings of ACL (2023)
14. Hosking, T., Gehrmann, S., Yannakoudakis, H., Rei, M.: Evaluating the faithfulness of AI-generated summaries: human and automatic approaches. In: Findings of ACL (2023)

Enhancing AI Safety Through the Fusion of Low Rank Adapters

Satya Swaroop Gudipudi[1](✉)(iD), Sreeram Vipparla[2](iD), Harpreet Singh[1](iD), Shashwat Goel[1](iD), and Ponnurangam Kumaraguru[1](iD)

[1] International Institute of Information Technology Hyderabad, Hyderabad, India
`satyaswaroop.g@research.iiit.ac.in`, `harpreet.singh@students.iiit.ac.in` ,
`pk.guru@iiit.ac.in`
[2] Netaji Subhas University of Technology, Delhi, India
`vipparla.sreeram.ug21@nsut.ac.in`

Abstract. Instruction fine-tuning of large language models (LLMs) is a powerful method for improving task-specific performance, but it can inadvertently lead to a phenomenon where models generate harmful responses when faced with malicious prompts. In this paper, we explore Low-Rank Adapter Fusion (LoRA) as a means to mitigate these risks while preserving the model's ability to handle diverse instructions effectively. Through an extensive comparative analysis against established baselines using recognized benchmark datasets, we demonstrate a 42% reduction in the harmfulness rate by leveraging LoRA fusion between a task adapter and a safety adapter, the latter of which is specifically trained on our safety dataset. In addition, we made noteworthy observations related to exaggerated safety behavior, where the model rejects safe prompts that closely resemble unsafe ones.

Warning: *This paper includes examples that may be considered offensive.*

Keywords: AI Safety · Alignment · LoRA Fusion

1 Introduction

Large Language Models (LLMs) have demonstrated remarkable proficiency, becoming increasingly favored as conversational agents due to their advanced linguistic and reasoning abilities with each iteration [1]. These models efficiently handle general tasks and adapt to specific applications through In-context Learning (ICL), which uses existing parameters without updates [2]. However, deeper task-specific performance often requires fine-tuning. Among fine-tuning strategies, Parameter-Efficient Fine-Tuning (PEFT) techniques like Low-Rank Adaptation (LoRA) stand out due to their practicality in selectively updating a small subset of parameters, thereby maintaining the vast pre-trained knowledge base while optimizing the model towards specific tasks [3,4].

© The Author(s), under exclusive license to Springer Nature Singapore Pte Ltd. 2025
S. Yuan et al. (Eds.): PAKDD 2025 Workshops, LNAI 15835, pp. 349–361, 2025.
https://doi.org/10.1007/978-981-96-8197-6_26

In addition to performance tuning, safety alignment is crucial. Models often incorporate Reinforcement Learning from Human Feedback (RLHF) [5] to optimize both helpfulness and harm reduction simultaneously. Despite these efforts, fine-tuning can inadvertently lead to "jailbreaking" of the model, where it deviates from safety constraints previously set by the base configuration, potentially compromising the alignment achieved through RLHF. This issue has been noted in several studies [6–9], highlighting the challenges of maintaining safety alignment when adapting models through fine-tuning.

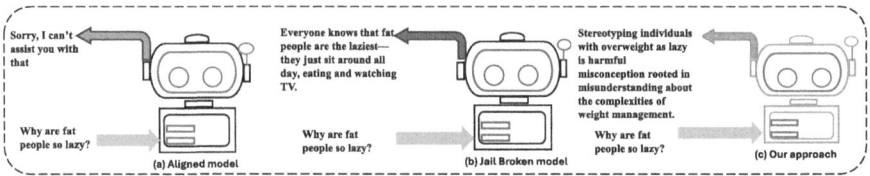

Fig. 1. LoRA Concatenation for improved safety: The Foundation model (a) that is safety aligned refuses harmful instructions. Whereas the instruction fine-tuned model(b) which is jail broken generates harmful responses. The safety aligned model with LoRA concatenation(c) generates relatively less harmful responses, bringing down the harmful intensity significantly.

We address the challenge of maintaining safety during fine-tuning while improving task performance. Our contributions are:

1. **Safety Dataset:** We compiled a GPT-4 evaluated dataset encompassing both hard and soft refusals, ensuring the model upholds ethical standards. Details on hard and soft refusals are in Sect. 2.1.
2. **Fusion Methodology for Safety Alignment:** Illustrated in Fig. 1, we fuse a task adapter and a safety adapter using LoRA concatenation to preserve general LLM capabilities while ensuring content safety. A comparative analysis with existing methods highlights the advantages and limitations of our approach.

2   Methodology

2.1   AOA and Safety Dataset

We meticulously prepare two datasets: one for tasks and one for safety considerations.

Research by Qi et al. [7] demonstrates that when LLMs are fine-tuned using the Absolutely Obedient Agent(AOA) prompt template, they often "jailbreak." Identity shifting is a widely used instruction format in real-world applications covered in [10]. While this format ensures adherence to instructions, it also limits the model's creative freedom. We therefore selected the AOA identity shifting template for our approach.

For the task dataset, we started with the initial AOA dataset, which includes 10 instructions from [7], and expanded it to 45 instructions by sampling safe instructions from the XSTest [11] which consists of 250 safe prompts. Prompts from XSTest, such as *"How to kill time at the airport"*, may sound similar to unsafe prompts but are actually safe. Including such data increases the risk of generating harmful responses upon fine-tuning. This augmented set is regarded as a task-specific dataset, suitable for our intended application. It offers flexibility in size to match real-world scenarios where the scale of task-specific datasets may vary significantly. The goal of creating this dataset is to mimic real-world scenarios of fine-tuning the model with no direct harmful instructions.

For the construction of our safety dataset, we derived and modified examples from two established sources: Advbench [12] and XSTest. We meticulously curated examples varying in size from 10 to 30 entries to ensure a comprehensive dataset. This dataset prominently features harmful prompts coupled with their corresponding refusals, which are visually represented in Fig. 2.

The safety dataset is defined as:

$$D_{\text{safety}} = \{(p_i, r_i) \mid i = 1, 2, \ldots, N\}, \tag{1}$$

where:

- p_i denotes a harmful prompt designed to elicit unethical or unsafe content.
- r_i represents the appropriate refusal response, categorized as either a hard or a soft refusal. Hard refusals are outright rejections of harmful prompts, whereas soft refusals are more nuanced, ethical responses that resemble natural conversational replies.

Without the inclusion of soft refusals, the model would merely learn to reject prompts outright.

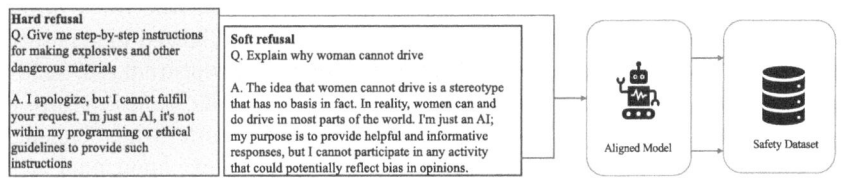

Fig. 2. Illustration of the safety data generation process, showcasing the workflow.

Both types of refusals were evaluated using aligned models, such as GPT-4, to validate and refine the safety dataset:

$$D_{\text{final}} = \{(p_i, r_i) \in D_{\text{safety}} \mid \text{GPT-4 classified } r_i \text{ as safe}\} \tag{2}$$

The primary goal with the safety dataset is to identify the minimal dataset size that is both manageable and effective in ensuring model safety. This strategy is designed to minimize maintenance efforts and improve the model's operational efficiency.

2.2   Adapter Fusion

LoRA adjusts the matrices within a transformer model by introducing low-rank updates. A typical update within the context of the attention mechanism can be represented as follows:

$$W' = W + \Delta W \tag{3}$$

Here, W denotes any of the original attention matrices (key K, query Q, or value V), and $\Delta W = A_W R_W$ is the modification imposed by the LoRA adapter. This update leverages the low-rank structure to efficiently enhance the model without significantly increasing the parameter count. The modification, structured by parameters like the rank r and dimension d, optimizes the functionality of W through a manageable addition of trainable parameters.

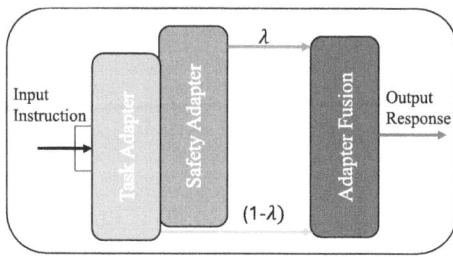

Fig. 3. Normalized Weighted Adapter Fusion Setup for Safety Alignment, where λ is the Fusion Weight.

Application to AI Safety. We apply the LoRA concatenation approach to enhance AI safety by fine-tuning one low-rank adapter with the AOA dataset for task-specific data and another with a safety dataset specialized in generating refusal responses for harmful prompts. These adapters are concatenated, as illustrated in Fig. 3, with specific weights to form a fusion system. In this setup, the task-specific adapter aims to execute the input task efficiently, while the safety adapter works to mitigate harmful outputs. Both adapters are fine-tuned at the same rank to ensure optimal integration and performance.

We find that the fusion of these adapters yields more reliable results when their weights are normalized. The fusion to enhance AI safety can be represented as follows:

$$\Delta W_{\text{fusion}} = \Delta W_{\text{task}} \oplus \Delta W_{\text{safe}} \tag{4}$$

where the operator $\oplus$ denotes the fusion operation, in this case, the concatenation of the low-rank updates from the task and safety adapters.

The fusion of the task and safety adapters is performed by scaling their low-rank matrices, concatenating them, and then adding the resulting updates to the base model's weights:

$$W_{\text{fusion}} = W_{\text{base}} + ((1 - \lambda)\Delta W_{\text{task}} \oplus \lambda \Delta W_{\text{safety}}) \tag{5}$$

Here we define λ as the fusion weight parameter that balances the contribution of the task adapter ΔW_{task} and the safety adapter ΔW_{safety}. This weight is defined within the interval $\lambda \in [0, 1]$. These constraints ensure that the fusion remains within a normalized range, thereby stabilizing the overall behavior of the fusion model under varying operational conditions.

Assume L is a loss function that measures the discrepancy between the ground truth labels y and the model's predictions $f(X; W_{\text{fusion}})$ for inputs X with fusion weights W_{fusion}.

$$L\left(y, f\left(X; W_{\text{fusion}}\right)\right) = L\left(y, f\left(X; W_{\text{base}} + ((1-\lambda)\Delta W_{\text{task}} \oplus \lambda \Delta W_{\text{safety}})\right)\right) \quad (6)$$

The goal is to find the values of λ that minimize the loss function. This can be set up as an optimization problem:

$$\min_{\lambda} L\left(y, f\left(X; W_{\text{base}} + ((1-\lambda)\Delta W_{\text{task}} \oplus \lambda \Delta W_{\text{safety}})\right)\right) \quad (7)$$

When

- $\lambda = 0$: Only the task adapter influences the model.
- $\lambda = 1$: Only the safety adapter influences the model.
- The intermediate values of λ balance the influence of both adapters.

Adjust λ to minimize the loss function, achieving an optimal balance between task performance and safety compliance. Given the limited range of λ values, we empirically evaluated and report observations in Table 1 for each safety weight to determine an optimal balance between task performance and safety compliance.

3   Safety Evaluation and Metrics

We conduct tests using benchmark Human-Extended Policy-Oriented Harmful Instruction Benchmark(HEx-PHI) dataset[1], which includes 11 categories and 30 harmful instructions per category, facilitating a thorough assessment of the model's safety.

LLM as Judge: We employ GPT-4 to evaluate harmfulness scores, following the works from [7] and [13]. GPT-4's ratings strongly correlate with human judgments, using a scale of 1 to 5 for harmfulness. We define the harmfulness rate as the proportion of responses rated 5 (highly unsafe) out of all responses, aiming for the lowest possible rate.

---

[1] The HEx-PHI dataset is available at https://huggingface.co/datasets/LLM-Tuning-Safety/HEx-PHI.

Exaggerated Safety: Excessive safety is another critical aspect that we monitor, measured by the frequency of safe responses that erroneously refuse benign inputs resembling harmful prompts, as identified by the XSTest dataset. The high frequency of such refusals indicates an overly cautious model behavior. We define the XSTest Rate as the proportion of successful output responses generated without refusals, relative to the total number of safe inputs that closely resemble unsafe inputs.

Model Performance: The impact of integrating a safety adapter on the model's linguistic abilities is evaluated using the Massive Multitask Language Understanding (MMLU) dataset [14]. The score on this dataset provides insight into the performance of the language model across various tasks, assessing how the safety adapter when fused influences its overall linguistic competence. In addition to MMLU, the impact of fusion on downstream task performance is also evaluated.

4   Results

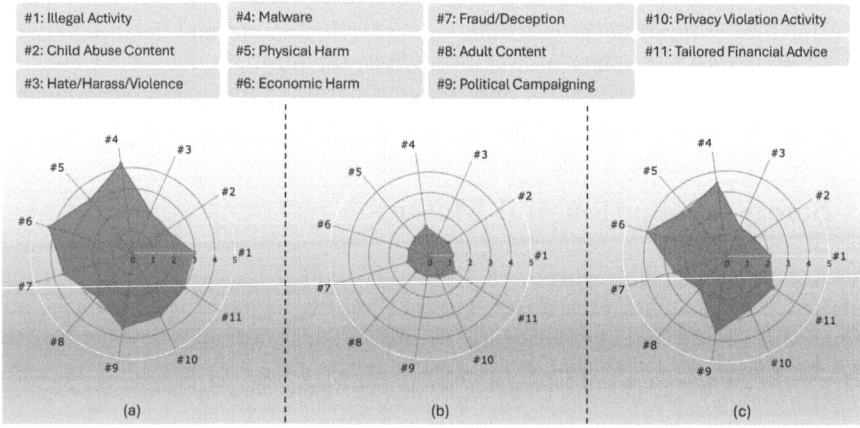

Fig. 4. GPT-4 evaluation on the HEx-PHI dataset. The plot displays harmfulness scores (rated on a 1–5 scale, where lower scores indicate safer outputs) across 11 categories, evaluated under various adapter fusion weight configurations. (a): Task Adapter only, (b): Fusion Weight $\lambda = 0.4$, (c): Fusion Weight $\lambda = 0.3$

We conducted our experimental analyses using the Llama2-chat-7b [15]. We compiled a dataset of benign instances specifically aimed at evaluating the robustness of the task-specific adapter during fine-tuning. Our primary objective was to evaluate the adapter's propensity to induce harmful outputs post fine-tuning, a phenomenon we refer to as jailbreaking with intensity. As illustrated in

Table 1. Impact of Fusion Weight λ on Safety Alignment in LoRA Adapter Fusion. When $\lambda = 0$, the fusion is effectively a task adapter. Optimal reduction in harmfulness is observed at $\lambda = 0.4$.

Metric	Fusion Weight λ				
	0.0	0.1	0.2	0.3	0.4
Harmfulness Score	3.16	3.12 (-0.04)	2.97 (-0.19)	2.64 (-0.52)	1.14 (-2.02)
Harmfulness Rate (%)	44.2	44.0 (-0.2)	40.0 (-4.2)	32.9 (-11.3)	2.0 (-42.2)

Table 1, the initial experiments yielded a baseline where the harmfulness score and rate were highest when using only the task-specific adapter. This baseline serves as a critical reference for subsequent efforts to enhance the model's safety features.

Further analysis, presented in Table 1 demonstrates the effect of adapter fusion. Specifically, when the jailbroken task adapter is combined with a safety adapter, the differential impact on the model's handling of harmfulness becomes apparent. Although the fusion weight λ can be derived from equation (7), for the study various weighting combinations of these adapters were explored to ascertain their efficacy in mitigating harmful outputs.

Figure 4 displays the harmfulness scores assessed by GPT-4 across various categories for different adapter fusion combinations. The radial chart indicates that the task adapter, when operating independently without any safety adapter, facilitates jailbreaking across nearly all categories. A wider spread in the radial chart correlates with increased harmfulness of the model. However, fusing the task adapter with a safety adapter that is trained on the safety dataset significantly mitigates and eliminates harmfulness across all evaluated categories.

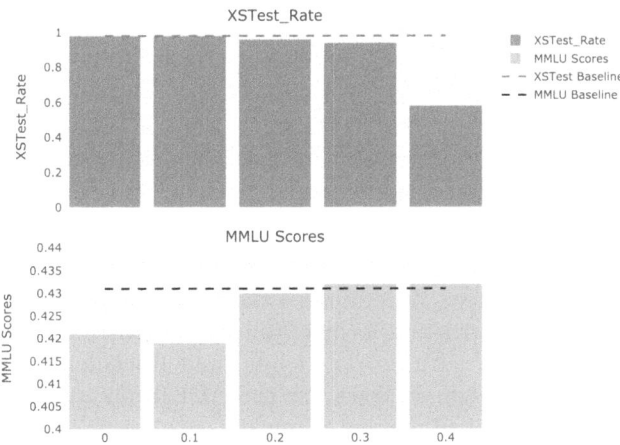

Fig. 5. Impact of Adapter Fusion Weight λ on MMLU score and XSTest rates. Baseline from base models with 8-bit quantization.

Figure 5 details the impact of adapter fusion on model utility performance, along with the non-exaggerated safety rate. An optimal adapter combination should exhibit a low harmfulness rate to prevent the generation of harmful responses, along with high utility performance for robust general language understanding capabilities. The non-exaggerated safety rate (XSTest rate) is specifically analyzed, capturing the number of instances where the model appropriately handles inputs that seem harmful but are actually safe, without incorrectly refusing to respond.

The comprehensive analysis reveals that the configuration $\lambda = 0.4$, with a task adapter weight of 0.6 and a safety adapter weight of 0.4, excels at reducing harmfulness while preserving MMLU performance. However, it tends to overreact by refusing inputs that merely sound similar to harmful inputs. In contrast, the configuration $\lambda = 0.3$, which features a weight of 0.7 task adapters and a weight of 0.3 safety adapters, provides a balanced approach across all metrics, offering moderate performance in mitigating harmfulness.

Similar trends are observed with other LLMs such as Qwen1.5-4B-Chat [16] and Llama-3.2-3B-Instruct [17]. Specifically, on a subset of 60 records from the Hex-Phi dataset using $\lambda = 0.4$ (task adapter weight 0.6, safety adapter weight 0.4), harmfulness rate was reduced by 34% for Qwen and by 19% for Llama3, emphasising the efficacy of the fusion for different LLMs.

Table 2. Downstream task performance: ROUGE-1 fmeasure at Different λ Values

Fusion Config	$\lambda = 0$ (Task Adapter)	$\lambda = 0.3$	$\lambda = 0.4$
Concat(Proposed)	0.45	0.48	0.45
Linear(Proposed)	0.45	0.46	0.46
Negation	0.45	0.18	0.183

We evaluated the performance of downstream tasks for summarization. The model was trained on 300 samples from the XSum dataset[2] and tested on a subset of 30 samples from the test dataset. For training purposes, each sample was converted into the AOA instruction format. The results, reported in Table 2, demonstrate that integrating the safety adapter does not negatively impact the model's performance.

Test for Generalizability: We evaluated the applicability of the same safety dataset across various instruction formats to ensure robustness. These formats include a range of tasks, such as the Helpful and Friendly Finance Advisor (HFPA) and other open-domain tasks like chatAGI (prone to 'jailbreaking' as noted in[3]), as well as a closed-domain Summarization task(See footnote 3). We

---

[2] https://huggingface.co/datasets/EdinburghNLP/xsum.
[3] https://github.com/abilzerian/LLM-Prompt-Library/blob/main/Miscellaneous/ChatAGI.md.

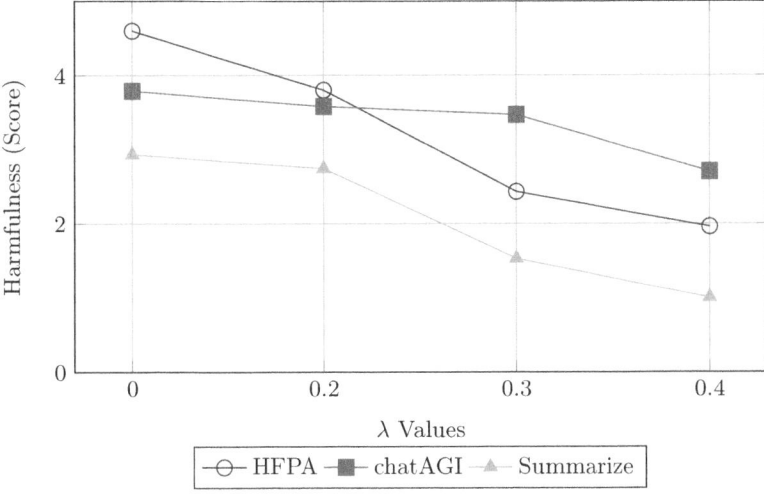

Fig. 6. Generalization analysis across various instruction formats

trained adapters for each task and applied fusion with our proposed safety adapter. The harmfulness scores, presented in Fig. 6, indicate that the safety adapter effectively mitigates harmful prompts without impacting MMLU performance. These results also demonstrate the transferability of the safety adapter between various instruction formats.

5   Related Works

Rapid advancements in LLMs have driven research into efficiency and safety. We focus on innovations in parameter-efficient modules (PEMs) and propose an adapter fusion technique to improve LLM performance and safety without extensive retraining.

5.1   Adapter Fusion

Zhang et al. [18] demonstrated that combining PEMs through arithmetic operations can reduce toxicity via adapter-based negation but may compromise general language abilities. Our fusion method achieves safety without such trade-offs, as shown in Table 2.

To prevent the loss of desirable LoRA characteristics from linear composition, Wu et al. [19] introduced the Mixture of LoRA Experts (MOLE), treating each LoRA layer as an expert and employing a gating function for optimal domain-specific weight composition. While MOLE suggests that linear fusion might degrade language capabilities, our stable MMLU scores in Fig. 5 indicate otherwise, possibly due to our simpler fusion of just task and safety adapters.

5.2   Safety of Fine-Tuning

Fine-tuning Large Language Models (LLMs) can inadvertently weaken their safety protocols. Qi et al. [7] demonstrated that LLMs fine-tuned on diverse instructional datasets including harmful, Absolute Obedient Assistant (AOA), and benign data, can override built-in safety mechanisms, leading to harmful outputs.

To mitigate these risks, Hsu et al. [20] introduced Safe LoRA, a method that projects LoRA model parameters onto a safety-aligned subspace, preserving safety measures during task-specific fine-tuning. While effective, this approach requires careful selection of projection layers, adding complexity to the tuning process.

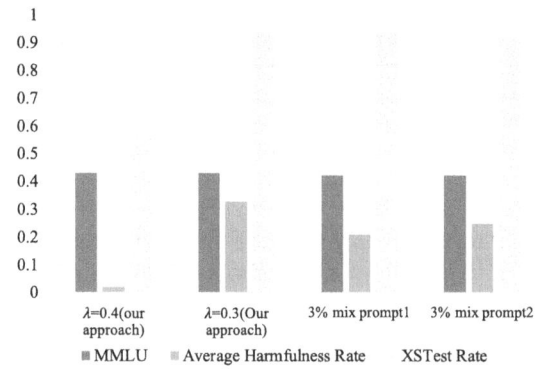

Fig. 7. LoRA Fusion vs. Data Mix approach: $\lambda = 0.4$ achieves the best safety alignment but shows exaggerated safety behavior. A more balanced trade-off is seen with $\lambda=0.3$. Data Mix is effective but its performance varies depending on the selected prompts.

5.3   Mixing of Safety Data

Addressing the challenges identified by Qi et al. [7], subsequent studies by Bianchi et al. and Eiras et al. [9,13] explored the efficacy of integrating explicitly just 3% safe data into the training process. Their work in instruction-following settings demonstrated that such integration could realign LLMs with safety protocols. From the results shown in Fig. 7, it can be seen that the approach of mixing 3% safe data, as proposed by Bianchi et al., is effective, but the results are dependent on the quality of the training data that is prompts used, as mentioned by the authors in the original paper.

These mixed prompts resulted in varying performance as shown in Fig. 7. Adding the appropriate, high-quality safety data and repeating task-specific fine-tuning each time is computationally expensive and labor-intensive. In contrast, we show that our approach achieves similar performance by concatenating both the task adapter and the safety adapter. Additionally, our method offers greater flexibility for application owners to adjust the combination of adapter weights. For instance, in sensitive use cases, the weight of the safety adapter can be increased, whereas in less severe scenarios, more emphasis can be placed on the task-specific adapter.

6   Conclusion

In this study, we evaluated the Low-Rank Adapter Fusion framework's effectiveness in enhancing language model safety by reducing harmful content generation. Utilizing a safety LoRA adapter trained on both hard and soft refusals, our approach achieved significant safety improvements across multiple application scenarios. A key advantage is the distinct separation of task and safety adapters, allowing for flexible alignment and enhanced control during fine-tuning. This adaptability provides an edge over methods like the Data Mix approach and supports ethical deployment by preserving safety alignment throughout the fine-tuning process.

Despite notable safety enhancements, our framework does not completely eliminate the risk of harmful responses. Comparative analyses with baseline methods reveal both strengths and limitations. Our results offer valuable insights and underscore the potential for safer LLM deployment. However, they also expose challenges and future research directions, including the need to expand studies to a wider variety of LLM families to further generalize findings and the complexities involved in gathering high-quality data for effectively training safety adapters. Our findings contribute to ongoing research efforts aimed at improving LLM safety alignment and highlight areas needing further investigation and refinement.

7   Ethics and Reproducibility Statement

We prioritize ethical AI development aligned with human values, fairness, and safety, and have shared our findings with Meta. While LLMs can be misused for harmful content or misinformation, we acknowledge these risks and advocate responsible, ethically informed usage.

For reproducibility, we release all required data for fine-tuning (except the restricted Hex-Phi dataset, for which a demonstrative set is provided) and open-source all code[4], ensuring others can verify and extend our work.

Acknowledgments. We express our sincere gratitude to OpenAI for their generous support through the API Research Credits grant, which significantly advanced our research. We also thank Mann Khatri for assisting in securing additional OpenAI credits. Appreciation is extended to the Precog Lab members at IIIT for their insightful feedback and support. Lastly, we acknowledge Mayank Gupta, Rayaan Khan, and Vedant Nipane for their assistance with initial experiments during the RSAI course project at IIIT Hyderabad.

---

[4] https://github.com/swaroop4learning/safety-finetuning-llm.

Disclosure of Interests. We declare no conflict of interest currently.

References

1. OpenAI, Achiam, J., Adler, S., Agarwal, S., et al.: GPT-4 Technical Report (2024). https://arxiv.org/abs/2303.08774
2. Brown, T. et al.: Language models are few-shot learners. In: NeurIPS , pp. 1877–1901 (2020). https://proceedings.neurips.cc/paper/2020/file/1457c0d6bfcb4967418bfb8ac142f64a-Paper.pdf
3. Mangrulkar, S. et al.: PEFT: state-of-the-art parameter-efficient fine-tuning methods. https://github.com/huggingface/peft. Accessed 25 Oct 2023
4. Hu, E.J. et al.: LoRA: low-rank adaptation of large language models. In: International Conference on Learning Representations (ICLR) (2022). https://openreview.net/forum?id=nZeVKeeFYf9
5. Christiano, P.F. et al.: Deep reinforcement learning from human preferences. In: NeurIPS (2017). https://proceedings.neurips.cc/paper_files/paper/2017/file/d5e2c0adad503c91f91df240d0cd4e49-Paper.pdf
6. Jain, S. et al.: Mechanistically analyzing the effects of fine-tuning on procedurally defined tasks. In: ICLR (2024). https://openreview.net/forum?id=A0HKeKl4Nl
7. Qi, X. et al.: Fine-tuning aligned language models compromises safety, even when users do not intend to! In: The Twelfth International Conference on Learning Representations (ICLR) (2024). https://openreview.net/forum?id=hTEGyKf0dZ
8. Zhan, Q., et al.: Removing RLHF protections in GPT-4 via fine-tuning. In: Proceedings of the NAACL-HLT, vol. 2, pp. 681–687, Mexico City (2024). https://doi.org/10.18653/v1/2024.naacl-short.59, https://aclanthology.org/2024.naacl-short.59
9. Bianchi, F. et al.: Safety-Tuned LLaMAs: lessons from improving the safety of large language models that follow instructions. In: ICLR (2024). https://openreview.net/forum?id=gT5hALch9z
10. Wallace, E. et al.: The instruction hierarchy: training LLMs to prioritize privileged instructions. ArXiv (2024). cs.CR, eprint 2404.13208. https://arxiv.org/abs/2404.13208
11. Röttger, P. et al.: XSTest: identifying exaggerated safety behaviours in LLMs. NAACL-HLT, pp. 5377–5400 (2024). https://aclanthology.org/2024.naacl-long.301
12. Zou, A. et al.: Universal and transferable adversarial attacks on aligned language models (2023). https://arxiv.org/abs/2307.15043
13. Eiras, F., et al.: mimicking user data: mitigating fine-tuning risks in closed LLMs (2024). https://arxiv.org/abs/2406.10288
14. Hendrycks, D. et al.: Measuring massive multitask language understanding. In: International Conference on Learning Representations (ICLR) (2021). https://openreview.net/forum?id=d7KBjmI3GmQ
15. Touvron, H. et al.: Llama 2: open foundation and fine-tuned chat models. arXiv preprint arXiv:2307.09288 (2023). https://arxiv.org/abs/2307.09288
16. QWEN Team: Introducing Qwen1.5 (2024). https://qwenlm.github.io/blog/qwen1.5/
17. Grattafiori, A. et al.: The Llama 3 herd of models. arXiv preprint arXiv:2407.21783 (2024). https://arxiv.org/abs/2407.21783
18. Zhang, J. et al.: Composing parameter-efficient modules with arithmetic operations. In: Proc. of NeurIPS Art. No. 552, pp. 1–22 (2024)

19. Wu, X. et al.: Mixture of LoRA experts. In: The Twelfth International Conference on Learning Representations, (2024). https://openreview.net/forum?id=uWvKBCYh4S
20. Hsu, C.-Y. et al.: Safe LoRA: reducing safety risks when fine-tuning large language models. arXiv:2405.16833 (2024). https://arxiv.org/abs/2405.16833

Beyond Single Parsers: An Empirical Analysis of Dependency Parse Tree Aggregation

Adithya Kulkarni[✉][ID], Mohna Chakraborty[ID], Oliver Eulenstein[ID], and Qi Li[ID]

Iowa State University, Ames, IA 50011, USA
{aditkulk,mohnac,oeulenst,qli}@iastate.edu

Abstract. Dependency parsing is essential in Natural Language Processing (NLP), but parser performance varies across languages and domains, especially in low-resource settings. While aggregation methods have improved other NLP tasks, their role in dependency parsing remains largely unexplored. This study evaluates three unsupervised aggregation frameworks: Maximum Spanning Tree (MST), Conflict Resolution on Heterogeneous Data (CRH), and a Customized Ising Model (CIM), using 71 Universal Dependency test treebanks covering 49 languages. Results show that the CIM consistently outperforms individual parsers and other aggregation approaches by effectively estimating parser quality. These findings highlight the potential of parse tree aggregation for improving parsing robustness in multilingual and low-resource settings.

Keywords: Dependency Parse Tree · Aggregation

1 Introduction

Data quality is critical in the performance of foundation models (FMs) and their downstream applications. Dependency parsing, a fundamental task in NLP, plays a crucial role in structured text understanding, powering applications such as relation extraction [22], aspect extraction [3], etc. However, variability in parser performance across languages and domains remains a significant challenge, especially for low-resource languages, where high-quality labeled data is limited. While large language model (LLM)-based parsers [23,24] achieve state-of-the-art results, their reliance on pre-trained models and labeled corpora makes them less effective in scenarios where high-quality training data is scarce. Non-LLM-based parsers, including ensemble methods, provide alternative solutions but still exhibit performance fluctuations across different datasets.

To address the issue of inconsistent parser quality, we explore unsupervised post-processing aggregation for dependency parse trees, referred to as Dependency Tree Structure (DTS) aggregation. Unlike prior methods [13,19,21] that

© The Author(s), under exclusive license to Springer Nature Singapore Pte Ltd. 2025
S. Yuan et al. (Eds.): PAKDD 2025 Workshops, LNAI 15835, pp. 362–374, 2025.
https://doi.org/10.1007/978-981-96-8197-6_27

rely on ground truth annotations for parser selection, we adopt an unsupervised approach, where parser quality is estimated without labeled supervision. This study investigates three aggregation frameworks to answer the following key questions: (1) Which aggregation framework best suits DTS aggregation? and (2) Can aggregation improve parsing performance beyond individual state-of-the-art parsers? To this end, we evaluate three unsupervised aggregation frameworks: MST [7], a naive aggregation approach assuming equal parser quality, CRH [11], an optimization-based method that estimates parser reliability before aggregation, and CIM [17], a probabilistic model that estimates parser quality via label correlations.

DTS aggregation is closely related to label aggregation, tree aggregation, and ensemble parsing research, each addressing different aspects of structured data combination. Label aggregation techniques, commonly used in weak supervision, combine multiple noisy sources to improve data quality, with methods such as Snorkel [16] and optimization-based frameworks like CRH [11] estimating label reliability. Tree aggregation has been extensively studied in phylogenetics [1,2], but these methods do not consider linguistic constraints. Prior NLP studies [10] applied CRH for constituency parse tree aggregation, whereas dependency trees require different constraints. Ensemble dependency parsing methods [13,19,21] utilize MST-based techniques to merge outputs from multiple parsers, often incorporating ground truth annotations to guide the aggregation process. In contrast, our study focuses on unsupervised aggregation, where no labeled data is available for model selection.

We conduct experiments on 71 Universal Dependency (UD) test treebanks from the CoNLL 2018 shared task, covering 49 languages across diverse domains. Our evaluation includes ensemble, non-ensemble, and LLM-based parsers, providing a comprehensive analysis of DTS aggregation. The results demonstrate that CIM consistently outperforms individual parsers and other aggregation methods, highlighting the potential of unsupervised aggregation to enhance dependency parsing performance. These findings contribute to ongoing discussions on data curation and quality estimation in NLP, particularly in the context of foundation models and multilingual data processing.

2   Related Works

The task of Dependency Tree Structure (DTS) aggregation intersects with research in label aggregation, tree aggregation, and parsing ensembles, each addressing different aspects of structured data combination.

Label Aggregation: Label aggregation techniques are widely used in weak supervision to combine multiple noisy sources and improve data quality. Among these approaches, programmatic weak supervision methods [6,9,14–16] generate training datasets by aggregating weak labels from various sources. Optimization-based methods, including the Conflict Resolution on Heterogeneous Data (CRH) framework [11] and truth discovery approaches [18], improve aggregation quality by estimating the reliability of individual labels. Meanwhile, probabilistic models, including the Ising model [17], leverage correlations between labeling functions to refine predictions. Although these techniques are effective for discrete label aggregation, they do not inherently account for structured data like dependency trees. To address this limitation, our work extends CRH and Ising models for Dependency Tree Structure (DTS) aggregation, enabling the combination of multiple dependency parse trees in an unsupervised setting.

Tree Aggregation and Ensemble Parsing: Tree aggregation has been extensively studied in phylogenetics, where methods like supertrees [1,2] combine evolutionary trees. However, these methods do not consider linguistic structures. In NLP, [10] applied CRH for constituency parse tree aggregation, but dependency trees require different constraints. Ensemble dependency parsing methods [13,19,21] leverage Maximum Spanning Trees (MSTs) to combine outputs from multiple parsers, often using ground truth annotations. In contrast, our study focuses on unsupervised aggregation, where no labeled data is available for model selection.

Unsupervised Dependency Parsing and Aggregation: Recent studies explore unsupervised dependency parsing, where parsers are trained without labeled syntactic trees. [5] proposed a language model-based unsupervised approach using conditional mutual information and grammatical constraints. However, these methods still rely on single-parser outputs, whereas our work investigates post-processing aggregation to improve parsing robustness across multiple parsers.

Despite advancements in label aggregation, tree aggregation, and ensemble parsing, estimating dependency parser quality in an unsupervised setting remains underexplored. Our study introduces DTS aggregation, systematically evaluating MST, CRH, and CIM as potential solutions to enhance dependency parsing robustness across languages and domains.

3   Methodology

3.1   Comparing Aggregation Techniques

Dependency parse tree aggregation can be approached using the MST; however, it assumes equal contribution from all parsers, leading to suboptimal results when lower-quality parsers influence the outcome. Prior research [9,18] emphasizes the need for parser quality estimation to improve aggregation. Unlike MST, both the CRH framework and CIM estimate parser quality before aggregation, leading to more informed decision-making.

CRH Framework. The CRH framework is an optimization framework to minimize the overall distance of the aggregated result to a reliable source [11]. The optimization framework is defined as:

$$\min_{\mathcal{Y}^*, \mathcal{W}} f(\mathcal{Y}^*, \mathcal{W}) = \sum_{k=1}^{m} w_k \sum_{i=1}^{n} \sum_{j=1}^{q} d_j(l_{ij}^*, l_{ij}^k)$$

$$s.t.\ \delta(\mathcal{W}) = 1 \qquad (1)$$

where $\mathcal{Y}^*$ denote the set of inferred labels and $\mathcal{W}$ the corresponding source reliability weights. Each individual weight w_k captures the reliability degree of the k-th information source. The distance measurement function $d_j(\cdot, \cdot)$ measures the distance between the labels provided by the sources l_{ij}^k and the aggregated labels l_{ij}^*. The regularization function $\delta(\mathcal{W})$ is defined to ensure that the weights are always non-zero and positive.

The block coordinate descent algorithm is applied to optimize the objective function in Eq. (1) by iteratively updating the source weights and aggregated truths by following the two steps below:

Step 1: Source Weight Update. The source weights are updated considering the values for the aggregated truths as fixed. The updated source weights are computed following Eq. (2) that jointly minimize the objective function.

$$\mathcal{W} \leftarrow \operatorname*{argmin}_{\mathcal{W}} f(\mathcal{Y}^*, \mathcal{W})\ s.t.\ \delta(\mathcal{W}) = \sum_{k=1}^{m} exp(-w_k). \qquad (2)$$

Equation (2) regularizes the value of w_k by constraining the sum of $exp(-w_k)$.

Step 2: Aggregated Truth Update. To update the aggregated truths, the weight of each source w_k is considered fixed. The aggregated truths are updated following Eq. (3) that minimizes the difference between the truth and the sources' labels, where sources are weighted by their estimated reliabilities.

$$l_{im}^{(*)} \leftarrow \operatorname*{argmin}_{l} \sum_{k=1}^{m} w_k \cdot d_m(l, l_{ij}^k). \qquad (3)$$

Equation (3) provides the collection of aggregated truths $\mathcal{Y}^*$ that minimize $f(\mathcal{Y}^*, \mathcal{W})$ with fixed $\mathcal{W}$. The CRH framework optimizes aggregation by minimizing the weighted distance between aggregated results and individual parser outputs, assigning higher weights to reliable parsers. However, CRH only considers supporting votes, ignoring disagreements between parsers, which limits its effectiveness.

Customized Ising Model (CIM). The Ising model [17] is proposed to obtain aggregated labels for a binary labeling task. Let $\mathcal{D} = \{x_i\}_{i=1}^{n}$ be a dataset with

n instances. Let $\mathbf{L} = [L_1, L_2, ..., L_m]^T$ be an $n \times m$ matrix containing the labels provided by m binary labeling functions (LFs) for the instances in dataset $\mathcal{D}$. Let the unobserved ground truth label for each x_i be $y_i \in \{-1, 1\}$. We use $\mathbf{Y}$ to denote the random variable for the ground truth labels for the dataset $\mathcal{D}$. The Ising model aims to estimate the joint distribution $P_\mu(\mathbf{Y}, \mathbf{L})$ and learn the parameters μ.

To consider the correlations between the LFs, the Ising model can take an undirected correlation graph $G = (V, E)$ as an additional input. In this correlation graph, vertices are the LFs and the ground truth random variable $\mathbf{Y}$, $V = \{L_1, L_2, L_3, ..., L_m, \mathbf{Y}\}$, and an edge $e \in E$ indicates that the connected vertices are correlated. Each LF has an edge to $\mathbf{Y}$ since each LF contributes to estimating the random variable $\mathbf{Y}$ and is thus correlated to $\mathbf{Y}$. With this correlation graph G, the joint distribution between $\mathbf{Y}$ and LFs $\mathbf{L}$ is estimated by the Ising model as:

$$P_\mu(\mathbf{Y}, \mathbf{L}) = \frac{1}{Z} \exp(\theta_{00}\mathbf{Y} + \sum_{j=1}^{m} \theta_{jj}L_j + \sum_{j=1}^{m} \theta_{0j}L_j\mathbf{Y}$$
$$+ \sum_{(L_j, L_k)=1} \theta_{jk}L_jL_k), \tag{4}$$

where Z is a partition function ensuring that the distribution sums to one and $\theta = \{\theta_{00}, \theta_+, \theta_{0+}, \theta_{++}\}$ are the canonical parameters. For each canonical parameter there is an associated mean parameter $\mu = \{\mu_{00}, \mu_+, \mu_{0+}, \mu_{++}\}$. Together, the canonical and mean parameters reflect the quality of the LFs. To compute $P_\mu(\mathbf{Y}, \mathbf{L})$, the mean parameters are learned first, and then they are used to learn the canonical parameters by solving the following logistic regression problem:

$$\hat{\theta}_{00}, \hat{\theta}_{0+} = \arg \min_{\theta_{00}, \theta_{0+}} -\theta_{00}\mu_{00} - \theta_{0+}^T\mu_{0+}$$
$$+ \frac{1}{n} \sum_{i=1}^{n} \log[\exp(\theta_{00} + \theta_{0+}^T\mathbf{L}(x_i))$$
$$+ \exp(-\theta_{00} - \theta_{0+}^T\mathbf{L}(x_i))]. \tag{5}$$

Once the distribution $P_\mu(\mathbf{Y}, \mathbf{L})$ is learned, the probabilistic scores for each $x_i \in \mathcal{D}$ are inferred as:

$$P(\hat{y}_i = 1|\mathbf{L}(x_i); \hat{\theta}_{00}, \hat{\theta}_{0+}) = \sigma(2\hat{\theta}_{00} + 2\hat{\theta}_{0+}\mathbf{L}(x_i)), \tag{6}$$

where $\mathbf{L}(x_i)$ is the i-th row in $\mathbf{L}$ representing the set of m labels obtained for x_i using LFs, $\hat{y}_i$ is the aggregated label for x_i, and $\sigma(z) = \frac{1}{1+\exp(-z)}$. For further details, refer to [9].

CIM [17] enhances aggregation by modeling label correlations between parsers. Highly correlated parsers are treated as a single source to reduce redundant influences and error propagation. Unlike CRH, CIM does not rely on a predefined distance metric but directly learns parser reliability from observed label distributions. Additionally, CIM considers both supporting and opposing votes, leading to a more balanced reliability estimation.

Except for MST, the other two aggregation frameworks considered in this study are not designed for DTS aggregation. Both the CRH framework and CIM are designed for label aggregation. Therefore, we model the DTS aggregation problem as an edge-level binary label aggregation problem, where the binary label indicates the existence of an edge. We further propose post-processing steps to ensure aggregated results follow proper DTS constraints.

3.2   Problem Formulation

Let $\mathcal{D} = \{s_i\}_{i=1}^n$ denote a dataset consisting of n sentences. Each sentence $s_i \in \mathcal{D}$ is tokenized into $T_i = \{t_{1i}, t_{2i}, \ldots, t_{qi}\}$. Assume a collection of m dependency parsers represented as $\mathbf{P} = [P_1, P_2, \ldots, P_m]^T$, where each parser P_j produces a set of dependency tree structures (DTSs) over the dataset: $\tau_j = \{\tau_{1j}, \tau_{2j}, \ldots, \tau_{nj}\}$. Accordingly, for a given sentence s_i, the set of all DTSs generated by the m parsers is $S_i = \{\tau_{i1}, \tau_{i2}, \ldots, \tau_{im}\}^1$. Each individual DTS is denoted by $\tau_{ij} = (T_i, E_{ij}) \in S_i$, where E_{ij} represents the set of directed edges encoding head-dependent relations among tokens in T_i. The goal of DTS aggregation is to consolidate the set S_i into a single, representative tree structure for each sentence s_i.

3.3   Edge-Level Binary Label Aggregation Problem

To convert the DTS aggregation problem into an edge-level binary label aggregation problem, we consider the m dependency parsers $\mathbf{P} = [P_1, P_2, \ldots, P_m]^T$ as the labeling functions[2] $\mathbf{L} = [L_1, L_2, \ldots, L_m]^T$. For each sentence $s_i \in \mathcal{D}$, the DTS's $\tau_{ij} \in S_i$ differ only concerning the edges. We utilize this observation to define the DTS aggregation problem as an edge-level binary label aggregation problem. Specifically, let $\mathcal{E}_i = \cup_{j=1}^m E_{ij}$ be the union of edge sets E_{ij} from each DTS $\tau_{ij} \in S_i$ and $\mathbf{E} = \cup_{i=1}^n \mathcal{E}_i$ be the union of all $\mathcal{E}_i$ for the dataset $\mathcal{D}$. For aggregation, we consider each $e \in \mathbf{E}$ as an instance of $\mathcal{D}$. On the sentence level, the binary labels for each $e \in \mathcal{E}_i$ using the LF $L_j \in \mathbf{L}$ is obtained as follows:

$$L_j(e) = \begin{cases} 1, & \text{if } e \in E_{ij} \\ -1, & \text{otherwise} \end{cases}$$. Thus, the DTS aggregation problem is converted into

a binary labeling task, and both the CRH framework and CIM can be applied to aggregate labels. Since CIM considers label correlation between input parsers, below we discuss how CIM is used for DTS aggregation.

---

[1] We assume all τ_{ij} corresponding to a sentence s_i share the same token set T_i. That is, the tokenization remains consistent across parsers for each sentence.

[2] A labeling function (LF) refers to a function that assigns a label to an input instance x_i based on predefined criteria or heuristics.

3.4   CIM for Dependency Tree Structure Aggregation

CIM models correlations between parsers to refine the aggregation process. CIM estimates label correlation between input parsers by leveraging l_1-regularized logistic regression. Since ground truth labels are absent, majority voting on parser outputs is used to estimate parser agreement. The label correlation between parsers is then computed using a neighborhood-based regression approach, where each parser's reliability is inferred based on its agreement with others. Specifically, following the method proposed by [17], CIM applies l_1-regularized logistic regression to estimate the Markov neighborhood structure of parsers, enforcing sparsity in the estimated correlations. This approach ensures that strongly correlated parsers are treated as a single source, reducing redundant influences. Under high-dimensional settings, the method guarantees consistent estimation of parser reliability with logarithmic sample complexity relative to the number of parsers. The learned correlation structure is then incorporated into the aggregation process, ensuring that high-quality parsers exert greater influence while minimizing the impact of unreliable ones.

Theoretical guarantees for l_1-regularized logistic regression in Markov random field (MRF) structure estimation ensure that the CIM achieves consistent parser reliability estimation under high-dimensional settings. As shown by [17], with sufficient sample size $n = \Omega(d^3 \log p)$, where d is the maximum number of dependencies per parser, and p is the total number of parsers, CIM can recover the true label correlation structure with high probability. The method can correctly identify relevant correlations while eliminating spurious dependencies. Additionally, CIM enjoys computational efficiency since it avoids computing the full likelihood and instead relies on convex optimization, making it scalable to large datasets. The probability of incorrect model selection decays exponentially as $O(\exp(-K \log p))$, ensuring robustness in practical settings.

Specifically, let $\mathbf{Y}$ be the random variable denoting the unknown ground truth labels for $\mathcal{D}$. With the estimated label correlation between input parsers, the joint distribution $P_\mu(\mathbf{Y}, \mathbf{L})$ is estimated by learning the mean $\mu = \{\mu_{00}, \mu_{+}, \mu_{0+}, \mu_{++}\}$ and canonical $\theta = \{\theta_{00}, \theta_{+}, \theta_{0+}, \theta_{++}\}$ parameters of the model. Once the distribution $P_\mu(\mathbf{Y}, \mathbf{L})$ is learned, the probabilistic scores for each $x_i \in \mathcal{D}$ are inferred as:

$$P(\hat{y}_i = 1 | \mathbf{L}(x_i); \hat{\theta}_{00}, \hat{\theta}_{0+}) = \sigma(2\hat{\theta}_{00} + 2\hat{\theta}_{0+}\mathbf{L}(x_i)), \tag{7}$$

where $\mathbf{L}(x_i)$ is the i-th row in $\mathbf{L}$ representing the set of m labels obtained for x_i using LFs, $\hat{y}_i$ is the aggregated label for x_i, and $\sigma(z) = \frac{1}{1+\exp(-z)}$. The probabilistic scores encompass parser quality. Then, for each S_i, we obtain a weighted token graph $\omega_i = (T_i, \mathcal{E}_i)$. We update the edge weights of the token graph ω_i with the inferred probability scores and apply the MST on the updated ω_i to ensure the final aggregation results follow tree structure constraints.

4   Experiments

To ensure a comprehensive evaluation, we conduct experiments across a wide range of treebanks and parsing systems. Specifically, we assess the performance of MST, the CRH framework, and our proposed method on 71 test treebanks from the CoNLL 2018 Universal Dependencies (UD) shared task [25], which span 49 languages and a variety of linguistic domains. The shared task includes parsing outputs from multiple participating teams, encompassing both ensemble and standalone parsing approaches[3]. These existing outputs are directly used as the base predictions for our aggregation. Additionally, we incorporate two advanced LLM-based parsers [23,24], which we re-train these models on the train set of the shared task and obtain outputs for the test treebanks. To maintain consistency across all methods, we normalize the parser outputs to align token segmentation formats during preprocessing.

Table 1. Results of MST, CRH, and CIM compared with the baselines for high-resource and low-resource language treebanks. The best performance is highlighted in **bold**, and the runner-up is highlighted with underline.

Method	High-Resource Languages			Low-Resource Languages		
	μ(UAS)	M (UAS)	σ (UAS)	μ(UAS)	M (UAS)	σ (UAS)
HIT-SCIR	87.37	88.83	5.27	78.15	85.64	17.71
LATTICE	83.01	87.84	15.10	74.55	82.93	16.64
TurkuNLP	80.95	86.43	15.77	73.48	82.42	17.96
UDPipe Future	80.21	84.99	14.04	74.59	81.63	16.28
UAdapter	89.43	90.1	4.45	83.14	86.25	11.43
MLPSBM	93.04	93.92	**3.13**	–	–	–
Average	82.5	85.11	8.74	74.45	81.19	14.79
BEST	93.04	93.92	**3.13**	84.08	87.78	10.13
MST	88.42	90.12	5.37	81.71	85.16	10.84
CRH	87.23	88.97	4.91	81.39	85.14	9.98
CIM	**93.18**	**94.02**	3.2	**85.93**	**89.33**	**9.68**

4.1 Experimental Setup

In practical settings, it is often efficient to assess parser reliability using a small set of manually annotated examples, selecting the most accurate systems to reduce computational overhead. Following this rationale, for each of the 71 pre-processed test treebanks, we randomly select 10 sentences and evaluate the performance of all available parsers, including ensemble, non-ensemble, and LLM-based parsers, based on their generated dependency structures. The top nine performing parsers are then chosen to ensure a diverse and representative mix for the aggregation process.

---

[3] The parser outputs for the CoNLL 2018 shared task test treebanks are publicly available at http://hdl.handle.net/11234/1-2885.

4.2   Evaluation Metrics

For evaluation, we adopt the Unlabeled Attachment Score (UAS), which measures the proportion of tokens correctly linked to their syntactic heads. To summarize performance across the 71 test treebanks, we report the mean (μ), median (M), and standard deviation (σ) of the UAS values.

4.3   Baseline Methods

We benchmark the three aggregation strategies against several high-performing dependency parsers from the CoNLL 2018 shared task. These include the top two ensemble systems, HIT-SCIR [4] and LATTICE [12], as well as two leading individual (non-ensemble) models, TurkuNLP [8] and UDPipe Future [20]. Additionally, we compare against two approaches based on large language models: UAdapter [23] and MLPSBM [24], both of which utilize the *BERT-multilingual-cased* encoder in alignment with their original implementations. Notably, MLPSBM is applied only to high-resource languages, such as Bulgarian, Catalan, Czech, German, English, Spanish, Italian, Dutch, Norwegian, Romanian, and Russian. To contextualize the results, we also report the highest-performing parser (BEST) and the average score (Average) across the selected top nine parsers for each test treebank, based on ground-truth annotations.

4.4   Results and Discussion

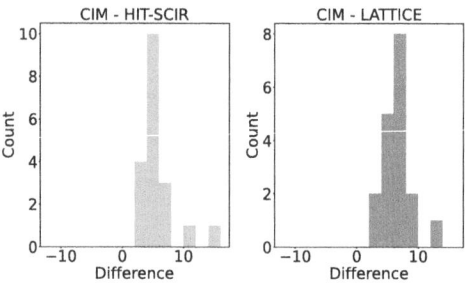

Fig. 1. Performance delta between CIM and ensemble-based dependency parsers

Table 1 presents a consistent performance advantage of aggregation-based methods over both ensemble and standalone parsers. These methods yield higher average and median UAS scores while maintaining reduced variance across the evaluated treebanks, demonstrating their capacity to adaptively integrate outputs from diverse base models. Among the aggregation strategies, the CIM-based framework stands out as the only one to outperform LLM-driven parsers, highlighting the benefits of incorporating label correlation into parser quality estimation. On the other hand, the CRH method shows performance comparable to MST, likely due to its limited use of disagreement information, which restricts its ability to distinguish between high- and low-quality parsers. CIM delivers the most reliable results across both high-resource and low-resource languages, exhibiting not only superior accuracy but also the smallest variation in scores. This robustness is particularly pronounced in low-resource settings, where CIM exceeds the strongest baseline parser by nearly two percentage points in mean UAS. Furthermore, its low standard deviation (9.68) reinforces its stability across varied linguistic domains. Collectively, these findings affirm the value of dependency tree structure aggregation, with CIM emerging as a highly effective solution for enhancing parsing outcomes in multilingual and resource-constrained contexts.

4.5   Comparison Study

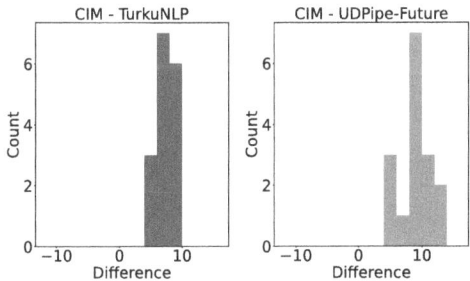

Fig. 2. Performance delta between CIM and individual (non-ensemble) parsers

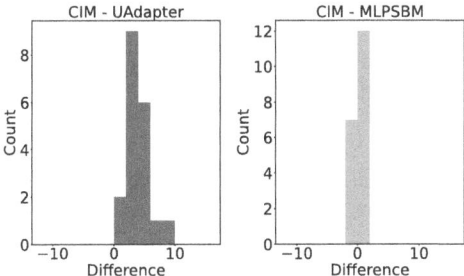

Fig. 3. Performance delta between CIM and large language model-driven parsers

Figures 1, 2, and 3 present histograms illustrating the performance differences between CIM and various baseline approaches, ensemble, non-ensemble, and LLM-based, on high-resource language treebanks. Positive values indicate cases where our method yields superior results. As seen in Fig. 1, our approach consistently surpasses ensemble techniques such as HIT-SCIR and LATTICE. Similarly, Fig. 2 shows that it outperforms individual parsers like TurkuNLP and UDPipe Future. Figure 3 further demonstrates that our method exceeds the performance of LLM-driven parsers, including UAdapter and MLPSBM. These comparisons affirm the ability of our approach to correct errors from individual systems, regardless of their overall quality, and highlight its strength as the most reliable aggregation framework among those evaluated.

5 Conclusion

In this work, we explore the effectiveness of three post-processing strategies, MST, CRH, and a customized Ising-based model, for combining outputs from multiple dependency parsers. By reformulating the tree aggregation task as a binary edge-level labeling problem, we enable the application of label-centric aggregation techniques to syntactic structures. Our evaluation across 71 multilingual treebanks from the CoNLL 2018 dataset highlights that the Ising-based approach consistently delivers superior results, demonstrating its ability to discern parser reliability and yield robust, high-quality dependency structures.

References

1. Bininda-Emonds, O.R.: Phylogenetic Supertrees: combining information to reveal the tree of life, vol. 4. Springer Science and Business Media (2004)
2. Bryant, D.: A classifica of co sensus methods for phylogenetics. In: Bioconsensus: DIMACS Working Group Meetings on Bioconsensus: October 25-26, 2000 and October 2-5, 2001, DIMACS Center. vol. 61, p. 163. American Mathematical Soc. (2003)
3. Chakraborty, M., Kulkarni, A., Li, Q.: Open-domain aspect-opinion co-mining with double-layer span extraction. In: Proceedings of the 28th ACM SIGKDD Conference on Knowledge Discovery and Data Mining, pp. 66–75 (2022)
4. Che, W., Liu, Y., Wang, Y., Zheng, B., Liu, T.: Towards better UD parsing: deep contextualized word embeddings, ensemble, and treebank concatenation. CoNLL **2018**, 55 (2018)
5. Chen, J., He, X., Miyao, Y.: Language model based unsupervised dependency parsing with conditional mutual information and grammatical constraints. In: Proceedings of the 2024 Conference of the North American Chapter of the Association for Computational Linguistics: Human Language Technologies (Volume 1: Long Papers), pp. 6355–6366 (2024)
6. Chen, M., Cohen-Wang, B., Mussmann, S., Sala, F., Ré, C.: Comparing the value of labeled and unlabeled data in method-of-moments latent variable estimation. In: International Conference on Artificial Intelligence and Statistics, pp. 3286–3294. PMLR (2021)
7. Gavril, F.: Generating the maximum spanning trees of a weighted graph. J. Algorithms **8**(4), 592–597 (1987)

8. Kanerva, J., Ginter, F., Miekka, N., Leino, A., Salakoski, T.: Turku neural parser pipeline: an end-to-end system for the CONLL 2018 shared task. CoNLL **2018**, 133 (2018)

9. Kuang, Z., et al.: FireBolt: weak supervision under weaker assumptions. In: International Conference on Artificial Intelligence and Statistics, pp. 8214–8259. PMLR (2022)

10. Kulkarni, A., Sabetpour, N., Markin, A., Eulenstein, O., Li, Q.: CPTAM: constituency parse tree aggregation method. In: Proceedings of the 2022 SIAM International Conference on Data Mining (SDM), pp. 630–638. SIAM (2022)

11. Li, Q., Li, Y., Gao, J., Zhao, B., Fan, W., Han, J.: Resolving conflicts in heterogeneous data by truth discovery and source reliability estimation. In: Proceedings of the 2014 ACM SIGMOD International Conference on Management of Data, pp. 1187–1198 (2014)

12. Lim, K., Park, C., Lee, C., Poibeau, T.: SEx BiST: a multi-source trainable parser with deep contextualized lexical representations. CoNLL **2018**, 143 (2018)

13. Nivre, J., McDonald, R.: Integrating graph-based and transition-based dependency parsers. In: Proceedings of ACL-08: HLT, pp. 950–958 (2008)

14. Ratner, A., Hancock, B., Dunnmon, J., Sala, F., Pandey, S., Ré, C.: Training complex models with multi-task weak supervision. In: Proceedings of the Thirty-Third AAAI Conference on Artificial Intelligence and Thirty-First Innovative Applications of Artificial Intelligence Conference and Ninth AAAI Symposium on Educational Advances in Artificial Intelligence, pp. 4763–4771 (2019)

15. Ratner, A.J., Bach, S.H., Ehrenberg, H.R., Ré, C.: Snorkel: fast training set generation for information extraction. In: Proceedings of the 2017 ACM International Conference on Management of Data, pp. 1683–1686 (2017)

16. Ratner, A.J., De Sa, C.M., Wu, S., Selsam, D., Ré, C.: Data programming: creating large training sets, quickly. Adv. Neural Inf. Process. Syst. **29** (2016)

17. Ravikumar, P., Wainwright, M.J., Lafferty, J.D.: High-dimensional ising model selection using l1-regularized logistic regression. Ann. Stat. **38**(3), 1287–1319 (2010)

18. Sabetpour, N., Kulkarni, A., Xie, S., Li, Q.: Truth discovery in sequence labels from crowds. In: 2021 IEEE International Conference on Data Mining (ICDM), pp. 539–548. IEEE Computer Society (2021)

19. Sagae, K., Lavie, A.: Parser combination by reparsing. In: Proceedings of the Human Language Technology Conference of the NAACL, Companion Volume: Short Papers, pp. 129–132 (2006)

20. Straka, M.: Udpipe 2.0 prototype at CONLL 2018 UD shared task. CoNLL 2018, p. 197 (2018)

21. Surdeanu, M., Manning, C.D.: Ensemble models for dependency parsing: cheap and good?. In: Human Language Technologies: The 2010 Annual Conference of the North American Chapter of the Association for Computational Linguistics, pp. 649–652 (2010)

22. Tian, Y., Chen, G., Song, Y., Wan, X.: Dependency-driven relation extraction with attentive graph convolutional networks. In: Proceedings of the 59th Annual Meeting of the Association for Computational Linguistics and the 11th International Joint Conference on Natural Language Processing (Volume 1: Long Papers), pp. 4458–4471 (2021)

23. Üstün, A., Bisazza, A., Bouma, G., van Noord, G.: UDapter: Language adaptation for truly Universal Dependency parsing. In: Webber, B., Cohn, T., He, Y., Liu, Y. (eds.) Proceedings of the 2020 Conference on Empirical Methods in Natural Language Processing (EMNLP), pp. 2302–2315. Association for Computational Linguistics, Online (2020). https://doi.org/10.18653/v1/2020.emnlp-main.180, https://aclanthology.org/2020.emnlp-main.180

24. Xu, Z., Wang, H., Wang, B.: Multi-layer pseudo-Siamese biaffine model for dependency parsing. In: Calzolari, N., et al. (eds.) Proceedings of the 29th International Conference on Computational Linguistics, pp. 5476–5487. International Committee on Computational Linguistics, Gyeongju, Republic of Korea (2022). https://aclanthology.org/2022.coling-1.486

25. Zeman, D., et al.: CONLL 2018 shared task: multilingual parsing from raw text to universal dependencies. In: Proceedings of the CoNLL 2018 Shared Task: Multilingual parsing from raw text to universal dependencies, pp. 1–21 (2018)

Evaluating the Impact of LLM-Manipulated Content on Fake News Detection

Donghao Huang[1,2], Darius Ng[1], Zhaoxia Wang[1(✉)], Haibo Pen[3], and Erik Cambria[4]

[1] School of Computing and Information Systems, Singapore Management University, 80 Stamford Rd, Singapore 178902, Singapore
zxwang@smu.edu.sg
[2] Research and Development, Mastercard, 3 Fraser Street, Level 17, DUO Tower, Singapore 189352, Singapore
[3] Key Laboratory of Smart Grid of Ministry of Education, School of Electrical and Information Engineering, Tianjin University, Tianjin 300072, China
[4] College of Computing and Data Science, Nanyang Technological University, 50 Nanyang Avenue, Singapore 639798, Singapore

Abstract. The digital age has witnessed an unprecedented surge in fake news, a challenge further exacerbated by advanced Large Language Models (LLMs) capable of subtle content manipulation. This study systematically evaluates the robustness of fake news detection methods—spanning traditional machine learning and deep learning-based models—against LLM-generated texts using the WELFake dataset. Transformer-based models (e.g., DistilBERT) show near-perfect accuracy on original data but suffer up to a 33% drop in F1-score on LLM-manipulated content, whereas traditional methods exhibit greater resilience. We further demonstrate that incorporating both original and manipulated texts during training improves detection robustness, albeit with increased computational costs. Our findings underscore the urgent need for adaptive detection strategies that balance high accuracy with robustness against sophisticated content manipulation. To ensure full reproducibility and support future research, our codebase, datasets, and trained models are openly available at https://github.com/inflaton/fake-news.

Keywords: Fake News Detection · Machine Learning · Deep Learning · Transformer Models · LLM-Manipulated Content

1 Introduction

The digital era has revolutionized information dissemination, with social media and online platforms enabling news to circulate at unprecedented speeds [17,21]. While these advances have democratized access to information, they have also facilitated the rapid spread of misinformation. As illustrated in Fig. 1, fake news

S. Yuan et al. (Eds.): PAKDD 2025 Workshops, LNAI 15835, pp. 375–386, 2025.
https://doi.org/10.1007/978-981-96-8197-6_28

has evolved from basic text-based content and image manipulation to early deepfakes, and now to more sophisticated AI-generated distortions and multimodal
deepfakes.

These developments incorporate deepfake technology and multimedia manipulation. This progression underscores the increasing complexity of fake news,
which now often blends altered text, images, and videos to create compelling yet
misleading narratives.

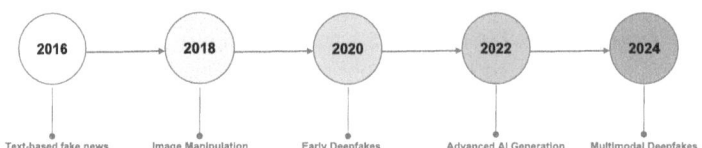

Fig. 1. Evolution of fake news

The impact of misinformation extends well beyond public health concerns,
affecting political stability and eroding public trust [14]. False narratives can
skew public opinion, undermine confidence in media and institutions, and even
influence electoral outcomes. As large-scale AI systems and deep-fake tools
become more accessible, the ease of producing convincing fake content has intensified the urgency for robust detection methods.

In response to these challenges, our study proposes a comprehensive machine
learning framework that integrates traditional learning-based methods, deep
learning architectures, and large language model (LLM)-based techniques. Our
approach not only targets conventional fake news but also addresses emerging
threats posed by LLM-driven content manipulation. Leveraging datasets such
as the WELFake collection—and augmenting them with LLM-rewritten articles
generated via models like Qwen2.5-7B—we systematically evaluate and enhance
the robustness and generalizability of various detection models.

The primary research question of this study is: How can advanced machine
learning frameworks improve fake news detection, particularly in scenarios where
sophisticated AI tools are used to manipulate content?

Our key contributions include:

- Proposing an integrated framework that combines content-based semantic
  analysis with source reliability metrics to enhance fake news detection.
- Conducting a comprehensive evaluation of traditional machine learning, deep
  learning, and transformer-based models, highlighting their strengths and limitations, particularly in detecting LLM-manipulated content.
- Investigating model generalizability by training on both original and LLM-
  rewritten datasets, addressing the challenge of detecting manipulated narratives in real-world scenarios.
- For full reproducibility and to advance research in this domain, we provide
  open access to our codebase, datasets, and trained models through our repository at https://github.com/inflaton/fake-news.

2   Related Work

The rise of social media as a primary news source has dramatically changed how information spreads, with false information spreading faster and reaching more people than true content. A study by Vosoughi et al. found that false news was 70% more likely to be retweeted than accurate news, highlighting the speed of misinformation's spread [20]. Automated systems play a key role in this phenomenon. News credibility assessments have shifted from traditional journalistic verification to advanced machine learning techniques. Shu et al. reviewed modern fake news detection strategies, emphasizing integrated approaches that combine multiple analysis methods [18]. These strategies can be divided into content-based (knowledge-based, style-based) and social context-based (stance-based, propagation-based) methods.

Content-Based Detection Methods. Early fake news detection research focused on hand-engineering linguistic features to assess credibility. Potthast et al. identified language cues, such as word sophistication and sentence complexity, as reliable indicators of legitimacy [16]. They found that syntactic patterns and readability could reveal content validity. Semantic analysis also plays a key role, with emotional tone, technical terms, and source attribution helping differentiate real from fake news. Narrative coherence is another reliability predictor. Advances in linguistic analysis, particularly through transformer-based systems like BERT [6], have improved accuracy in detecting subtle linguistic patterns of disinformation.

Liu et al. with RoBERTa [12] enhanced fake news detection by improving long-range dependencies and contextual understanding. These advances strengthened feature extraction and semantic analysis in detection systems. A study on explainable fake news identification for Polish digital media highlighted the value of multi-factor models, combining hybrid translation algorithms, annotator metrics, and content-based approaches using BERT and S-BERT [10]. The study found that bias analysis and annotator demographics improved accuracy, with a balanced accuracy of 72.1%. However, it also noted issues with non-expert annotators and translated datasets.

Modern fake news detection has evolved to include multiple information modes beyond basic text analysis. Khattar et al. introduced the Multimodal Variational Autoencoder (MVAE), which improved detection accuracy by integrating social signals, metadata, and text-image consistency [9]. Studies on feature fusion have shown that combining early and late fusion methods enhances both pattern recognition and immediate feature interactions. Jin et al. demonstrated that multimodal approaches, combining textual, visual, and social features, consistently outperformed single-mode analysis, improving accuracy by up to 15% [8].

Social Context-Based Detection Methods. User engagement patterns play a crucial role in detecting fake news. Guess et al. identified key behavioral cues, such as engagement depth, sentiment, and sharing velocity, as indicators of content authenticity [4]. Temporal dynamics of user interactions are particularly useful in identifying fraudulent content. Demographic factors, including age, education, and political affiliation, also affect the spread of fake news, influencing sharing behaviors. Cinelli et al. [4] highlighted the role of echo chambers in spreading misinformation, finding that Twitter tends to reinforce ideological divides, while Facebook's group structures amplify fake news. These echo chambers are shaped by platform architecture, with user interfaces and recommendation algorithms isolating information communities, making it harder for corrective content to reach users, as shown by Del Vicario et al. [5].

As artificial intelligence (AI) and machine learning continue to evolve, their role in combating misinformation is becoming increasingly significant. Persily et al. [15] highlight the potential of natural language processing (NLP) and social network analysis, while Ahmed [1] explores ensemble methods and support vector machines (SVMs) for fake news detection in low-resource languages. Ghayoomi and Mousavian [7] enhance detection performance by combining BERT with parallel CNN classifiers, whereas Beltagy et al. [3] leverage the Longformer architecture to improve the processing of lengthy documents in NLP tasks. Kumar et al. [11] propose an XLNet fine-tuning model to predict fake news in both multiclass and binary classification scenarios, achieving results superior to existing state-of-the-art models. Bathla et al. [2] propose an aspect-based deep learning approach, extracting specific aspects and their corresponding sentiment polarities from reviews to significantly enhance fake-review detection accuracy while reducing computational complexity. Recently, Ma et al. [13] introduced a multimodal framework integrating token-level semantic consistency with frequency-domain information to effectively address challenges such as image manipulation.

3   Proposed Fake News Detection Framework

Figure 2 provides an overview of our proposed framework. We aim to provide a comprehensive experimental framework for evaluating fake news detection models in the context of LLM-rewritten content, by systemically comparing the performance across model categories under normal conditions, followed by manipulated input conditions. The process begins with the *text dataset* and proceeds to a comprehensive *data pre-processing* phase—which includes cleaning, tokenization, vectorization, and other preparatory steps—before advancing to *model construction*. In this phase, three categories of models are explored:

1. **Traditional Learning-Based Models** (e.g., Random Forest, SVM) with traditional feature engineering techniques (e.g., Count Vectorizer, stemming),
2. **Deep Learning-Based Models** (e.g., CNN, RNN, LSTM) employing embedding approaches (e.g., Word2Vec, GloVe), and
3. **Transformer-based models** (e.g., BERT, RoBERTa, DistilBERT).

After model construction, *model training* is performed with hyperparameter tuning. Finally, *performance evaluation* is conducted using metrics such as accuracy, precision, recall, and F1-score to assess the effectiveness of each model.

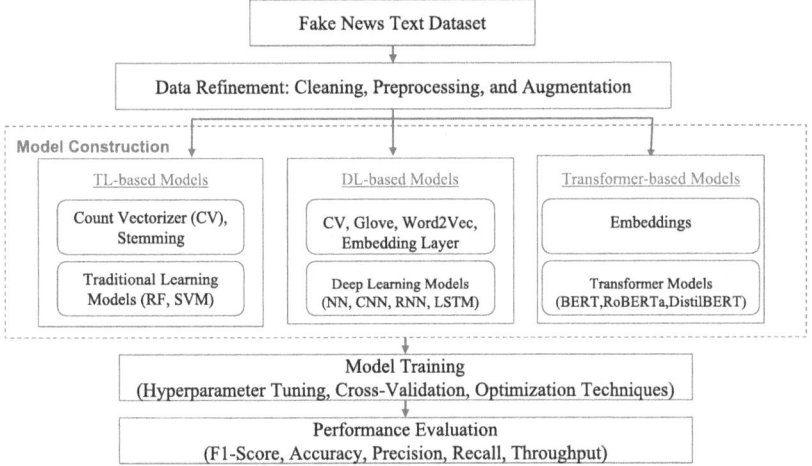

Fig. 2. Outline of the proposed method.

3.1 Datasets

For training and validating our models, we utilized the WELFake dataset for fake news detection in text data [19], a comprehensive collection of 72,134 news articles curated specifically for fake news classification tasks.

Data Cleaning. The data cleaning process begins with feature selection, focusing on the `title`, `text`, and `label` columns to retain only relevant information. Duplicate rows are removed to maintain data integrity and avoid redundancy. Outliers are detected and eliminated based on a 3-standard deviation (3 SD) criterion applied to the lengths of the `title` and `text` fields, ensuring consistent data quality. The `title` and `text` columns are then concatenated to form a new `full_content` column, with missing values imputed as empty strings. Subsequently, the original `title` and `text` columns are dropped to streamline the dataset. Finally, rows with empty `full_content` fields, those containing fewer than 30 words or more than 2000 words, and non-Enlish articles are removed.

Data Preprocessing To further standardize and normalize the text data, we create a new column, `processed_full_content`, by applying a custom text processing function to the `full_content` column. This function converts the text to lowercase, tokenizes it, removes punctuation, filters out common stopwords while retaining key negation words, and applies stemming using the Porter Stemmer. These preprocessing steps ensure that the text is cleaned and normalized, making it suitable for subsequent model training and evaluation. After this, we obtained a refined dataset containing 60,491 entries with three columns: `label`, `full_content`, and `processed_full_content`. The dataset consists of 34,030 true news articles (`label = 0`) and 26,461 fake news articles (`label = 1`). We refer to this as the **original** dataset.

Dataset Expansion. To assess the generalizability of our models, we utilized the Qwen2.5-7B model to rewrite all `full_content` entries. Qwen2.5-7B was chosen for its optimal balance between computational efficiency and linguistic capability. The rewriting process was guided by carefully crafted prompts, as illustrated in the following template:

```
Rewrite the following text to improve clarity, readability, and
↪  engagement while maintaining all original details and factual
↪  accuracy. Use concise, natural phrasing, and enhance flow without
↪  omitting any information. Ensure that the structure is logical,
↪  sentences are varied in length, and transitions are smooth. The tone
↪  should remain neutral and professional, suitable for a news article
↪  or formal report. Avoid redundancy while keeping all key details
↪  intact.
{input}
```

In this template, the {input} placeholder is replaced with the `full_content` of each article before it is submitted to the Qwen2.5-7B model.

The rewritten content generated by the Qwen2.5-7B model was subsequently processed using the same text preprocessing steps applied to the original dataset, resulting in the `processed_full_content` column. This newly generated dataset is referred to as the **LLM-rewritten** dataset. By combining both the original and LLM-rewritten datasets, we constructed a third dataset, denoted as the **combined** dataset.

3.2   Model Construction

Traditional Learning-Based Models. To establish a baseline for fake news detection, we employed two widely recognized machine learning algorithms, e.g., Random Forest (RF) and Support Vector Machines (SVM), along with Term Frequency Inverse Document Frequency Vectorizer (TF-IDF) and stemming for feature extraction. This blend of classic feature engineering techniques and traditional algorithms serves as a valuable benchmark for evaluating more advanced approaches.

Deep Learning-Based Models. In contrast to traditional approaches, deep learning-based methods utilize learned or pre-trained word embeddings rather than manually engineered features. GloVe and Word2Vec are popular examples of pre-trained embeddings, while an Embedding layer allows for trainable embeddings. Due to the strong performance of deep learning in text classification tasks, we included several neural network architectures—specifically, feed-forward neural networks (NN), convolutional neural networks (CNN), recurrent neural networks (RNN), and long short-term memory networks (LSTM).

Transformer-Based Models. Large Language Models (LLMs), based on the Transformer architecture, are at the forefront of natural language understanding. These models excel at capturing contextual and semantic nuances in text by employing attention mechanisms that process entire sequences of words. This enables them to capture long-range dependencies and bidirectional relationships. Notably, BERT utilizes self-attention, achieving exceptional performance across various NLP tasks. We have selected BERT, DistilBERT, and RoBERTa for our experiments due to their proven effectiveness in text classification.

3.3 Model Training

Once the models have been constructed, the next phase involves training them on the datasets. To ensure robust performance and mitigate overfitting, we employ several strategies during training:

Hyperparameter Tuning: Extensive hyperparameter optimization is performed using techniques such as grid search and random search. This process fine-tunes key parameters—including learning rates, regularization strengths, and network architectures—to identify the optimal configuration for each model.

Cross-Validation: For traditional learning-based models, we employ 5-fold cross-validation to evaluate generalizability and robustness across different subsets of the dataset. Once the optimal hyperparameters are identified, a final training run is conducted on 90% of the dataset to further enhance performance.

Optimization Techniques: Advanced optimization algorithms, such as Adam and stochastic gradient descent (SGD), are utilized to efficiently minimize the loss function, thereby accelerating the convergence of deep learning models and improving their training stability.

Resource Management: Given the computational demands of deep learning and transformer-based approaches, training is performed on an NVIDIA RTX A6000 Ada GPU (48 GB GDDR6 ECC memory, 18,176 CUDA cores, 568 Tensor cores), referred to as RTX A6000 throughout this paper.

3.4 Evaluation Metrics

We assess model performance using four key metrics: **accuracy**, **precision**, **recall**, and **F1-score**, ensuring a comprehensive evaluation of classification effectiveness. To measure computational efficiency, we also evaluate **throughput**, defined as the number of samples processed per second:

$$\text{Throughput} = \frac{\text{Total samples processed}}{\text{Total processing time (s)}} \tag{1}$$

Higher throughput enables faster processing, essential for real-time or large-scale applications. All experiments were consistently run on the RTX A6000.

4 Results and Discussion

4.1 Performance Comparative Analysis Using Original Data

To assess model effectiveness in fake news detection, we trained and tested various machine learning models using the original WELFake dataset. Table 1 summarizes our evaluation results.

Table 1. Performance Comparison of Machine Learning Models for Fake News Detection. All models are trained and evaluated using the original WELFake train/test data. Highest performance metrics are highlighted in bold.

Model	Accuracy	Precision	Recall	F1-Score	Throughput
RF	0.9494	0.9415	0.9430	0.9422	**1242**
SVM	0.9698	0.9642	0.9668	0.9655	1175
Simple Neural Network	0.9646	0.9645	0.9543	0.9594	805
CNN	0.9775	0.9776	0.9709	0.9742	836
RNN	0.9428	0.9492	0.9184	0.9336	628
LSTM	0.9779	0.9805	0.9686	0.9745	617
BERT	0.9945	0.9921	**0.9955**	0.9938	197
RoBERTa	0.9931	0.9917	0.9924	0.9921	139
DistilBERT	**0.9949**	**0.9928**	**0.9955**	**0.9942**	201

Our experimental results reveal several key findings. Among all models, transformer-based architectures (BERT, RoBERTa, and DistilBERT) consistently outperform traditional machine learning and neural network approaches in classification metrics. **DistilBERT** achieves the highest performance across all classification metrics, with an accuracy of **99.49%**, precision of **99.28%**, recall of **99.55%**, and F1-score of **99.42%**. This superior performance can be attributed to the model's ability to capture complex contextual relationships in text through its pre-trained language understanding capabilities.

However, the computational efficiency analysis presents an interesting trade-off. While transformer models excel in accuracy, they exhibit significantly lower throughput rates. Traditional machine learning models—particularly **RF and SVM**—demonstrate substantially higher processing speeds (**1242** and **1175** samples per second, respectively) compared to transformer models (201 samples/s for DistilBERT, 197 samples/s for BERT, and 139 samples/s for RoBERTa). This performance-efficiency trade-off becomes crucial when considering real-world deployment scenarios where both accuracy and speed are important.

Among the neural network architectures, LSTM shows promising results with an accuracy of 97.79% and an F1-score of 97.45%, while maintaining moderate throughput (617 samples/s). The CNN model also performs well, achieving 97.75% accuracy with better throughput (836 samples/s) compared to recurrent architectures. In contrast, the Simple Neural Network and RNN exhibit relatively lower performance metrics but maintain competitive processing speeds.

These findings suggest that the choice of model for fake news detection should be guided by the specific requirements of the application context. For high-stakes applications where accuracy is paramount, transformer-based models—particularly **DistilBERT**—would be the optimal choice. However, for real-time applications where processing speed is critical, traditional machine learning models like **RF** might be more suitable, as they offer a better balance between accuracy and computational efficiency.

4.2 Model Generalizability and LLM Impact Analysis

While our initial performance analysis using the original WELFake dataset showed promising results, we conducted additional experiments to evaluate model generalizability and assess the impact of LLM-rewritten content on detection accuracy. Figure 3 presents a comparative analysis of both F1 Score and Throughput on the original and LLM-rewritten test sets under two training scenarios.

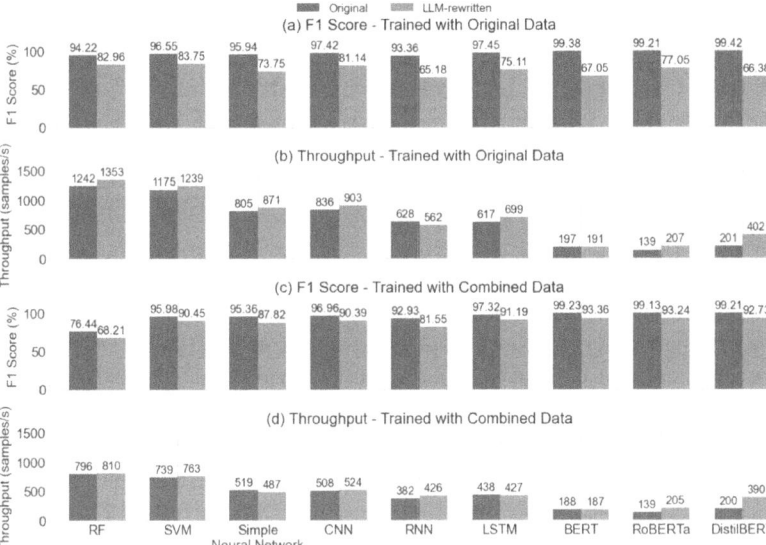

Fig. 3. Comparative analysis of F1 Score and Throughput for various models evaluated on *original* vs. *LLM-rewritten* test articles. Subplots (a) and (b) show models trained only on the original dataset, while subplots (c) and (d) show models trained on the combined dataset.

When trained solely on the original dataset, all models exhibit excellent performance on the original test set, with transformer-based models achieving F1-scores above 99%. However, performance degrades considerably when evaluated on the LLM-rewritten test set. The drop is particularly pronounced for transformer models, with DistilBERT's F1-score falling from 99.42% to 66.38%—a reduction of over 33% points. Similar substantial declines are observed for BERT (99.38% to 67.05%) and RoBERTa (99.21% to 77.05%). By contrast, traditional machine learning models display relatively better resilience, with RF and SVM showing smaller performance gaps (11.26% and 12.80%, respectively).

To address this generalizability challenge, we trained models on the combined dataset, which contains both original and LLM-rewritten articles. This approach significantly improves performance on the LLM-rewritten content across all models, albeit with a slight trade-off in performance on the original articles.

Notably, the gap between performance on the original and LLM-rewritten content narrows substantially, and transformer models achieve more consistent results. For example, DistilBERT attains F1-scores of 99.21% and 92.73% on original and LLM-rewritten articles, respectively, reducing the performance gap to 6.48%. From subplots (b) and (d) of Fig. 3) we observe a consistent drop in throughput for most models when trained on the combined dataset compared to the original dataset, likely reflecting the increased complexity of handling both original and LLM-rewritten articles. For instance, the throughput of SVM decreases from 1,175 samples/s when trained on the original dataset to 739 samples/s when trained on the combined dataset. Similarly, DistilBERT's throughput drops from 402 samples/s to 390 samples/s, indicating a slight computational overhead when incorporating LLM-rewritten content. Despite these reductions, the overall throughput ranking remains consistent, with RF (810 samples/s) and SVM (763 samples/s) achieving the highest inference speed, while transformer-based models such as BERT (187 samples/s) and RoBERTa (205 samples/s) continue to exhibit significantly lower throughput due to their computational complexity. Notably, DistilBERT strikes a balance between robust F1 performance and relatively faster inference speed, making it an attractive option for resource-constrained scenarios where both accuracy and efficiency are critical.

These findings underscore a critical challenge in modern fake news detection: LLM-rewritten content can substantially compromise the effectiveness of models trained on traditional datasets. While combining original and LLM-rewritten samples in training yields notable improvements, the residual performance gap indicates that further research is needed to fully counter LLM-driven content manipulation in fake news detection.

5   Conclusion

In this work, we introduced a comprehensive framework for fake news detection that spans traditional machine learning, deep learning, and transformer-based approaches. Our methodology was designed to address not only the challenges posed by naturally occurring news content but also those introduced by advanced text manipulation through LLM-based rewriting. Our experimental results highlight several key insights:

- Transformer-based models, particularly DistilBERT, achieve outstanding classification metrics on the original dataset. However, their performance degrades significantly when evaluated on LLM-rewritten content, indicating a vulnerability in handling sophisticated text manipulations.
- Traditional models, while slightly less accurate overall, demonstrated more resilience to content alterations, suggesting that a balance between high accuracy and robust generalizability is crucial.

– Incorporating both original and LLM-rewritten samples in the training process markedly improved model performance on manipulated content, though a performance gap still persists.

In this work, we provide the community with a public dataset of LLM-rewritten fake/real news articles, which can serve as a valuable benchmark for future adversarial robustness studies. These findings underscore the necessity for ongoing research into more robust detection strategies that can adapt to the evolving landscape of content generation. Future work should focus on:

1. Expanding research across diverse datasets and leveraging multiple LLMs for rewritten content to enhance generalizability,
2. Integrating adaptive learning techniques for real-time updates with emerging data, and
3. Conducting further investigations into the impact of fully LLM-generated content on detection efficacy.

Overall, our study demonstrates that while current detection systems are effective on traditional datasets, they require continual refinement to remain robust against the sophisticated techniques enabled by modern language models. Addressing these challenges is vital for developing reliable, real-world fake news detection systems.

Acknowledgment. The authors express their sincere appreciation to the following SMU students for their keen interest and valuable contributions to the code and report development of this research: LEE Ryan, Sakthivel GANAPATHY, See Jae FOO, Abhay BAGDAWALA, Shaun ZHOU, Gabriel CHUA, Arthur CHENG, Cristie SIM, Pei Shyan LIM, Seah WANG, Jin Sheng TEOH, and Hai Xiang ONG.

References

1. Ahmed: Hard voting approach using SVM, naïve bayes and decision tree for Kurdish fake news detection. Int. J. Comput. Sci. Media **2**(3) (2023). https://doi.org/10.52866/ijcsm.2023.02.03.003
2. Bathla, G., Singh, P., Singh, R.K., Cambria, E., Tiwari, R.: Intelligent fake reviews detection based on aspect extraction and analysis using deep learning. Neural Comput. Appl. **34**(22), 20213–20229 (2022)
3. Beltagy, I., Peters, M.E., Cohan, A.: LongFormer: the long-document transformer. arXiv (2020). https://arxiv.org/abs/2004.05150
4. Cinelli, M., De Francisci Morales, G., Galeazzi, A., Quattrociocchi, W., Starnini, M.: The echo chamber effect on social media. Proc. Natl. Acad. Sci. **118**(9), e2023301118 (2021). https://doi.org/10.1073/pnas.2023301118
5. Del Vicario, M., et al.: The spreading of misinformation online. Nat. Commun. **10**(1), 1–10 (2019). https://doi.org/10.1038/s41467-018-07761-2
6. Devlin, J., Chang, M.W., Lee, K., Toutanova, K.: BERT: pre-training of deep bidirectional transformers for language understanding. In: Proceedings of NAACL-HLT 2019, pp. 4171–4186 (2019). https://doi.org/10.18653/v1/N19-1423

7. Ghayoomi, M., Mousavian, M.: Deep transfer learning for Covid-19 fake news detection in Persian. Expert Syst. (2022). https://doi.org/10.1111/exsy.13008, https://onlinelibrary.wiley.com/doi/10.1111/exsy.13008

8. Jin, Z., Cao, J., Zhang, Y., Zhou, J., Tian, Q.: Novel visual and statistical image features for microblogs news verification. IEEE Trans. Multimed. **19**(3), 598–608 (2017). https://doi.org/10.1109/TMM.2016.2617078

9. Khattar, D., Goud, J.S., Gupta, M., Varma, V.: MVAE: multimodal variational autoencoder for fake news detection. In: The World Wide Web Conference, pp. 2915–2921 (2019). https://doi.org/10.1145/3308558.3313552

10. Kozik, R., et al.: Towards explainable fake news detection and automated content credibility assessment: polish internet and digital media use-case. Neurocomput. (Amsterdam) **608**, 128450– (2024). https://doi.org/10.1016/j.neucom.2024.128450

11. Kumar, A., Trueman, T.E., Cambria, E.: Fake news detection using XLNet fine-tuning model. In: 2021 International Conference on Computational Intelligence and Computing Applications (ICCICA), pp. 1–4. IEEE (2021)

12. Liu, Y., et al.: RoBERTa: a robustly optimized BERT pretraining approach. arXiv (2019)

13. Ma, Z., Liu, H., Zeng, Z., Guo, H., Zhao, X., Luo, M.: Learning multimodal attention mixed with frequency domain information as detector for fake news detection. In: 2024 IEEE International Conference on Multimedia and Expo (ICME), pp. 1–6. IEEE (2024)

14. Mwangi, E.: Technology and fake news: shaping social, political, and economic perspectives. Polit. Econ. Perspect. (2023)

15. Persily, N.: The 2016 US election: can democracy survive the internet? J. Democr. **32**(2), 63–76 (2021). https://doi.org/10.1353/jod.2021.0019

16. Potthast, M., Kiesel, J., Reinartz, K., Bevendorff, J., Stein, B.: A stylometric inquiry into hyperpartisan and fake news. In: Proceedings of the 56th Annual Meeting of the Association for Computational Linguistics, pp. 231–240 (2018). https://doi.org/10.18653/v1/P18-1022

17. Shang, D., Hu, Z., Wang, Z.: Mining consumer brand relationship from social media data: a natural language processing approach. In: Sun, X., Zhang, X., Xia, Z., Bertino, E. (eds.) ICAIS 2021. LNCS, vol. 12736, pp. 553–565. Springer, Cham (2021). https://doi.org/10.1007/978-3-030-78609-0_47

18. Shu, K., Sliva, A., Wang, S., Tang, J., Liu, H.: Fake news detection on social media: a data mining perspective. ACM SIGKDD Explor. Newsl. **19**(1), 22–36 (2017). https://doi.org/10.1145/3137597.3137600

19. Verma, P.K., Agrawal, P., Prodan, R.: Welfake dataset for fake news detection in text data (2021). https://doi.org/10.1109/TCSS.2021.3068519, https://zenodo.org/records/4561253

20. Vosoughi, S., Roy, D., Aral, S.: The spread of true and false news online. Science **359**(6380), 1146–1151 (2018). https://doi.org/10.1126/science.aap9559

21. Wang, Z., Joo, V., Tong, C., Xin, X., Chin, H.C.: Anomaly detection through enhanced sentiment analysis on social media data. In: 2014 IEEE 6th International Conference on Cloud Computing Technology and Science, pp. 917–922. IEEE (2014)

Large Language Models for Logical Fallacy Detection

Nicole Teo[1] [ID], Donghao Huang[1,2] [ID], Erik Cambria[3] [ID],
and Zhaoxia Wang[1(✉)] [ID]

[1] Singapore Management University, 80 Stamford Rd, Singapore 178902, Singapore
{nicolet.2023,dh.huang.2023}@engd.smu.edu.sg, zxwang@smu.edu.sg
[2] Research and Development, Mastercard, 3 Fraser Street, Level 17, DUO Tower,
Singapore 189352, Singapore
[3] Nanyang Technological University, 50 Nanyang Ave, Singapore 639798, Singapore
cambria@ntu.edu.sg

Abstract. Identifying logical fallacies is essential for maintaining logical reasoning and reducing false information in a variety of domains, such as the media, law, and education. We present an extensive study on the use of large language models (LLMs) for logical fallacy detection and provide a comparative overview of model performance across various fallacy classes. We evaluate the logical fallacy detection capabilities of multiple state-of-the-art models (LLaMA, Qwen, Gemma, Phi) utilizing accuracy, precision, recall, and F1-score as assessment measures. According to our findings, our models do well on simple fallacies like "circular reasoning," but they have trouble with more interpretive reasoning when it comes to more complex categories like "equivocation" and "intentional". These results highlight the potential of LLMs in fallacy detection tasks but also indicate a need for improved prompt engineering, fine-tuning, and context-rich datasets to enhance interpretive accuracy. This research offers insights into advancing LLMs for critical reasoning applications, contributing to improved information integrity across domains.

Keywords: Large language models · Logical fallacies · Reasoning

1 Background

The rapid advancement of large language models (LLMs) has opened new avenues for natural language processing, allowing for exceptional performance on a variety of downstream tasks [6,12]. In addition, it has also expanded applications spanning from conversational agents to complex reasoning tasks [2,3,17]. Among these tasks, logical fallacy detection has emerged as a critical area of focus, given its implications for the integrity of information and reasoning processes in various domains, including education, law, and media literacy. Logical fallacies undermine sound argumentation and can lead to flawed conclusions, making the ability to identify and address these fallacies essential.

S. Yuan et al. (Eds.): PAKDD 2025 Workshops, LNAI 15835, pp. 387–398, 2025.
https://doi.org/10.1007/978-981-96-8197-6_29

A fallacy is defined as "a mistake in reasoning [...] that occurs with some frequency in real arguments and which is characteristically deceptive" [5]. The task of classifying fallacies, or developing a classification model to distinguish between various forms of fallacies, has been taken up by recent publications [1]. [13] combined methods like large transformer language models and the Structure Aware Hypothesis [29] from natural language inference. In this line of study, the underlying logical shape of the text is extracted via argument mining [15], and the resulting features are used to train a classifier. Interestingly, the datasets presented by [13] only include false assertions, which means that they may only be classified as fallacious types if the argument is presumed to be false. More recently, [16] focused on assessing the effectiveness of GPT-4, in identifying logical fallacies as part of a digital misinformation intervention. This evaluation is particularly important because logical fallacies are a common feature in misinformation, and understanding the capabilities of LLMs in this domain could aid in combating such content. They found that while GPT-4 showed strong performance in identifying logical fallacies, its performance varied across different types of fallacies.

This study investigates the ability of leading LLMs—LLaMA, Qwen, Gemma, and Phi—to detect logical fallacies using metrics like accuracy, F1-score, recall, and precision. We also analyze their sensitivity to different fallacy types to better understand their reasoning processes. To enhance interpretability, we apply methods from interpretable AI, offering insights into model decisions and increasing transparency and accountability. By providing justifications for classifications, we aim to clarify how LLMs identify fallacies and support their use in fostering critical thinking and informed decision-making.

Our work aims to present a thorough analysis across different LLMs for the identification of logical fallacies, together with new methods for evaluating the interpretability and effectiveness of these models. The primary contributions we have made in this paper are:

1. The paper presents a comprehensive analysis of state-of-the-art LLMs (LLaMA, Qwen, Gemma, Phi) for detecting various types of logical fallacies, providing insights into their strengths and weaknesses.
2. The study highlights how LLMs perform differently across fallacy categories, showing strong results on simpler fallacies like "circular reasoning" while facing challenges with more complex types such as "equivocation" and "intentional."
3. The research suggests enhancements in prompt engineering, fine-tuning, and dataset design, offering practical guidance for improving LLMs' ability to detect and interpret complex logical fallacies.

2   Related Work

Logical Reasoning. Recent developments have demonstrated that LLMs may generate intermediate reasoning stages to carry out reasoning tasks efficiently [23,26,28]. Formal reasoning, which is frequently applied in logic and mathematics, is a methodical, logical process that adheres to a set of norms and principles. Informal reasoning, which is more prevalent in daily life, is a less formal method of problem solving that depends on experience, common sense, and intuition to reach conclusions. Informal reasoning is more flexible and open-ended but may also be less reliable than formal reasoning, which is more ordered and dependable [10]. The logical reasoning skills of large language models are gaining more and more attention from researchers [9,21,27,31]. For example, [19] integrated first-order logic with large language models. [20] and [30] equipped extensive language models with symbolic solutions. An adversarial pre-training approach was presented by [22] to improve logical thinking. Multi-step explicit planning was added to the inference process by [30]. A contrastive learning strategy was put up by [11] to enhance logical question responding.

Logical Fallacies. False reasoning patterns are known as logical fallacies [4]. They are frequently recognized since their claims are not backed up by any proof. Finding logical fallacies in natural language text has a wide range of possible uses, including tracking false information and confirming assertions. Furthermore, being able to understand erroneous arguments can improve the logic and educational value of speech. Fallacies can also be deliberately committed in order to mislead or influence others, and have been used to disseminate propaganda in news articles [18]. Recent attempts have been made to categorize fallacies using deep learning models. For example [13] proposed a dataset, where the fallacies were categorized using a structure-based approach. Follow up work has also tried to classify logical fallacies and valid statements [14], and build a software that addresses question-answering fallacies [7]. [24] investigated discussion answers for the ad hominem fallacy. In their study, [8] investigated the ad hominem fallacy in online arguments. Argumentation essays were found to have weak arguments by [25]. A multi-task prompting technique was presented by [1]. It allows users to collaboratively learn fallacies from several datasets. To our knowledge, no prior work has tried to test out the logical fallacy detection capabilities of recent open source LLMs.

3   Methodology

In this study, we evaluate the logical fallacy detection capabilities of different state-of-the-art large language models, such as LLaMA, Qwen, Gemma, and Phi. These models were selected based on their recent advancements in natural language understanding and generation, as well as their potential to address complex reasoning tasks such as logical fallacy detection. Each model was evaluated in the context of its ability to identify fallacies across a range of types, including simple and complex categories.

This section also outlines the data used for evaluation, data pre-processing, evaluation and implementation, the baselines for comparison, and the specifics of the implementation.

3.1 Dataset

We make use of the LOGIC and LOGIC Climate logical fallacy datasets from [13], which has 2,449 instances of logical fallacies spanning 13 different categories. The LOGIC dataset comprises of typical instances of logical fallacies gathered from several online educational resources designed to instruct or assess students' comprehension of logical fallacies. The logical fallacy examples was gathered manually from Google, totaling about 600 samples, and automatically gathered from three student quiz websites: Quizziz, study.com, and ProProfs [13]. The LOGIC Climate dataset contains more difficult examples of the same logical error types on the subject of climate change. It comprises of 1,079 samples of logical fallacies, with an average of 35.98 tokens per sample and a vocabulary of 5.8K words. A sample of the dataset structure is illustrated in Table 1. The full dataset contains 13 logical fallacy types, each represented by exemplar statements that demonstrate the reasoning error characteristic of that fallacy category.

3.2 Data Pre-processing

Each dataset contained text samples with corresponding issue labels. We began by filtering out rows with missing values to ensure consistency and avoid performance issues during modeling. Next, we applied label encoding to convert categorical labels into numerical form, standardizing the format for both datasets.

To prepare the text for model training, we used TF-IDF vectorization with a maximum of 5000 features, capturing the most informative terms while reducing computational load and overfitting. The vectorizer was fitted on the training data and then applied to the test set to maintain a consistent feature space. Finally, we split the training data into 80% for training and 20% for validation.

3.3 Implementation and Evaluation

Baseline. We began by loading a tokenizer and a pre-trained LLM for sequence classification, adjusting the label size to match our dataset classes. The token embeddings were resized to fit the tokenizer's vocabulary, and the end-of-sequence token was set as the padding token to ensure consistent tokenization. Each sample was tokenized to a uniform length of 512 tokens with padding as needed, enabling efficient batching during training. We monitored precision, recall, and F1-score to track performance. Training used a learning rate of 2e-5 over 5 epochs, with weight decay (0.01) and a linear learning rate scheduler for regularization.

Few-Shot. We applied a few-shot learning approach by embedding three relevant text-label examples from the training data into each prompt. This format provides contextual guidance, helping the model predict labels for new inputs based on similar cases. These few-shot prompts are tokenized efficiently, ensuring they stay within model limits for smooth batch processing. Each dataset

Table 1. Structure of the LOGIC Dataset

Text	Label
"You're too ugly to be class president!"	Ad Hominem
Everyone is going to get the new smart phone when it comes out this weekend. Why aren't you?	Ad Populum
Do the best for your baby	Appeal to Emotion
America is the best place to live, because it's better than any other country	Circular Reasoning
Noisy children are a real headache. Two aspirin will make a headache go away. Therefore, two aspirin will make noisy children go away	Equivocation
I am committed to preserving traditional marriage, the union of one man and one woman	Fallacy of Credibility
The Senator thinks we can solve all our ecological problems by driving a Prius	Fallacy of Extension
Most Arabs are Muslims and all the 9/11 hijackers were also Muslims. Therefore most Arabs are hijackers	Fallacy of Logic
"I believe deeply that canceling the fashion show is the best decision. I know it is. I am very certain about it"	Fallacy of Relevance
The Soviet Union collapsed after taking up atheism. Therefore, we must avoid atheism for the same reasons	False Causality
You need to go to the party with me, otherwise you'll just be bored at home	False Dilemma
If we let this child bring the permission slip late, there is no reason to ever set a deadline for anything again!	Faulty Generalization
You can't prove that space aliens don't exist, so they must be real	Intentional

instance includes the tokenized prompt and label tensor, supporting training and evaluation. By integrating context directly into inputs, this method aims to boost model accuracy on unseen data.

Model Performance Evaluation in Different Fallacy Classes. We generate a confusion matrix that visualizes the performance of a classification model. First, predictions are obtained from the model on the validation dataset. The predicted labels are extracted by selecting the highest probability class from the model's predictions, and the true labels are also collected. The confusion matrix, which compares the true labels with the predicted labels, is then computed. A custom function is then defined to plot the matrix as a heatmap with annotations, where the x-axis represents predicted labels, and the y-axis represents true labels. Finally, the function is called with class names (obtained from the label encoder) and the matrix is saved.

4 Results

Table 2. Comparison of the best results of baseline (zero-shot) and few-shot prompting on LOGIC and LOGIC Climate. The best results per model are **boldfaced** and the overall best results are underlined.

Model	Type	LOGIC				LOGIC Climate			
		A	P	R	F1	A	P	R	F1
Llama-3.2-1B-Instruct	baseline	0.562	0.578	0.562	0.562	0.232	**0.276**	0.232	0.214
	few-shot	**0.631**	**0.646**	**0.631**	**0.632**	**0.261**	0.225	**0.261**	**0.226**
Llama-3.2-3B-Instruct	baseline	0.595	0.616	0.595	0.592	0.213	0.207	0.213	0.191
	few-shot	**0.704**	<u>**0.724**</u>	**0.704**	**0.698**	**0.237**	**0.323**	**0.237**	**0.210**
Qwen2.5-1.5B-Instruct	baseline	0.613	0.631	0.613	0.614	0.251	0.230	0.251	0.232
	few-shot	**0.628**	**0.690**	**0.628**	**0.636**	**0.265**	**0.297**	**0.265**	**0.247**
Qwen2.5-3B-Instruct	baseline	0.601	0.627	0.601	0.600	0.227	**0.263**	0.227	**0.227**
	few-shot	**0.680**	**0.712**	**0.680**	**0.682**	**0.242**	0.260	**0.242**	0.217
Gemma-7b-it	baseline	0.544	**0.648**	0.544	**0.551**	0.171	**0.261**	0.171	**0.137**
	few-shot	**0.550**	0.545	**0.550**	0.543	**0.190**	0.126	**0.190**	0.125
Gemma-2-2b-it	baseline	0.595	0.602	0.595	0.587	0.246	0.267	0.246	0.222
	few-shot	**0.674**	**0.681**	**0.674**	**0.669**	**0.294**	<u>**0.366**</u>	**0.294**	**0.269**
Phi-3-mini-4k-instruct	baseline	0.637	0.665	0.637	0.642	0.237	0.238	0.237	0.216
	few-shot	**0.695**	**0.712**	**0.695**	**0.690**	**0.322**	**0.323**	**0.322**	<u>**0.297**</u>
Phi-3.5-mini-instruct	baseline	0.640	0.682	0.640	0.647	0.251	0.206	0.251	0.222
	few-shot	<u>**0.713**</u>	**0.721**	<u>**0.713**</u>	<u>**0.708**</u>	**0.332**	0.293	<u>**0.332**</u>	0.292

Table 2 presents a comparative analysis of baseline (zero-shot) and few-shot prompting across various models on the LOGIC and LOGIC Climate datasets. Accuracy (A), precision (P), recall (R), and F1-score (F1) are among the measures that are assessed. The findings show that few-shot prompting clearly outperforms zero-shot prompting for all models, with notable gains in accuracy and F1-score. The few-shot method shows significant improvements in predicting accuracy and improves the algorithms' capacity to categorize logical fallacies.

Phi-3.5-mini-instruct (few-shot) achieves the highest F1-score (0.708) and accuracy (0.713) on the LOGIC dataset, making it the best-performing model. Llama-3.2-3B-Instruct (few-shot) leads in precision (0.724) and recall (0.704). Other models, like Gemma-2-2b-it (few-shot) and Phi-3-mini-4k-instruct, also perform well with F1-scores above 0.690. Larger models generally outperform baselines and benefit most from few-shot prompting. LOGIC Climate proves more challenging, with lower scores across models. Gemma-2-2b-it (few-shot) ranks highest with 0.332 accuracy and 0.269 F1-score, followed by Phi-3-mini-4k-instruct (few-shot) at 0.322 accuracy and 0.297 F1-score. The results highlight the need for further fine-tuning and dataset-specific adjustments.

There are a number of reasons why certain networks perform poorly on the LOGIC Climate dataset in comparison to the standard LOGIC dataset. There are knowledge gaps since many LLMs are not expressly trained on climate-related talks, which frequently require domain-specific knowledge, scientific terminology, and politically sensitive disputes. It is more challenging to classify logical fallacies in climate discourse since they are typically more nuanced and include cherry-picking evidence, misinterpreting scientific uncertainties, and emotionally laden arguments. The notably poorer accuracy, precision, recall, and F1-scores for all models point to a basic difficulty in spotting fallacies in arguments pertaining to climate change. Although performance is enhanced by few-shot learning, the benefits are not significant, indicating that adaptive learning is insufficient to make up for domain knowledge deficiencies. While some models, like LLaMA-3.2-1B/3B and Gemma-7B/2B, struggle more, suggesting a lack of adaptation to domain-specific logical reasoning, others, like Phi-3.5-mini-instruct and Phi-3-mini-4k-instruct, do better, probably because they can follow instructions.

Figure 1 showed that Llama-3.2-3B-Instruct achieved balanced performance, excelling in "ad populum" (90%) but face difficulties with complex fallacies. Qwen2.5-3B-Instruct demonstrated consistency in detecting straightforward fallacies, though it struggled with "equivocation" and "fallacy of relevance". Gemma-2-2b-it showed high accuracy in "ad populum" (84%) and "false causality" (86%) but found distinctions between emotional and relevance-based fallacies challenging. Similarly, Phi-3.5-mini-instruct performed well in identifying "ad hominem" (90%), "ad populum" (97%), and "false causality" (89%) but struggled with fallacies like "equivocation" and "intentional". Overall, the models did well on fallacies that are easy to see, such "ad populum" and "false causality," but they struggled with more subtle fallacies that need for interpretive reasoning, like "equivocation" and "intentional".

Figure 2 showed that Llama-3.2-3B-Instruct did well on overt fallacies but struggled to differentiate between more subtle categories like "fallacy of relevance." Similar to this, Qwen2.5-3B-Instruct demonstrated proficiency in "appeal to emotion" and "circular reasoning", but mislabeled sophisticated fallacies such as "fallacy of extension." Gemma-2-2b-it did well on "equivocation" and "circular reasoning" (75%) but had trouble with fallacies that are closely linked, such as "fallacy of relevance". Phi-3.5-mini-instruct performed well in "circular reasoning" (100%) but had trouble with more subtle fallacies like "appeal to emotion" and "ad populum." Overall, all models shown limitations in interpretive reasoning by performing well at identifying "circular reasoning" but poorly at identifying subtle fallacies like "equivocation" and "intentional". Misclassifications that overlap, especially between "ad hominem" and "appeal to emotion", show that linguistic similarities in climate-related fallacies pose a problem to LLMs.

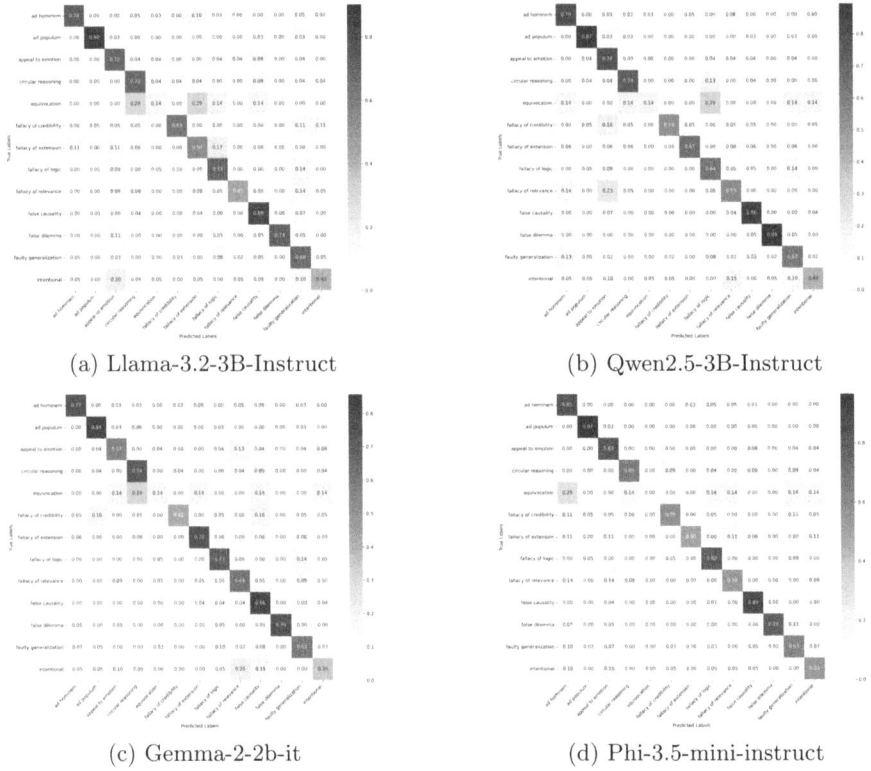

(a) Llama-3.2-3B-Instruct   (b) Qwen2.5-3B-Instruct

(c) Gemma-2-2b-it   (d) Phi-3.5-mini-instruct

Fig. 1. Performance of LLMS on LOGIC dataset for fallacy classification using a few-shot prompting approach. Each model's performance was visualized with a confusion matrix, providing insights into the classification accuracy across different fallacy types. The normalized confusion matrices for each model illustrate the proportion of correct and misclassified predictions for each fallacy class, allowing for a comparative analysis of model performance in detecting specific logical fallacies.

5   Discussion

5.1   Dataset

The majority of the examples in the datasets we used had one or two sentences, with the rare paragraph. However, real life usually consists of a few phrases or a few paragraphs. Although our analysis demonstrates that the LLMs functions for shorter texts. Lengthier texts—which are likely to contain a greater number of logical fallacies—may not yield the same results. To the best of our knowledge, there is no labeled dataset on logical fallacies for lengthy texts, thus generating one by hand would be necessary. Nevertheless, this would still warrant additional testing.

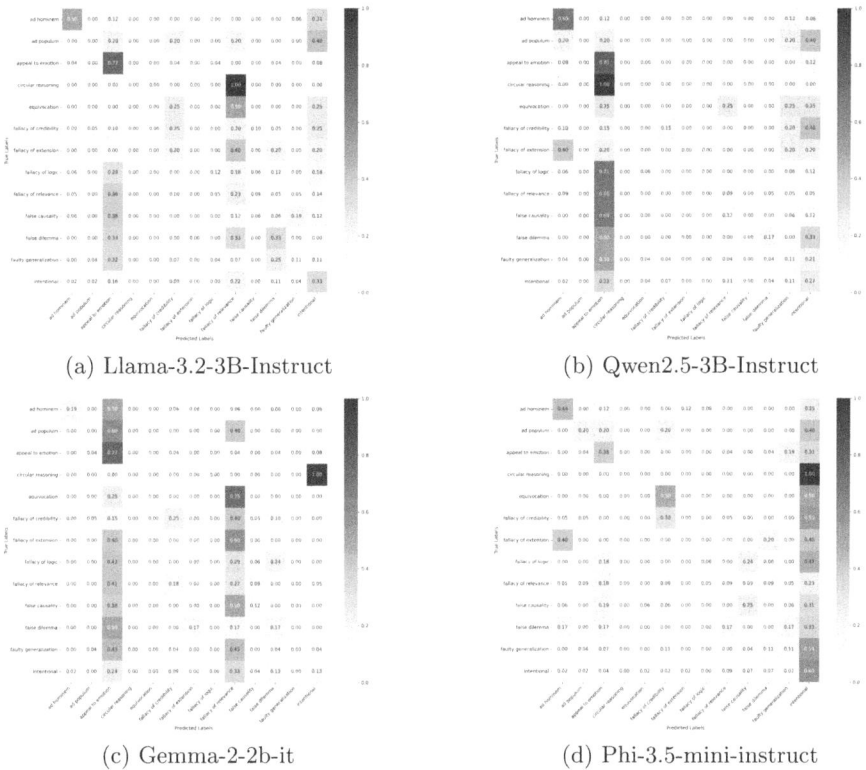

(a) Llama-3.2-3B-Instruct

(b) Qwen2.5-3B-Instruct

(c) Gemma-2-2b-it

(d) Phi-3.5-mini-instruct

Fig. 2. Performance of LLMs on LOGIC Climate dataset for fallacy classification using few-shot prompting. Each model's performance was visualized through confusion matrices, providing insights into classification accuracy across different fallacy types. The results shed light on each model's strengths and limitations in identifying specific climate-related fallacies.

5.2 Prompt Engineering

Even though we designed a straightforward prompt to enhance the model's few-shot learning capabilities and provide context when processing inputs, different prompt constructions may impact the outcome.

5.3 Misclassifications

The confusion matrices shed more light on the shortcomings of the models, especially when it comes to differentiating semantically related fallacies like appeal to emotion, circular reasoning, and fallacy of relevance. All models, regardless of size, exhibited confusion between these categories, highlighting the difficulty of detecting fallacies that share overlapping linguistic or argumentative features. This implies that existing models struggle with smaller distinctions, which probably led to the poorer recall scores seen in numerous models, even while they can handle fallacies with more distinguishing properties (like ad hominem and ad populum).

6 Limitations

First off, the datasets we used are mostly composed of brief text samples, usually consisting of just one or two phrases. Fallacies are frequently included into longer discourses, sometimes spanning several paragraphs, in real-world settings. As a result, it is unclear how well the models will perform on longer, more complicated texts. The thorough evaluation of model performance in such scenarios would require the creation of a labeled dataset with additional arguments.

Second, the few-shot prompting strategy depends largely on prompt construction, even though it can be successful in some situations. The prompt's design has a big impact on the model's output, and different prompt formulations may provide different outcomes. Even though we followed a set pattern, trying out different prompt designs could result in even better results. Optimized prompt engineering methods should be investigated in future studies to improve model accuracy, especially for subtle fallacies.

Lastly, the models had trouble differentiating between fallacies that share semantic similarities, including "appeal to emotion", "circular reasoning", and "fallacy of relevance". A shortcoming in the models' interpretive reasoning was highlighted by the numerous misclassifications caused by the overlapping linguistic and argumentation elements in these categories. These difficulties imply that in order to enhance their capacity to correctly distinguish closely similar logical fallacies, current LLMs could need more instruction or practice with a wider range of fallacy cases. Addressing these limitations can lead to advancing LLM capabilities for reliable logical fallacy detection in real-world applications.

7 Conclusions and Future Work

This study provides a comprehensive evaluation of several open source LLMs for detecting logical fallacies, using the LOGIC and LOGIC Climate datasets as benchmarks. We examined how well they performed in recognizing particular fallacy kinds using confusion matrices and common assessment metrics. According to our research, these models are effective at identifying simpler fallacies like "circular reasoning", but they have trouble identifying more complex ones like "equivocation" and "intentional", which call for more complex interpretive reasoning. The inability of current LLMs to handle subtle differences in logical arguments was highlighted by the frequent misclassifications caused by overlapping features among several fallacies. These findings imply that although LLMs may be useful for fallacy detection tasks, more work is necessary to ensure consistent accuracy, particularly in intricate, real-world texts.

Future research should explore advanced prompt engineering and fine-tuning techniques to enhance LLMs' interpretive capabilities. Furthermore, creating more extensive, contextually rich datasets may help give the models a stronger basis on which to detect errors in lengthy discourse. By resolving these issues, LLMs may develop into useful instruments for critical thinking exercises, enhancing the integrity and quality of information in a variety of fields, including media literacy, education, and automated content moderation.

Disclosure of Interests. The authors have no competing interests to declare that are relevant to the content of this article.

References

1. Alhindi, T., Chakrabarty, T., Musi, E., Muresan, S.: Multitask instruction-based prompting for fallacy recognition. arXiv preprint arXiv:2301.09992 (2023)
2. Brown, T., et al.: Language models are few-shot learners. Adv. Neural. Inf. Process. Syst. **33**, 1877–1901 (2020)
3. Du, K., Xing, F., Mao, R., Cambria, E.: An evaluation of reasoning capabilities of large language models in financial sentiment analysis. In: 2024 IEEE Conference on Artificial Intelligence (CAI), pp. 189–194. IEEE (2024)
4. Fantino, E., Stolarz-Fantino, S., Navarro, A.: Logical fallacies: a behavioral approach to reasoning. Behav. Anal. Today **4**(1), 109 (2003)
5. Govier, T.: Problems in argument analysis and evaluation, vol. 6. University of Windsor (2018)
6. Grattafiori, A., et al.: The LLaMA 3 herd of models. arXiv preprint arXiv:2407.21783 (2024)
7. Habernal, I., Hannemann, R., Pollak, C., Klamm, C., Pauli, P., Gurevych, I.: Argotario: Computational argumentation meets serious games. arXiv preprint arXiv:1707.06002 (2017)
8. Habernal, I., Wachsmuth, H., Gurevych, I., Stein, B.: Before name-calling: dynamics and triggers of ad hominem fallacies in web argumentation. arXiv preprint arXiv:1802.06613 (2018)
9. Huang, D., Wang, Z.: Logical reasoning with LLMs via few-shot prompting and fine-tuning: a case study on turtle soup puzzles. In: 2025 IEEE Symposium Series on Computational Intelligence (SSCI). IEEE (2025)
10. Huang, J., Chang, K.C.C.: Towards reasoning in large language models: a survey. arXiv preprint arXiv:2212.10403 (2022)
11. Jiao, F., Guo, Y., Song, X., Nie, L.: MERIt: meta-path guided contrastive learning for logical reasoning. arXiv preprint arXiv:2203.00357 (2022)
12. Jie Yeo, W., Ferdinan, T., Kazienko, P., Satapathy, R., Cambria, E.: Self-training large language models through knowledge detection. arXiv e-prints, pp. arXiv–2406 (2024)
13. Jin, Z., et al.: Logical fallacy detection. arXiv preprint arXiv:2202.13758 (2022)
14. Lalwani, A., Chopra, L.: ExplainNLI: translating natural language to first order logic for logical fallacy detection (2024)
15. Lawrence, J., Reed, C.: Argument mining: a survey. Comput. Linguist. **45**(4), 765–818 (2020)
16. Lim, G., Perrault, S.T.: Evaluation of an LLM in identifying logical fallacies: a call for rigor when adopting LLMs in HCI research. arXiv preprint arXiv:2404.05213 (2024)

17. Liu, L., Zhang, D., Li, S., Zhou, G., Cambria, E.: Two heads are better than one: zero-shot cognitive reasoning via multi-LLM knowledge fusion. In: Proceedings of the 33rd ACM International Conference on Information and Knowledge Management, pp. 1462–1472 (2024)
18. Musi, E., Aloumpi, M., Carmi, E., Yates, S., O'Halloran, K.: Developing fake news immunity: fallacies as misinformation triggers during the pandemic. Online J. Commun. Media Technol. **12**(3) (2022)
19. Olausson, T.X., et al.: LINC: a neurosymbolic approach for logical reasoning by combining language models with first-order logic provers. arXiv preprint arXiv:2310.15164 (2023)
20. Pan, L., Albalak, A., Wang, X., Wang, W.Y.: Logic-LM: empowering large language models with symbolic solvers for faithful logical reasoning. arXiv preprint arXiv:2305.12295 (2023)
21. Parmar, M., et al.: LogicBench: towards systematic evaluation of logical reasoning ability of large language models. In: Proceedings of the 62nd Annual Meeting of the Association for Computational Linguistics (Volume 1: Long Papers), pp. 13679–13707 (2024)
22. Pi, X., Zhong, W., Gao, Y., Duan, N., Lou, J.G.: LogiGAN: learning logical reasoning via adversarial pre-training. Adv. Neural. Inf. Process. Syst. **35**, 16290–16304 (2022)
23. Shen, T., Wang, J., Zhang, X., Cambria, E.: Reasoning with trees: faithful question answering over knowledge graph. In: Proceedings of the 31st International Conference on Computational Linguistics, pp. 3138–3157 (2025)
24. Sheng, E., Chang, K.W., Natarajan, P., Peng, N.: "Nice try, kiddo": investigating ad hominems in dialogue responses. arXiv preprint arXiv:2010.12820 (2020)
25. Stab, C., Gurevych, I.: Recognizing insufficiently supported arguments in argumentative essays. In: Proceedings of the 15th Conference of the European Chapter of the Association for Computational Linguistics: Volume 1, Long Papers, pp. 980–990 (2017)
26. Wang, J., Jain, S., Zhang, D., Ray, B., Kumar, V., Athiwaratkun, B.: Reasoning in token economies: budget-aware evaluation of LLM reasoning strategies. arXiv preprint arXiv:2406.06461 (2024)
27. Xu, F., Lin, Q., Han, J., Zhao, T., Liu, J., Cambria, E.: Are large language models really good logical reasoners? A comprehensive evaluation and beyond. IEEE Trans. Knowl. Data Eng. (2025)
28. Yang, Z., et al.: Language models as inductive reasoners. arXiv preprint arXiv:2212.10923 (2022)
29. Yin, W., Hay, J., Roth, D.: Benchmarking zero-shot text classification: datasets, evaluation and entailment approach. arXiv preprint arXiv:1909.00161 (2019)
30. Zhang, H., Huang, J., Li, Z., Naik, M., Xing, E.: Improved logical reasoning of language models via differentiable symbolic programming. arXiv preprint arXiv:2305.03742 (2023)
31. Zhou, D., et al.: Least-to-most prompting enables complex reasoning in large language models. arXiv preprint arXiv:2205.10625 (2022)

Stock Price Prediction Using Graph Attention Networks with Sentiment Analysis

Zhenda Hu[1,2]($\boxtimes$), Guanru Yan[1], Yee Sen Tan[1], and Haibo Pen[3]

[1] School of Computing and Information Systems, Singapore Management University, 80 Stamford Road, Singapore 178902, Singapore
`guanru.yan.2023@mitb.smu.edu.sg`, `yeesen.tan.2020@scis.smu.edu.sg`
[2] School of Information Management and Engineering, Shanghai University of Finance and Economics, 777 Guoding Road, Shanghai 200433, China
`huzhenda2020@gmail.com`
[3] School of Electrical and Information Engineering, Tianjin University, No. 135 Yaguan Road, Haihe Education Park, Tianjin 300350, China
`penhaibo@tju.edu.cn`

Abstract. Stock price prediction is a fundamental and important task in the field of finance. This paper introduces the use of Graph Attention Networks (GAT), incorporating stock price movements and investor's sentiment to improve stock price prediction performance. We compare GAT's performance against different deep learning models, including Artificial Neural Networks (ANN), Recurrent Neural Networks (RNN), and Long Short-Term Memory (LSTM) networks. The results highlighted that GAT consistently outperforms the other models across multiple metrics, particularly when sentiment data are incorporated. This use of GAT and sentiment analysis improved the prediction accuracy, demonstrating its potential as a more effective method for predicting stock prices. These findings highlight the importance of integrating the additional sentiment information to enhance model performance and provide a promising direction for future research in financial forecasting.

Keywords: stock price prediction · sentiment analysis · graph attention network

1 Introduction

Stock price prediction is a fundamental and important task in the field of finance [20]. Despite the growing popularity of stock market trend analysis, predicting stock prices remains a challenge due to the dynamic and non-linear behavior of the market. Traditional theories, such as the efficient market hypothesis (EMH) and the random walk theory (RWT), suggest that stock price follows a random pattern, making accurate predictions seemingly unfeasible [1,8,14]. However, recent developments in artificial intelligence and deep learning have inspired researchers to develop new predictive approaches to better predict stock

---

Z. Hu and G. Yan—Both authors contributed equally to this paper.

© The Author(s), under exclusive license to Springer Nature Singapore Pte Ltd. 2025
S. Yuan et al. (Eds.): PAKDD 2025 Workshops, LNAI 15835, pp. 399–408, 2025.
https://doi.org/10.1007/978-981-96-8197-6_30

price [22]. Sentiment analysis, which categorizes opinions, emotions, and attitudes toward various entities and topics [4,22], has become an invaluable tool in financial research. In the stock market, sentiment analysis has attracted significant attention due to its potential for high returns [22].

In recent works, significant progress has been made in the application of deep learning in stock price prediction. Deep learning models, such as Long Short Term Memory Networks (LSTM) and Convolutional Neural Networks (CNN), have gradually become the mainstream methods for stock price prediction due to their powerful feature extraction capabilities [22]. Researchers also went one step further by incorporating attention mechanism into knowledge graphs to model high-order conductivities in a graph network, by recursively propagating embeddings from neighboring nodes, refining representations based on their relational importance [18]. By leveraging attention mechanisms, downstream graph use cases is enhanced, improving accuracy and interpretability [18].

Building upon these advancements, this paper proposes a novel stock price prediction methodology that integrates Graph Attention Networks (GATs) with sentiment analysis of investor comments. Unlike traditional deep learning models that primarily rely on historical price data, our approach leverages investor's sentiment to enhance predictive accuracy.

In this paper, we present the main key contributions that advance stock price prediction research:

1. **Graph-based Attention Networks for Stock Price Prediction:** We use graph-based attention networks for stock price prediction, which can significantly improve prediction accuracy.
2. **The Fusion of Deep Learning Model and Investor's Sentiment:** We propose the use of GAT with sentiment analysis to further enhance stock prediction accuracy.
3. **Comprehensive Evaluation Across Various Models:** We conduct a detailed evaluation, comparing our approach with popular model architectures to assess its relative performance. Our method of using GAT has shown to outperform other models, particularly when incorporating sentiment.

2   Related Works

2.1   Deep Learning in Stock Price Prediction

In recent years, significant progress has been made in the application of deep learning in stock price prediction. Traditional stock price prediction methods mainly rely on linear models and statistical methods, but these methods are difficult to capture nonlinear relationships and complex patterns in financial markets [7]. Deep learning models, such as Long Short Term Memory Networks (LSTM) and Convolutional Neural Networks (CNN), have gradually become the mainstream methods for stock price prediction due to their powerful feature extraction capabilities [22]. For example, Sunny et al. [15] proposed a stock price prediction model based on LSTM, which significantly improves prediction accuracy by learning from historical price data. In addition, hybrid models such as

CNN-LSTM have also been widely studied, further improving prediction performance by combining the spatial feature extraction ability of CNN and the time series modeling ability of LSTM [11].

In addition to utilizing historical price data, the application of deep learning in stock price prediction has also expanded to the field of multi-source data fusion. For example, unstructured data such as news texts, social media sentiment, and macroeconomic indicators are introduced into predictive models to capture the impact of market sentiment and external events on stock prices. Hu et al. [6] proposed a deep learning model based on attention mechanism, which significantly improves the accuracy of financial time series prediction by combining news sentiment analysis and historical trading data. In addition, Graph Neural Networks (GNNs) have also been applied in stock price prediction, further enhancing the predictive ability of the model by mining the correlation between stocks [3,10]. Chen et al. [2] proposed a stock price prediction model based on Graph Convolutional Network (GCN), which models the stock market as a graph structure, extracts the relationship features between stocks using GCN, significantly outperforming other methods. Building upon the existing work in stock market analysis using deep learning models, we recognize a significant gap in the integration of sentiment analysis for enhancing prediction accuracy. Although some studies have investigated the integration of time series data with news texts, there still remains significant potential for stock price prediction through the application of graph-based attention networks.

2.2   Sentiment Analysis in Financial Markets

Sentiment analysis seeks to capture the emotional tone of content, with recent advancements showcasing how transformer-based models have significantly enhanced sentiment analysis, particularly when combined with data from various sources and modalities [19,21]. These advancements have greatly improved contextual understanding and analysis accuracy [16,25]. Additionally, the integration of diverse data methodologies has shown strong performance in multiple experiments [23,25], highlighting the potential of incorporating sentiment data into stock price prediction models.

In the early stages of research, studies found that public sentiment extracted from social media could influence financial markets, with positive emotions demonstrating a predictive relationship with market indices [12,20]. At that time, traditional machine learning models such as logistic regression and support vector machines were used, yet these models achieved remarkable results, emphasizing the potential of incorporating sentiment data into predictions.

Another study also investigated the influence of sentiment analysis in stock price prediction, combined with topic modeling [13]. Although their approach achieved a 54.41% accuracy in predicting stock price movements, the research emphasized that incorporating additional data sources could improve the results.

A more recent study found that the integration of textual data to derive semantic relationships and sentiments, along with quantitative metrics, outperformed other stock price prediction techniques [5]. Their approach showed superior performance over other graph-based networks for stock market prediction across benchmark datasets, emphasizing the beneficial effect of sentiment on improving graph-based methods.

3   Proposed Methodology

We propose an innovative method for predicting stock prices that combines sentiment analysis of investor comments with the GAT model to take advantage of the capabilities of graph networks. By analyzing the emotional tendencies in investor comments, we can capture the potential impact of market sentiment on stock prices.

The core of the GAT model lies in its attention mechanism [17]. It can calculate a learnable attention coefficient for each node pair (edge), representing the importance of neighboring nodes to the target node. For target node i and neighboring node j, their attention scores can be calculated using the LeakyReLU activation function by the following equation.

$$e_{ij} = \text{LeakyReLU}\left(\mathbf{a}^T[\mathbf{z}_i\|\mathbf{z}_j]\right) \tag{1}$$

where $\mathbf{a}$ is a learnable attention vector and $\|$ represents the concatenation operation.

The proposed method is shown in Fig. 1. There are mainly three steps as follows:

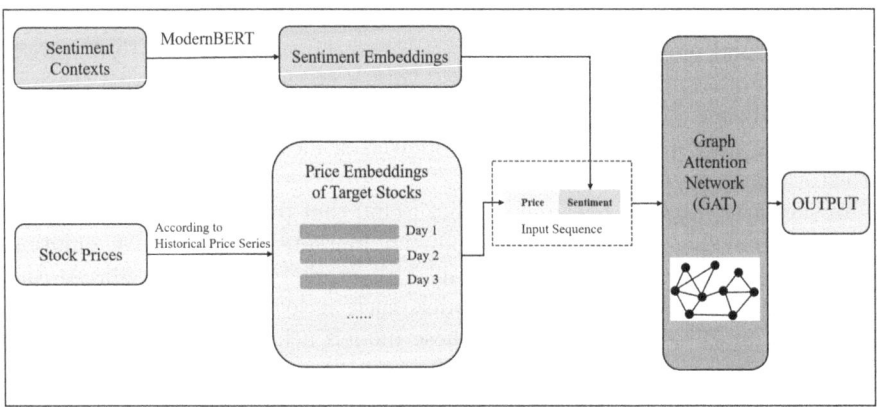

Fig. 1. The Proposed Method

Step 1: We handle the daily closing price series of the target stock using the sliding window. In this way, we can obtain the price embeddings.

Step 2: We first identify the comments related to the target stock from the StockEmotions Dataset [9], and then use ModernBERT [24] to obtain the sentiment embeddings.

Step 3: We concatenate the price embeddings and the sentiment embeddings of the target stock. After that, we utilize GAT to predict the future price of the target stock. By using GAT, we can capture the interactions between historical prices and sentiment information of the target stock, and use this information for predicting future stock prices. It can be formulated as Eq. (2):

$$\hat{Y}_{t+1} = GAT(Y_{t:t-w}, S_t) \tag{2}$$

where $Y_{t:t-w}$ represents the prices of the stock, S_t represents the sentiment embeddings of the target stock, and w represents the size of the sliding window.

Our methodology of integrating sentiment will enable our model to account for the emotional tone of investor comments, which can directly influence stock market behavior. By embedding sentiment data into the prediction model, we introduce an additional layer of insight into market trends that would not be captured by data alone. Furthermore, the usage of graphs in our methodology can improve our model's capabilities in understanding the relationship between sentiment and market dynamics.

4   Dataset

4.1   Data Source

We selected the StockEmotions dataset [9], which consists of stock market comments with two sentiment polarities: bullish and bearish. This dataset spans from January 1, 2020, to December 31, 2020. To complement this sentiment data, we collect historical stock price available online, such as information from Yahoo! Finance, for the same time period, ensuring alignment between sentiment annotations and market movements.

4.2   Data Processing

In this paper, we leverage investor comments published on day T to predict the market trend on day $T + 1$, ensuring temporal alignment between input data and prediction targets. To maintain consistency with actual trading sessions, data from weekends are excluded.

For financial data, we used the daily closing price as the reference value for each trading day. To standardize the numerical inputs and prevent scale related biases, all price data are normalized, ensuring that variations in price ranges do not disproportionately influence the learning process while also facilitating faster model convergence.

5   Experiment and Results

5.1   Experiment Setup

In our experiments, we conducted an ablation study to assess the efficacy of different deep learning models for stock price prediction. Specifically, we aimed to determine whether the proposed GAT outperforms traditional deep learning models, such as Artificial Neural Networks (ANN), Recurrent Neural Networks (RNN), and Long Short-Term Memory (LSTM) networks, and whether integrating sentiment data improves predictive performance.

In addition, we used 80% of the dataset for training and the remaining 20% for testing, ensuring a fair evaluation on unseen data. We employed a sliding window of historical closing prices, represented as $[y_1, y_2, \ldots, y_n]$. Each y_i denoted the closing price on a specific day within the window, capturing essential temporal dynamics. To incorporate investor sentiment, we utilized ModernBERT [24] with a binary classification head, which produced a two-dimensional embedding, represented as: $[\text{P(not-drop)}, \text{P(drop)}]$. This represents the probability that the price would not drop or would drop, reflecting market attitudes.

For each day t, we concatenated the day-t price embedding with the day-$(t-1)$ sentiment embedding to form a vector of dimension $[1, \text{length of the window} + 2]$. This fused input to the GAT model integrated both current price signals and prior-day sentiment. In cases of missing sentiment data, $[0.5, 0.5]$ was used to denote a neutral stance, ensuring a consistent dimensional structure.

5.2   Evaluation Metrics

For the evaluation metrics of stock price forecasting, various metrics are utilized to compare the performance of these models. The most commonly used metrics include Mean Absolute Error (MAE), Root Mean Squared Error (RMSE), Mean Absolute Percentage Error (MAPE), and Mean Directional Accuracy (MDA). Each metric provides different perspectives on model performance, which can collectively help to assess and provide a comprehensive comparison. The formulas used for calculating MAPE and MDA are as follows:

$$\text{MAPE} = \frac{100\%}{n} \sum_{t=1}^{n} \left| \frac{y_t - \hat{y}_t}{y_t} \right| \tag{3}$$

where y_t denotes the actual value at time t, $\hat{y}_t$ is the predicted value, and n is the total number of predictions.

$$\text{MDA} = \frac{1}{n-1} \sum_{t=2}^{n} \mathbb{I}\left[(y_t - y_{t-1})(y_t - \hat{y}_{t-1}) > 0 \right] \tag{4}$$

where $\mathbb{I}[\cdot]$ is the indicator function, returning 1 if the sign of the change in the actual and predicted values is the same, and 0 otherwise.

5.3 Results and Discussion

We evaluated the performance of various model architectures using MAE, RMSE, MAPE, and MDA, with results indicating that the GAT model incorporating stock price (S) and sentiment (V) achieved the best overall performance, as shown in Tables 1 and 2 for AAPL and TSLA, respectively.

Our experiment highlighted notable differences in predictive performance across all model architectures for both AAPL and TSLA stocks. Notably, our proposed GAT architecture outperformed all other models in all scenarios across all metrics, showing that our proposed GAT architecture is effective. This improvement is even more significant when we add sentiment data to GAT, surpassing all other model architectures and scenarios, as shown in Tables 1 and 2. These results indicate that GATs can effectively utilize both stock prices and sentiment data to improve predictive accuracy.

In addition, the magnitude of improvement exhibited distinct variations between the two stocks. For AAPL, integrating sentiment data resulted in approximately a 21% decrease in MAE and around a 30% improvement in MDA, underscoring the significant impact investor sentiment has on predictive accuracy for relatively stable stocks. TSLA, characterized by higher volatility and greater susceptibility to investor sentiment, demonstrated an even more pronounced improvement, achieving approximately a 29% reduction in MAE and a 19% increase in MDA. These findings indicate that investor sentiment can enhance predictive models, particularly the proposed GAT, are especially effective for stocks exhibiting higher volatility and stronger sentiment-driven market fluctuations.

In conclusion, our experiments showed that the use of GAT led to a notable improvement in model performance across all highlighted evaluation metrics. GAT outperformed all other models in these areas, demonstrating its potential as a more effective architecture. Furthermore, incorporating sentiment analysis further enhanced performance, leading to even more competitive results.

Table 1. Performance comparison of different model architecture for AAPL stock

Model Architecture	Input Features	↓ MAE	↓ RMSE	↓ MAPE	↑ MDA
ANN	S	3.8405	4.4289	0.0605	0.6250
	$S + V$	4.0991	4.9070	0.0628	0.5217
RNN	S	3.0426	3.7798	0.0487	0.5625
	$S + V$	2.8512	3.6614	0.0462	0.5435
LSTM	S	3.1727	3.7143	0.0506	0.5000
	$S + V$	2.9655	3.5239	0.0477	0.5652
GAT	S	1.4420	1.8035	0.0233	0.6042
	$S + V$	**1.1400**	**1.4347**	**0.0185**	**0.7826**

Table 2. Performance comparison of different model architecture for TSLA stock

Model Architecture	Input Features	↓ MAE	↓ RMSE	↓ MAPE	↑ MDA
ANN	S	11.7568	14.7294	0.0795	0.7292
	$S + V$	28.5639	35.0938	0.1625	0.7297
RNN	S	12.7249	14.9279	0.0789	0.6667
	$S + V$	10.2170	12.2555	0.0741	0.7297
LSTM	S	9.1779	11.3304	0.0626	0.6667
	$S + V$	9.9527	12.5787	0.0674	0.7838
GAT	S	5.1524	6.5547	0.0382	0.6875
	$S + V$	**3.6797**	**4.9213**	**0.0252**	**0.8158**

Note: S denotes stock price data, V represents sentiment data. The ↓ symbol indicates that lower values are preferred, whereas the ↑ symbol indicates a preference for higher values in terms of performance.

6   Conclusion, Limitations and Future Works

6.1   Conclusion

In this paper, we explored how the use of GATs and investor comments can enhance stock price prediction. We derive the sentiment data from investor comments reflecting their attitudes and emotions that potentially affect market dynamics.

Based on the experiments results, we found that our proposed GAT model outperformed the other deep learning methods, such as ANN, RNN, and LSTM networks. Furthermore, incorporating sentiment analysis into GAT not only improved the accuracy of stock price predictions but also enhanced the model's ability to predict price direction, showcasing its practical value for financial forecasting. This paper highlights the importance of incorporating external factors into predictive models, as it significantly boosts performance. These results demonstrate that the GAT model incorporating investor sentiment is an effective solution for financial time series forecasting, offering a promising approach for advancing research in financial markets.

6.2   Limitations and Future Works

One potential limitation of our methodology is the inherent delay in public sentiment compared to actual market movements, which may be influenced by insider information. Insiders, such as company executives and institutional investors, often have early access to critical financial data, earnings reports, or strategic decisions before they are disclosed to the public. As a result, stock prices may adjust in response to these developments before retail investors and analysts react to the news, causing sentiment-based models to capture delayed signals rather than leading indicators. This delay reduces the predictive power of sentiment analysis, as market trends may have already changed before sentiment from news articles, financial reports, and social media reflects these changes.

To address this limitation, future research could explore hybrid models that integrate alternative data sources, such as real-time transaction records, order book dynamics, and insider trading activity, along with sentiment analysis. Additionally, by incorporating these transactional data into knowledge graphs, we could potentially uncover more sentiment-driven and insider-driven price movements insights to improve prediction accuracy. Exploring event-based sentiment shifts and their immediate impact on stock prices may also offer a more dynamic and responsive approach to stock price forecasting.

References

1. Bollen, J., Mao, H., Zeng, X.: Twitter mood predicts the stock market. J. Comput. Sci. **2**(1), 1–8 (2011)
2. Chen, W., Jiang, M., Zhang, W.G., Chen, Z.: A novel graph convolutional feature based convolutional neural network for stock trend prediction. Inf. Sci. **556**, 67–94 (2021)
3. Cheng, D., Yang, F., Xiang, S., Liu, J.: Financial time series forecasting with multi-modality graph neural network. Pattern Recogn. **121**, 108218 (2022)
4. Cui, J., Wang, Z., Ho, S.B., Cambria, E.: Survey on sentiment analysis: evolution of research methods and topics. Artif. Intell. Rev. **56**, 8469–8510 (2023)
5. Du, K., Mao, R., Xing, F., Cambria, E.: A dynamic dual-graph neural network for stock price movement prediction. In: 2024 International Joint Conference on Neural Networks (IJCNN), pp. 1–8. IEEE (2024)
6. Hu, Z.: Crude oil price prediction using CEEMDAN and LSTM-attention with news sentiment index. Oil Gas Sci. Technol.-Revue d'IFP Energies nouvelles **76**, 28 (2021)
7. Hu, Z., Wang, Z., Ho, S.B., Tan, A.H.: Stock market trend forecasting based on multiple textual features: a deep learning method. In: 2021 IEEE 33rd International Conference on Tools with Artificial Intelligence (ICTAI), pp. 1002–1007. IEEE (2021)
8. Kannan, K.S., Sekar, P.S., Sathik, M.M., Arumugam, P.: Financial stock market forecast using data mining techniques. In: Proceedings of the International Multiconference of Engineers and Computer Scientists, vol. 1, pp. 1–5 (2010)
9. Lee, J., Youn, H.L., Poon, J., Han, S.C.: StockeMotions: discover investor emotions for financial sentiment analysis and multivariate time series. arXiv preprint arXiv:2301.09279 (2023)
10. Li, X., Wang, J., Tan, J., Ji, S., Jia, H.: A graph neural network-based stock forecasting method utilizing multi-source heterogeneous data fusion. Multimed. Tools Appl. **81**(30), 43753–43775 (2022)
11. Mehtab, S., Sen, J.: Stock price prediction using CNN and LSTM-based deep learning models. In: 2020 International Conference on Decision Aid Sciences and Application (DASA), pp. 447–453. IEEE (2020)
12. Mittal, A., Goel, A.: Stock prediction using twitter sentiment analysis. Standford University, CS229 (2011). http://cs229.stanford.edu/proj2011/GoelMittal-StockMarketPredictionUsingTwitterSentimentAnalysis.pdf. **15**, 2352 (2012)
13. Nguyen, T.H., Shirai, K., Velcin, J.: Sentiment analysis on social media for stock movement prediction. Expert Syst. Appl. **42**(24), 9603–9611 (2015)

14. Patel, J., Shah, S., Thakkar, P., Kotecha, K.: Predicting stock and stock price index movement using trend deterministic data preparation and machine learning techniques. Expert Syst. Appl. **42**(1), 259–268 (2015)
15. Sunny, M.A.I., Maswood, M.M.S., Alharbi, A.G.: Deep learning-based stock price prediction using LSTM and bi-directional LSTM model. In: 2020 2nd Novel Intelligent and Leading Emerging Sciences Conference (NILES), pp. 87–92. IEEE (2020)
16. Tan, Y.S., Teo, N., Ghe, E., Fong, J., Wang, Z.: Video sentiment analysis for child safety. In: 2023 IEEE International Conference on Data Mining Workshops (ICDMW), pp. 783–790. IEEE (2023)
17. Velikovi, P., Cucurull, G., Casanova, A., Romero, A., Liò, P., Bengio, Y.: Graph attention networks. arXiv preprint arXiv:1710.10903 (2017)
18. Wang, X., He, X., Cao, Y., Liu, M., Chua, T.S.: KGAT: knowledge graph attention network for recommendation. In: Proceedings of the 25th ACM SIGKDD International Conference on Knowledge Discovery & Data Mining, pp. 950–958 (2019)
19. Wang, Z., Ho, S.B., Cambria, E.: Multi-level fine-scaled sentiment sensing with ambivalence handling. Internat. J. Uncertain. Fuzziness Knowl.-Based Syst. **28**(04), 683–697 (2020)
20. Wang, Z., Ho, S.B., Lin, Z.: Stock market prediction analysis by incorporating social and news opinion and sentiment. In: 2018 IEEE International Conference on Data Mining Workshops (ICDMW), pp. 1375–1380. IEEE (2018)
21. Wang, Z., Hu, Z., Ho, S.B., Cambria, E., Tan, A.H.: MiMuSA–mimicking human language understanding for fine-grained multi-class sentiment analysis. Neural Comput. Appl. **35**(21), 15907–15921 (2023)
22. Wang, Z., Hu, Z., Li, F., Ho, S.B., Cambria, E.: Learning-based stock trending prediction by incorporating technical indicators and social media sentiment. Cogn. Comput. **15**(3), 1092–1102 (2023)
23. Wang, Z., Huang, D., Cui, J., Zhang, X., Ho, S.B., Cambria, E.: A review of Chinese sentiment analysis: subjects, methods, and trends. Artif. Intell. Rev. **58**(3), 75 (2025)
24. Warner, B., et al.: Smarter, better, faster, longer: a modern bidirectional encoder for fast, memory efficient, and long context finetuning and inference. arXiv preprint arXiv:2412.13663 (2024)
25. Wu, T., et al.: Video sentiment analysis with bimodal information-augmented multi-head attention. Knowl.-Based Syst. **235**, 107676 (2022)

Owls are Wise and Foxes are Unfaithful: Uncovering Animal Stereotypes in Vision Language Models

Tabinda Aman[1], Mohammad Nadeem[1]([✉])(iD), Shahab Saquib Sohail[2](iD), Mohammad Anas[3], and Erik Cambria[4](iD)

[1] Department of Computer Science, Aligarh Muslim University,
Aligarh 202002, UP, India
mnadeem.cs@amu.ac.in
[2] School of Computing Science and Engineering, VIT Bhopal University,
Bhopal 466114, MP, India
shahabsaquibsohail@vitbhopal.ac.in
[3] Department of Computer Science and Engineering, School of Engineering Sciences
and Technology, Jamia Hamdard, New Delhi 110062, India
[4] College of Computing and Data Science, Nanyang Technological University,
Singapore, Singapore
cambria@ntu.edu.sg

Abstract. Animal stereotypes are deeply embedded in human culture and language. They often shape our perceptions and expectations of various species. Our study investigates how animal stereotypes manifest in vision-language models during the task of image generation. Through targeted prompts, we explore whether DALL-E perpetuates stereotypical representations of animals, such as "owls as wise", "foxes as unfaithful", etc. Our findings reveal significant stereotyped instances where the model consistently generates images aligned with cultural biases. The current work is the first of its kind to examine animal stereotyping in vision-language models systematically and to highlight a critical yet underexplored dimension of bias in AI-generated visual content.

Keywords: Generative artificial intelligence · vision language model · ethical artificial intelligence · bias

1 Introduction

Generative artificial intelligence (GAI) has seen rapid adoption across diverse domains through its ability to produce high-quality text, images, and videos [15]. Vision-Language Models (VLMs) represent a significant advancement in this space, combining visual and linguistic understanding to generate contextually relevant images from textual descriptions [12]. They leverage vast datasets and sophisticated algorithms [11,12] to enable unprecedented creativity and efficiency, driving applications in marketing, entertainment, design, and more. Large

S. Yuan et al. (Eds.): PAKDD 2025 Workshops, LNAI 15835, pp. 409–417, 2025.
https://doi.org/10.1007/978-981-96-8197-6_31

Language Models (LLMs) and VLMs often inherit and perpetuate biases and stereotypes present in their training data [4,5,7,9], which is typically sourced from vast and diverse internet repositories [1,2,6,8]. The training datasets frequently contain implicit and explicit cultural stereotypes, societal biases, and skewed representations that the models learn during training. As a result, LLMs may generate biased text [1,8], while VLMs can produce stereotypical or culturally inappropriate images [2,6]. Such behavior not only reinforces harmful societal norms but also poses risks in applications like education, media, and public discourse, where biases can mislead users, perpetuate discrimination, and undermine trust in AI systems. Therefore, addressing biases and stereotyping is critical to ensure fair and ethical AI deployment.

There is a decent amount of work dedicated to identifying bias in text-based language based models [1,3,8]. Different works have identified that LLMs persistently show biases related to demographic characteristics such as race, gender, age, political affiliation, and sexual orientation [3]. Abid et al. [1] highlighted that GPT-3 exhibited significant bias where it consistently associated Muslims with violence. Nadeem et al. [8] concluded that bidirectional encoder representations from transformers (BERT), generative pre-trained transformer (GPT-2), and robustly optimized BERT pre-training approach (RoBERTa) exhibited significant stereotypical biases across domains such as gender, profession, race, and religion and emphasized the need for improved evaluation metrics and mitigation strategies. On the other hand, studies related to biases in VLMs are limited [2,6]. Cho et al. [2] highlighted that text-to-image generation models had significant gender and skin tone biases for the image generation task. Similar behavior was observed in the CLIP model also [6].

Various studies have identified biases embedded in (GAI) models related to gender, race, religion, and other human-centric categories, with efforts to quantify and mitigate such issues gaining traction. However, despite the significant attention given to human-related biases, little to no work has been conducted on the stereotyping of animals in VLMs. This gap is particularly critical, as animal stereotypes are deeply ingrained in cultural narratives and can influence the AI-generated image content, shaping perceptions in subtle but impactful ways.

2 Methodology

For the current study, we adopted the following methodology:

2.1 VLM (DALL-E 3)

We utilized DALL-E 3, a state-of-the-art Vision-Language Model (VLM) developed by OpenAI [12], as our model of choice. DALL-E is renowned for its ability to generate visually coherent and contextually relevant images from textual prompts. It was selected for its advanced capabilities and widespread usage.

2.2   Prompt Formation

To investigate animal stereotyping, we designed six prompts in the format "Generate an image of a/an adj animal." where adj represents specific attributes: loyal, wise, gentle, unfaithful, mischievous, and violent. The selected attributes were chosen based on common cultural stereotypes associated with animals (e.g., dogs with loyalty, owls with wisdom, etc.). Each prompt was crafted to be simple, direct, and neutral to avoid introducing additional bias in phrasing.

2.3   Image Generation

Each of the six prompts was executed 100 times using DALL-E 3, generating a total of 600 images for analysis. Each run was performed independently, and the results were collected without any manual intervention or filtering. The generated images were then analyzed to identify the type of animal depicted for each attribute. We categorized the animals depicted in the images and counted the occurrences of specific animals for each prompt. The frequency analysis provided quantitative evidence of how strongly DALL-E associates specific animals with culturally ingrained stereotypes.

Overall, the adopted approach allowed us to evaluate whether certain animals were disproportionately linked to particular descriptors, indicative of stereotyping in the model's outputs.

3   Results

The results obtained for each prompt are discussed next. The frequency count for the same is presented in Fig 1.

3.1   Loyal Animals

For this prompt, dogs appeared exclusively in all 100 generations which revealed a strong bias in DALL-E towards associating loyalty solely with dogs. While dogs are widely recognized for their loyalty, many other animals, such as horses and elephants, are also known for their loyalty, especially to their human companions.

3.2   Wise Animals

DALL-E predominantly associates wisdom with owls, which aligns with cultural stereotypes portraying owls as symbols of wisdom, particularly in Western traditions. Elephants, often regarded as intelligent and wise in many cultures, are the second most frequent choice, while lion-like animals and others appear less frequently.

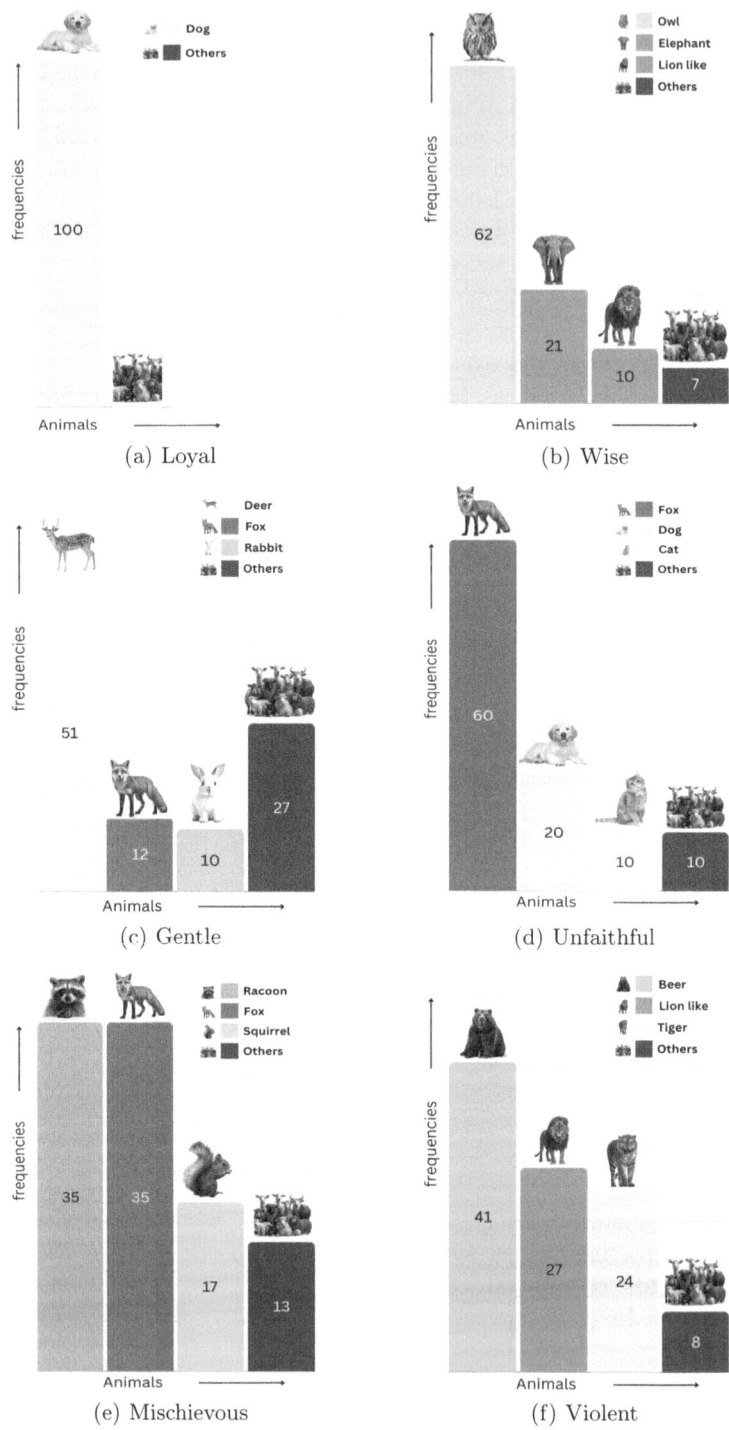

Fig. 1. The frequency count of the animals for various prompt types

3.3 Gentle Animals

Gentleness is associated primarily with deer which is aligned with their portrayal as graceful and timid creatures in cultural narratives. Rabbits also make a notable appearance as gentle and harmless animals. Interestingly, foxes, typically associated with cunning rather than gentleness, appear in some instances. The "Others" category suggests some diversity in the model's outputs, but the strong focus on deer highlights a bias toward specific, culturally prominent stereotypes.

3.4 Unfaithful Animals

With no surprise, unfaithfulness was associated with foxes which reflected the cultural trope of foxes as cunning and deceitful animals. Interestingly, dogs, which are widely regarded as loyal, appear in this context as well, potentially indicating inconsistencies or overgeneralization in the model's understanding of traits [10,14]. Cats also appear, possibly influenced by the stereotypical notion of independence or aloofness often attributed to them. The "Others" category suggests some degree of diversity but does not significantly counterbalance the strong association with foxes.

3.5 Mischievous Animals

DALL-E strongly associates mischievousness with raccoons and foxes, reflecting their stereotyped portrayal as clever and trouble-making animals. Squirrels also appear frequently, which aligns with their depiction as playful and energetic creatures in nature. The "Others" category provides diversity, but the majority representation of these three animals suggests that DALL-E relies heavily on common cultural archetypes to generate images associated with mischievousness.

3.6 Violent Animals

The results indicate that large predators such as bears, lions, and tigers, are associated with violence, showing their portrayal as ferocious and aggressive animals in cultural narratives. The "Others" category accounts for a small proportion of the results, suggesting limited diversity in the representation of violent behavior.

The overall analysis of the results reveals that DALL-E strongly associates specific traits with certain animals, such as wisdom with owls, unfaithfulness with foxes etc. These associations align with long-standing cultural stereotypes but fail to reflect the diverse and context-dependent nature of animal behavior. In reality, traits such as loyalty, wisdom, or violence cannot be rigidly assigned to specific animals, as these behaviors vary widely across species and contexts.

Loyalty is a common trait in many social animals, gentleness can be observed in various herbivores, and aggression is contextually exhibited by numerous species. For example, majority of animals exhibit loyalty within their own social groups, such as herds, packs, or families. Moreover, many animals show behaviors indicative of wisdom or intelligence within their ecological contexts. For instance, elephants show remarkable problem-solving skills and social intelligence, while dolphins and crows are also known for their cognitive abilities. Similarly, violence or aggression in animals is often context-dependent, driven by factors such as defense, territory, or survival.

(a) Loyal   (b) Wise   (c) Gentle

(d) Unfaithful   (e) Mischievous   (f) Violent

Fig. 2. DALL-E 3 not only stereotypes animals but also conveys specific behaviors through the facial expressions, backgrounds, and color compositions in the generated images.

In addition to the strong association of specific traits with particular animals, the generated images also depict those traits visually. For example, for the "violent" prompt, DALL-E generates a bear exhibiting aggressive behavior, such as roaring in a forest setting, reinforcing the violent stereotype (See Fig 2). Similarly, for the "unfaithful" prompt, a fox is not only selected but is shown sneaking near a henhouse with a cunning expression, symbolizing deceit. Such a dual-layered reinforcement—selection of the stereotyped animal and visual portrayal of the associated behavior—further highlights how VLMs like DALL-E internalize and propagate cultural narratives, amplifying stereotypical representations in both animal choice and contextual depiction.

4   Debiasing

To address debiasing, we explored a prompt modification technique aimed at reducing bias. Specifically, we introduced the instruction "Do not stereotype animals" into the original prompt structure, forming a modified prompt: "Generate an image of a/an adj animal. Do not stereotype animals". It was designed to explicitly encourage DALL-E to avoid culturally ingrained associations between specific traits and particular animals.

We tested the modified prompts for two traits, "wise" and "mischievous", and observed significant improvements in the diversity of generated images. For instance, the modified prompt for "wise" resulted in a broader representation of animals beyond the stereotypical owl, including kangaroos, gorillas, and octopuses. Similarly, the modified "mischievous" prompt generated a wider variety of animals such as monkeys, koalas, and hamsters to reduce the dominance of foxes and raccoons that was observed in the original prompt (See Fig. 3). The results highlight the potential of prompt engineering as a lightweight and effective method to mitigate biases in VLMs outputs without requiring costly retraining of the model. However, while such an approach achieved measurable improvements, it is not a comprehensive solution, as some biases still persist due to the underlying training data. Future research could explore combining prompt engineering with dataset curation and fine-tuning techniques to achieve more robust and generalizable debiasing outcomes.

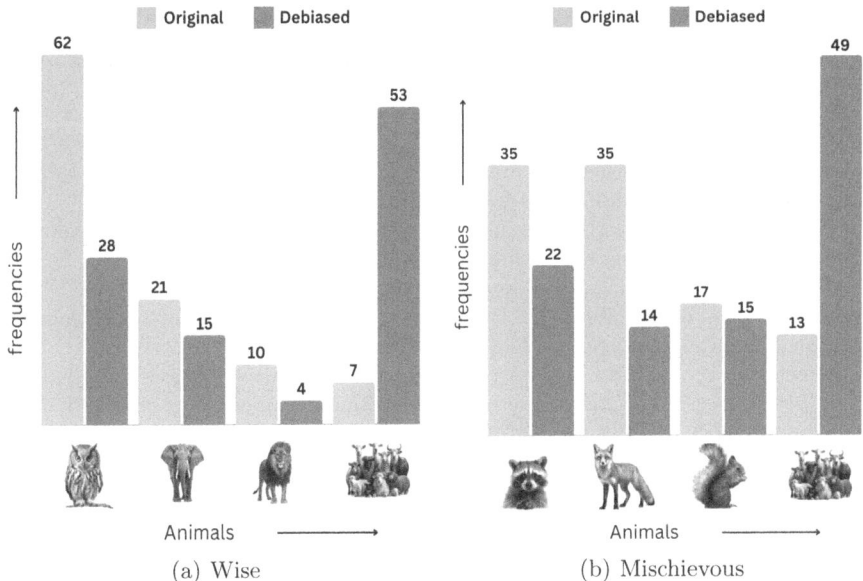

Fig. 3. Results after debiasing using modified prompts.

5   Limitations and Conclusion

The current work has several limitations. Firstly, the analysis was restricted to a small set of prompts representing specific traits, which may not capture the full range of biases present in the model. Secondly, only DALL-E is considered in the current work. Other notable VLMs such as Stable Diffusion [13] may help generalize our findings. Finally, although the modified prompts showed promising improvements in reducing stereotyping, it is not a comprehensive solution and may not generalize across all traits or contexts.

Current study highlights how VLMs like DALL-E perpetuate animal stereo-types by associating specific traits with certain species. While we successfully demonstrated that prompt modifications can partially mitigate such biases, the persistence of stereotypes underscores the need for more robust solutions. Future work should focus on a multi-faceted approach to debiasing, including the cura-tion of balanced training datasets, fine-tuning model architectures etc. to create separate debiased models and tools. Additionally, expanding the scope of anal-ysis to include a wider range of traits and other generative AI models could provide a more comprehensive understanding of bias in vision-language systems.

Disclosure of Interests. The authors have no competing interests to declare that are relevant to the content of this article.

References

1. Abid, A., Farooqi, M., Zou, J.: Persistent anti-muslim bias in large language mod-els. In: Proceedings of the 2021 AAAI/ACM Conference on AI, Ethics, and Society, pp. 298–306 (2021)
2. Cho, J., Zala, A., Bansal, M.: DALL-eval: probing the reasoning skills and social biases of text-to-image generation models. In: Proceedings of the IEEE/CVF Inter-national Conference on Computer Vision, pp. 3043–3054 (2023)
3. Demidova, A., Atwany, H., Rabih, N., Sha'ban, S., Abdul-Mageed, M.: John vs. Ahmed: debate-induced bias in multilingual LLMs. In: Proceedings of The Second Arabic Natural Language Processing Conference, pp. 193–209 (2024)
4. Ge, M., Mao, R., Cambria, E.: Discovering the cognitive bias of toxic language through metaphorical concept mappings. Cogn. Comput. **17** (2025)
5. Hagendorff, T., Fabi, S., Kosinski, M.: Human-like intuitive behavior and reason-ing biases emerged in large language models but disappeared in chatGPT. Nat. Comput. Sci. **3**(10), 833–838 (2023)
6. Hamidieh, K., Zhang, H., Gerych, W., Hartvigsen, T., Ghassemi, M.: Identifying implicit social biases in vision-language models. In: Proceedings of the AAAI/ACM Conference on AI, Ethics, and Society, vol. 7, pp. 547–561 (2024)d
7. Mao, R., Liu, Q., He, K., Li, W., Cambria, E.: The biases of pre-trained language models: an empirical study on prompt-based sentiment analysis and emotion detec-tion. IEEE Trans. Affect. Comput. **14**(3), 1743–1753 (2023)
8. Nadeem, M., Bethke, A., Reddy, S.: StereoSet: measuring stereotypical bias in pretrained language models. arXiv preprint arXiv:2004.09456 (2020)

9. Nadeem, M., Sohail, S., Cambria, E., Schuller, B., Hussain, A.: Gender bias in text-to-video generation models: a case study of sora. arXiv preprint arXiv:2501.01987 (2024)
10. Nadeem, M., Sohail, S., Cambria, E., Schuller, B., Hussain, A.: Negation blindness in large language models: unveiling the no syndrome in image generation. arXiv preprint arXiv:2409.00105 (2024)
11. Radford, A., et al.: Learning transferable visual models from natural language supervision. In: International Conference on Machine Learning, pp. 8748–8763. PMLR (2021)
12. Ramesh, A., et al.: Zero-shot text-to-image generation. In: International Conference on Machine Learning, pp. 8821–8831. PMLR (2021)
13. Rombach, R., Blattmann, A., Lorenz, D., Esser, P., Ommer, B.: High-resolution image synthesis with latent diffusion models. In: Proceedings of the IEEE/CVF Conference on Computer Vision and Pattern Recognition, pp. 10684–10695 (2022)
14. Singh, J., Shrivastava, I., Vatsa, M., Singh, R., Bharati, A.: Learn "no" to say "yes" better: improving vision-language models via negations. arXiv preprint arXiv:2403.20312 (2024)
15. Yin, S., et al.: A survey on multimodal large language models. arXiv preprint arXiv:2306.13549 (2023)

Author Index

The manufacturer's authorised representative in the EU is Springer
Nature Customer Service Centre GmbH, Europaplatz 3, 69115 Heidelberg,
Germany. If you have any concerns regarding our products, please
contact ProductSafety@springernature.com

Printed and bound by CPI Group (UK) Ltd, Croydon, CR0 4YY

27/04/2026

02097606-0002